Organismic Evolution

Organismic Evolution

Verne Grant
UNIVERSITY OF TEXAS AT AUSTIN

FOREWORD BY

George Gaylord Simpson

W. H. FREEMAN AND COMPANY
SAN FRANCISCO

Library of Congress Cataloging in Publication Data
Grant, Verne.
    Organismic evolution.

    Bibliography: p.
    Includes indexes.
    1.   Evolution.   I.   Title.
QH366.2.G68   575   76–54175
ISBN 0–7167–0372–6

Printed in the United States of America

1 2 3 4 5 6 7 8 9

Foreword

The present book gives an overall view of the most important things in our current knowledge of organic evolution. This foreword seeks to place the book within a comparative framework. To do so in full detail would require another book or several of them, but it can be done in an abbreviated way by focusing attention on some of the most significant features of studies of evolution in three periods: the first about a hundred years ago, the second now some fifty years past, and the third the present. For each period the focus can be further sharpened and the account further abbreviated by special attention to a general book on evolution characteristic for its time: first, Darwin's masterpiece, which, more than any other one work, started it all; second, a work by Lull that was the most popular general textbook on the subject fifty years ago; and third, the book now in your hands.

A hundred years ago Darwin was still alive, and was to be actively at work almost to the day of his death at the age of 73 in 1882. His master work, the long title of which is usually abbreviated to *The Origin of Species* or simply *The Origin,* had first appeared in 1859, and the last edition revised by Darwin (the sixth edition) had been published in 1872. In the 1870s the reality of evolution had been accepted by almost all competent and well-informed biologists. The two most notable exceptions were Louis Agassiz, but he died in 1873 and his followers were already evolutionists, and Richard Owen, who lived to the age of 88 in 1892 and whose late works hinted obscurely that he objected not so much to acceptance of evolution as to Darwin's particular ideas about its course and causes.

Thus a century ago anyone who studied biology under a qualified teacher was already introduced to evolution as a general element of that subject. As far as I know there were no courses explicitly devoted to evolution as a distinct topic, and although there were many publications on one aspect or another, there was no truly general text other than *The Origin* itself. Anyone who studied *The Origin* would learn first of all that evolution is the only rational explanation of a multitude of observed facts. Next he would learn that the Darwinian theory, properly speaking, ascribed evolutionary change primarily to natural selection—inadequately summed up in the expression "survival of the fittest" —and in lesser degree to inherited effects of the use and disuse of organs, to the effects of the environment, and to "variations which seem to us in our ignorance to arise spontaneously." Most later discussants combined Darwin's second and third factors as inheritance of acquired characters. His fourth factor entailed what we would call mutations, but for the most part mutations with larger and more obvious effects than are usual. The student would also observe that Darwin was quite positive on the role of natural selection and adduced strong evidence for it, but was less emphatic and had weaker evidence for his other three factors.

During the next fifty years, which bring us into the 1920s, many more factual and relevant observations were made. All indicated evolution as the only rational and judicious interpretation of the history of organisms, and opposition to this view was no longer possible on any scientific grounds, although it does continue even now in some nonscientific or antiscientific circles, as noted in the last part of the present book. Nevertheless scientific unanimity on that general point had not produced a definite consensus as to the specific factors or causes of evolution.

In the 1920s college courses in organic evolution were being given. That by Richard Swann Lull was among the most popular lecture courses at Yale. His textbook *Organic Evolution*, published in 1917 and in revised form in 1929, was probably the most widely used of the several then available, and may serve as indicative of the state of knowledge and of pedagogy in this subject fifty years ago. Only a small part of the book, less than 15 percent of the text, was devoted to the theories, principles, or causes ("mechanisms") of evolution. Here there are two striking differences between Lull's and almost any now current treatment, such as that of the present book: the multiplicity of theories given serious attention—some now considered worthy of no more than passing historical mention, at most—and the minimal involvement of genetics.

Lull's discussion of natural selection was brief and anecdotal, with the conclusion that ". . . whereas natural selection may be conceded to be a factor of importance, it is apparently not the only factor nor indeed the only important factor in the evolution of organic life." The inheritance of acquired characters was treated at almost equal length. Lull's verdict on that was that it was "unproved," but he added that "one can not help feeling" that in some unknown way the inheritance of acquired characters may occur. In the present book such inheritance, in the sense hitherto given it, is quite properly considered not even worthy of discussion.

Lull maintained Darwin's distinction between natural and sexual selection, with some emphasis on the latter. Sexual selection is now considered a real but somewhat minor special case of natural selection. Lull also singled out for consideration rectigradation and kinetogenesis. Both the terms and the concepts have been discarded and are now forgotten except by historians. Orthogenesis, lengthily and rather favorably discussed by Lull, is still named and briefly discussed but rejected in the present book.

The most important development bearing on evolutionary biology in the interval between a hundred years and fifty years ago was the rise of what is often, but somewhat misleadingly, called Mendelian genetics. Lull gave only an elementary and brief account of "Mendel's law" and concluded in his first edition (1917) that it "does not apply universally to all cases of inheritance." In the revision (1929) the account was even shorter but the conclusion then was that "Mendel's laws" (now in the plural) do "apply universally to all cases of inheritance." No clear connection was made between "Mendel's laws," or the then rapidly advancing science of genetics, and the explanation of evolution. A supposed connection then made by some geneticists was anti-Darwinian in that the control of evolution was presumed to be by mutations, especially those that were saltatory in their phenetic effects, whereas natural selection was given only a negative role, if any.

Here is the most marked difference between a textbook of the 1920s and those of today, and this reflects a sort of subrevolution in evolutionary biology supplementing and extending the Darwinian revolution of a century and more ago. In the present book the approach to basic evolutionary theory is primarily genetical. Natural selection, although it involves a broader interplay of factors, is consonant with the genetical approach and can be discussed in terms of causes and effects in genetic systems, which are now considered far more broadly than as simple Mendelism.

Grant has noted in this book that the new movement in evolutionary theory started at the end of the 1920s and in the 1930s in the work of a number of biologists, but most prominently, at first, in that of R. A. Fisher, an English statistician, of J. B. S. Haldane, an English biologist who was successively a professor of physiology, genetics, and biochemistry, and of Sewall Wright, an American geneticist. This new approach to evolutionary theory expanded in the 1940s as contributions came in from every branch of organismal biology, including most noticeably at that time population genetics, systematics, paleontology, and botany. The modified and extended evolutionary theory, or rather body of theory, thus became a synthesis from many sources and so is commonly called the synthetic theory, as it is in this book.

As in any science, or for that matter any subject at all, there are differences of opinion about parts of this body of theory, but its general approach and broader conclusions are now adopted by a great majority of organismal biologists. The synthesis also continues as new discoveries are made and new relationships studied. Grant suggests (in Chapter 17) that if the genetic phenomenon of induction proves to have a significant role in evolution this would go beyond the synthetic theory. I may add the suggestion that it would not remain outside the scope of that theory but would enlarge that scope.

There is a rather widespread human tendency, to which human scientists are also liable, to rediscover, enlarge, and embellish old ideas that seem new when expressed in a later vocabulary. A major part of Lull's text was devoted to adaptive radiation, a fruitful evolutionary concept that Lull ascribed to Henry Fairfield Osborn, although he did mention in passing and incorrectly that Lamarck had had the same idea under a different term, and correctly that Darwin had. In fact it was stated with perfect clarity and for the first time by Darwin, who noted in his autobiography the exact moment when that insight occurred to him. I bring this up here just to emphasize Grant's skill in this book at avoiding the confusion of new terms with new concepts and ideas. It is necessary to use the vocabulary now usual among evolutionists. For example, the term "character displacement" is current and is used as such, although this is another instance of new words and new examples for a phenomenon clearly noted by Darwin. In other instances, as the reader will find for himself, Grant quietly rejects or quite demolishes terminological fallacies.

Here the synthetic theory is not merely expounded. This book is by an active research contributor to that body of theory. The author does not hesitate to express his own opinions and to put forward new ideas. The book has personality.

Finally, it may surprise some readers of this extensive and intensive treatment that some aspects of current evolutionary studies are barely mentioned, or even omitted. To support that statement, and for no other purpose, a simple example should be given: there is no discussion here of mathematical models of faunal changes on islands and in other isolated communities. The point is that no single book can cover all of a subject as vast and complicated as organismal evolution. Darwin considered the first edition of *The Origin*, which ran to 490 text pages, a mere abstract of the book on evolution that he had prepared to write. A single book on this topic has to be highly selective. Darwin selected well, and so has Grant.

Tucson, Arizona   *George Gaylord Simpson*
December 31, 1976

Contents

Preface

This book deals with the processes that bring about evolutionary changes in organisms and with the principal factors that affect these processes. The title, *Organismic Evolution,* is meant to be taken literally, since the emphasis is on the evolution of organisms, particularly animals and plants, rather than on molecular evolution or primitive organic evolution or mathematical models of evolution.

Furthermore, the emphasis in this book is on principles rather than details, on fundamentals rather than current topics, and on well-chosen examples rather than catalogs of facts.

Organismic Evolution has grown out of a one-semester, senior-level course that I have been teaching since 1952. The classes typically contain juniors, seniors, and graduate students, and occasionally a professional biologist as a visitor. This is the range of readers I have had in mind in writing *Organismic Evolution.* In short, *Organismic Evolution* was designed and written as an advanced text.

Let me hasten to enter a caveat. Students do not particularly like textbooks that read and look like textbooks, but they do like good books, and I share the same predilection. My primary goal, then, has been to write a good general book on evolution, of potential interest to a wide range of readers; and, within this broader objective, the secondary goal was to shape the book to fit the specific needs of students.

The subject matter is organized into groups of chapters corresponding to three broad levels of evolutionary change: microevolution (Parts II and III), speciation (Part V), and macroevolution (Part VI). These parts can be read separately. Since macroevolution is underemphasized in many recent books on evolution, a special effort was made to give an adequate introduction to that subject here. Human evolution is dealt with in Part VII. The evolution theory has had in the past, and continues to have, implications for human thought and culture; these social aspects of evolution are considered briefly in Part VIII.

The chapters are short. Only by keeping the individual chapters short was I able to cover a very wide range of topics within the limits of one medium-sized book. The first corollary is that the chapters provide introductions to, not exhaustive treatises on, their respective topics. The second corollary is that a need will often exist for supplementary reading on a specific topic introduced in the text. I have attempted to meet this need in two ways: by providing lists of selected collateral readings at the end of most chapters, and by giving bibliographical citations in the body of the text.

The technical terminology of evolutionary biology is now quite extensive and contains a considerable amount of redundancy. I have grouped the technical terms into two categories for the purpose of this book. The first category consists of terms that are essential for an understanding of basic concepts. These terms are defined and illustrated in the text. In the second category I place those terms that I regard as unnecessary, but that other workers evidently do not, since the terms in question are being used currently in one school or another; consequently students will encounter them in readings, seminars, and lectures. These second-category terms are mentioned parenthetically in the text, usually as synonyms or near-synonyms of the primary terms.

The reader will find an overall list of technical terms—both the essential and the secondary ones—at the end of the book. This list is not a glossary, but rather a thesaurus or index. It gives the page numbers where the terms are either defined in context or placed in synonymy.

March 1976   *Verne Grant*
                                                                    UNIVERSITY OF TEXAS

Acknowledgments

I am very greatly indebted to Dr. George Gaylord Simpson of Tucson, Arizona, for critical reading of Chapters 26, 28, 29, 30, 31, 32, and 37. The criticisms, suggestions, and information provided by Dr. Simpson led to numerous corrections and improvements in this group of chapters. I also had the benefit of Dr. Simpson's comments concerning the chapter outline.

Dr. Robert Flake of the University of Texas critically read Chapters 6 and 14, making some helpful suggestions.

Dr. Theodore Downs of the Los Angeles County Museum of Natural History provided information from his published and unpublished studies of fossil horses; this information is incorporated in Chapter 26.

The entire manuscript was read by Mrs. Karen A. Grant with a critical eye for ambiguities and prolixities. Mrs. Grant also did a great deal of the proofreading.

My secretary, Ms. Helen Barler, typed the manuscript with care and accuracy, hunted up special items in the library, and skillfully handled the secretarial side of book production.

Mr. John Painter of W. H. Freeman and Company took a serious interest in the book from the start, and has been helpful in every way possible throughout the course of production. Mr. Fred Raab, also of W. H. Freeman and Company, edited the manuscript with understanding and skill.

Let me take this opportunity to express my sincere gratitude to all of these individuals.

Several book publishers kindly granted permission to use previously published illustrations in this book. The credit notes are given in the captions. The cooperation of the various publishing firms is gratefully acknowledged.

V.G.

Organismic Evolution

PART

I

INTRODUCTION

1

The Problem

Introduction

The world of living organisms exhibits several general features that have always aroused feelings of wonder in mankind. The first of these general phenomena is the great structural complexity of organisms. The second feature is the apparently purposive or adaptive nature of many of the characteristics of these organisms. The third striking general feature is the existence of a tremendous diversity of forms of life. The problem of biological complexity and adaptation is thus compounded by the fact that there are many diverse kinds of organisms with these properties in the world.

The questions evoked by these phenomena are obvious: How have complex organisms come into being? What forces have molded their adaptive characteristics? How has organic diversity originated and how is it maintained? To which can be added the special but relevant questions: What is the place of mankind in the organic world, and what is the ancestry of man?

In all ages man has sought intellectually satisfying answers to these questions. In pre-scientific societies the explanations have taken the form of myths, some of which have been carried over into the world religions. The scientific explanations are embodied in the theory of evolution. Before entering into our discussion of evolutionary theory, however, let us outline the problems that this theory has to explain in somewhat greater detail.

Organic Diversity

There are some 3700 species of mammals and 8600 species of birds in the Recent fauna. About 20,000 species of Recent fishes have been described. The living vertebrates as a collective whole comprise about 42,000 known species.

The number of known Recent species rises to greater heights in some other dominant major groups: to about 107,000 in the molluscs, 286,000 in the flowering plants, and 750,000 in the insects. Estimates of the species diversity in the various major groups of organisms are summarized in Table 1.1. The table shows that the total number of known species of organisms in the modern world is approximately 1.5 million.

The Herculean task of taxonomic exploration and description is well advanced in the birds and mammals, but is far from finished in most other groups. Substantial numbers of marine invertebrates, flowering plants, and other groups remain to be described taxonomically. Ichthyologists estimate that the number of Recent fish species described is about 20,000, but that the total number of Recent fishes, described and undescribed, is close to 40,000. The unfinished business in insect taxonomy bulks larger still. Entomologists suggest that the described species of insects (ca. 750,000) represent only a small fraction, perhaps one-fifth or one-tenth, of the insect species that actually exist in the modern world.

By making some conservative assumptions regarding the proportion of known to unknown species in the various major groups, it is possible to arrive at rough estimates of the total existing species diversity. We have previously arrived at an estimate of at least 4.5 million species of modern organisms (Grant, 1963, pp. 81–82).

Two independent estimates have been made of the number of species, both living and extinct, that have existed on earth throughout geological time. The ranges of the two estimates overlap broadly. We conclude tentatively that the total organic diversity throughout the history of life is in the neighborhood of one to several billion species (see Simpson, 1952; Grant, 1963).

Adaptation

Many of the hereditary characters of organisms conform to some feature of their normal environment in such a way as to benefit the organism. Such characters are adaptive.

The eighteenth-century naturalist Buffon described many adaptive characters

TABLE 1.1

Estimated numbers of described Recent species in various kingdoms and major groups of organisms. [Figures for animal kingdom from Mayr (1969, pp. 11–12); those for other kingdoms from Grant (1963, p. 80).]

Kingdom and group	Approximate number of described species	
Animal kingdom		
Chordates	43,000	
Arthropods	838,000	
Molluscs	107,250	
Echinoderms	6,000	
Segmented worms	8,500	
Flatworms	12,700	
Nematodes and relatives	12,500	
Coelenterates	5,300	
Bryozoans and relatives	3,750	
Sponges	4,800	
Miscellaneous small groups	2,100	
Total		*1,043,900*
Plant kingdom		
Flowering plants	286,000	
Gymnosperms	640	
Ferns and fern allies	10,000	
Bryophytes	23,000	
Green algae	5,280	
Brown and red algae	3,400	
Total		*328,320*
Fungus kingdom		
True fungi	40,000	
Slime molds	400	
Total		*40,400*
Protistan kingdom		
Protozoans, plant flagellates, diatoms	30,000	*30,000*
Moneran kingdom (prokaryotes)		
Blue-green algae	1,400	
Bacteria	1,630	
Total		*3,030*
Viruses	200	*200*
Grand total		*1,445,850*

of birds, fish, and other animals in the *Histoire Naturelle*. A typical statement is quoted here to illustrate the historical usage (Buffon, 1770, 1808):

> As to the external structure of Birds, it is peculiarly adapted for swiftness of motion; it is . . . designed to rise in the air. . . . Wide adaption of means to ends [appears] in the configuration of the feathered race. . . .

Darwin's stock example of adaptation was the woodpecker. One statement in *The Origin of Species* will suffice (Darwin, 1859, Ch. 6):

> Can a more striking instance of adaptation be given than that of a woodpecker for climbing trees and for seizing insects in the chinks of the bark?

One can think of the whole adaptive character combination of woodpeckers in this connection: the chisel bill, the strengthened head bones and head muscles, the extensile tongue with barbed tip, the feet with sharp-pointed toes pointing forward and backward for clinging to a vertical surface, and the stout tail for propping up the body while clinging in a vertical position.

Another classical example of adaptations is found in the types of feet in different groups of birds. This example is summarized graphically in Figure 1.1. Here we see the climbing foot of a woodpecker (Figure 1.1C). Along with it are the perching foot of a warbler, the grasping foot of an owl, the walking and scratching foot of a quail, the wading foot of a heron, and the webbed swimming foot of a duck.

It is useful to distinguish between general adaptations and special adaptations (Simpson, 1953a, Ch. 6). The former fit the organism for life in some broad zone of the environment; the latter are specializations for some particular way of life. Thus the wing of birds is a general adaptation, while the chisel bill and clinging foot of woodpeckers are special adaptations. General adaptations are among the distinguishing characteristics of major groups of organisms.

General and special adaptations are certainly ubiquitous in the living world. Whether *all* characters are adaptive, or whether some non-adaptive characters occur with any substantial frequency, is another and moot question. The pendulum of biological opinion has swung back and forth on this question for over a century. In the recent past, some morphologists and taxonomists have questioned the adaptive significance of many supposedly trivial morphological characters, such as those used in diagnostic taxonomic keys. Today certain schools of biochemists and biomathematicians are suggesting that some protein variations and changes are non-adaptive.

The question as to the common occurrence of non-adaptive characters can never be settled conclusively, since it involves the proof of a universal negative,

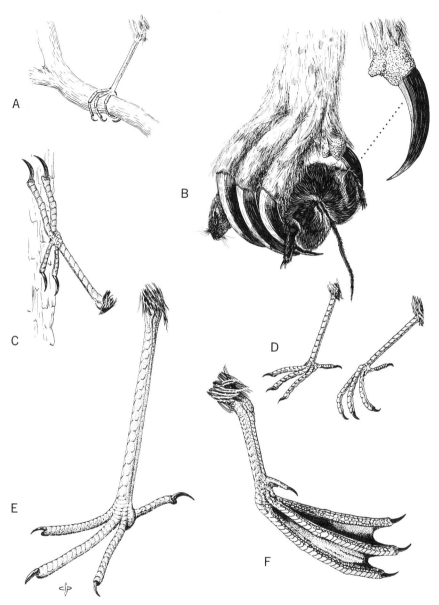

FIGURE 1.1

Different adaptive types of bird feet. Drawings not to same scale. (A) Perching foot (Audubon warbler, *Dendroica auduboni*). (B) Grasping foot with strong talons (Horned owl, *Bubo virginianus*). (C) Climbing foot (Acorn woodpecker, *Melanerpes formicivorus*). (D) Walking and scratching foot (California quail, *Lophortyx californicus*). (E) Wading foot (Green heron, *Butorides virescens*). (F) Swimming foot (Pintail duck, *Anas acuta*). (From V. Grant, *The Origin of Adaptations,* copyright 1963, Columbia University Press, New York; reproduced by permission.)

but some guiding thoughts can be mentioned. It is possible or even probable that some non-adaptive characters occur, and evolutionary mechanisms exist for establishing such characters. But their positive identification is very uncertain. Many human observers have a tendency to label as non-adaptive any character whose adaptive properties are not obvious to them. Such anthropocentric interpretations of nature can be misleading.

It would not be obvious at first glance that the difference between colored onions and white onions is adaptive. However, red and yellow onions are resistant to a fungus disease, a smudge *(Colletotrichum cincinans),* whereas white onions are susceptible to the same disease organism. The disease resistance is due to the presence in the onion bulbs of catechol and protocatechuic acid, which are toxic to smudge, and these compounds are associated with red or yellow pigments in the bulbs (Jones et al., 1946; Walker and Stahmann, 1955; Levin, 1971).

Evolutionary Explanations

The salient phenomena of organic complexity, organic diversity, and adaptiveness are explained scientifically as products of the process of evolution. Now there are two types of evolutionary explanation: the historical and the causal. The first traces the phylogenetic sequence leading up to the observed end result, whereas the second investigates the causal mechanisms involved. Both approaches are valid and both are necessary. This book is concerned primarily with the causes of evolution, but phylogenetic evidence is frequently and necessarily brought into the picture.

We have to deal with evolutionary phenomena at three broad levels. These are: evolutionary changes within populations (microevolution), evolution of races and species (speciation), and evolution of major groups (macroevolution). The three levels call for different methods of research, which in turn produce different types of evidence.

In this book we will discuss microevolution, speciation, and macroevolution separately and in that sequence. The order of presentation is important from a logical standpoint, since macroevolutionary interpretations, to be considered valid, must be consistent with microevolutionary findings.

Collateral Readings

Dobzhansky, Th. 1970. *Genetics of the Evolutionary Process.* Columbia University Press, New York. Chapter 1.

Mayr, E. 1970. *Populations, Species, and Evolution.* Harvard University Press, Cambridge, Mass. Chapter 1.

Simpson, G. G. 1944. *Tempo and Mode in Evolution.* Columbia University Press, New York. Reprinted by Hafner Publishing Co., New York, 1965. Chapter 6.

Simpson, G. G. 1953. *The Major Features of Evolution.* Columbia University Press, New York. Chapters 6, 7.

PART

II

MICROEVOLUTION

2

The Breeding Population

Introduction

Evolutionary change is a phenomenon of populations and population systems. Evolution in its most elementary form takes place within local breeding populations. This is microevolution. We can regard the local breeding population as the field for microevolution. It is necessary, therefore, to begin our survey of the processes of microevolution by considering the general characteristics of populations.

We cannot offer a precise, formal definition of the local breeding population that is applicable as a yardstick in all cases. This is due to the wide range of actual conditions. Some local populations are indeed sharply delimited. In other cases the local population has no definite boundary lines of its own, but is instead a component of a larger and more inclusive population system. It is useful to think of populations and population systems as forming a hierarchy from the randomly mating group to the species (more will be said of these units later). The local breeding population is, then, as its name implies, a population unit of local extent in this more or less continuous hierarchy.

Population Structure

The breeding population is a reproductive unit. In sexual organisms it is a community of individuals linked together by bonds of mating and parenthood (Dobzhansky, 1950). In asexual organisms the parenthood bonds remain present but the cross-linkages among individuals due to mating bonds are greatly reduced. We hesitate to say that mating bonds are absent entirely in asexual organisms, since many normally asexual organisms have some alternative, parasexual method of reproduction or have occasional reversions to sexuality. The population is usually an interbreeding group, whether the interbreeding is regular or occasional, and in all cases it is a reproductive entity.

The population is also an ecological unit. The individuals composing it are genotypically similar in their ecological tolerances, and occupy a definite area in some particular ecological niche or habitat.

Actual populations appear in a variety of sizes and shapes. Population size, reckoned as the number of adult breeding individuals in any generation, can vary from one or a few to many millions of individuals. Population structure has three main components: spatial configuration, breeding system, and migration rate. (The last of these will be discussed in Chapter 6.)

As regards the spatial characteristics of populations, three extreme conditions and their various intermediate conditions can be recognized. The extremes are: (1) large continuous populations, (2) small colonial populations (or populations conforming to the island model), and (3) linear populations.

Large continuous populations are exemplified by plains grasses, covering areas scores or hundreds of miles wide. Organisms with a colonial population system exist in a series of scattered, disjunct, and often small populations. Examples are furnished by terrestrial organisms in island archipelagoes, freshwater forms in series of lakes, mountaintop inhabitants in mountainous country, and organisms restricted to a particular soil or rock substrate that has a spotty distribution. Linear populations develop along rivers, coastlines, and similar habitats that are long and more or less continuous in one dimension but short and restricted in the other.

The various intermediate conditions are common. A large population may be continuous in some parts of its area but interrupted and semi-continuous in other parts. Likewise, the colonies belonging to an island system may be only semi-isolated rather than completely isolated. In the next section we will describe a concrete example of population structure in the giant sequoia *(Sequoiadendron giganteum)*, which exhibits a variety of conditions ranging from isolated colonies in the north to an interrupted forest belt in the south.

As regards the breeding system, the extremes are wide outcrossing and self-fertilization. Common intermediate conditions are: outcrossing between close neighbors; inbreeding by systems other than self-fertilization, such as brother-sister matings in animals; and mixtures of outcrossing and selfing, as in hermaphroditic but self-compatible flowering plants.

The spatial configurations and breeding systems are combined in numerous ways to give a great diversity of actual population structures. Thus a large continuous population may be composed of wide outcrossers, as in many wind-pollinated plains grasses, but it may also be composed of narrow outcrossers or inbreeders. Similarly, a small isolated colony may consist of either outcrossers or inbreeders. The type of population structure affects the variation pattern of the population, as will be brought out later.

Populations of Giant Sequoia

The giant sequoia *(Sequoiadendron giganteum)*, a wind-pollinated, outcrossing, coniferous tree, occurs in pine-fir forest at middle elevations (5000 to 8000 feet) on the west slope of the Sierra Nevada in California (Figure 2.1). Its distribution area forms a narrow band about 250 miles long (Figure 2.2A). Within this overall area, the giant sequoia occurs in a series of discrete and more or less disconnected populations (Figure 2.2B).

There are some 33 local populations of giant sequoia. At the turn of the century these had the sizes indicated in Table 2.1. The smaller populations are known as groves and the larger ones as forests. The forests had been extensively lumbered by the time of the surveys summarized in Table 2.1. The census figures given in the table thus refer to the size of the previous, undisturbed natural populations. The table reveals a wide range in population size, from groves of a few trees to forests consisting of thousands of individuals.

The northern and southern parts of the species area differ in population structure. In the northern part of its range, the giant sequoia usually exists in small, widely separated groves. Gaps of 10 to 50 miles between groves are not uncommon. In the southern area large forests are, or were, more common and the gaps between them narrower. In some cases the gaps are bridged by scattered trees so that the populations are linked into a semi-continuous belt.

The wide gaps between the northern groves correspond to valleys that were occupied by glaciers during the last glacial period. Presumably the giant sequoia formed larger and more continuous populations in the central Sierra Nevada prior to the last glaciation, was decimated and fragmented in that area by the glacial climate, and has been unable to regain its lost territory since. The populations in the southern Sierra were less severely affected by glaciation (Sudworth, 1908; Axelrod, 1959).

FIGURE 2.1
Giant sequoia (*Sequoiadendron giganteum*).

Polymorphism

Polymorphism is defined as the coexistence of two or more discontinuous segregating forms in a population, where the frequency of the rare type is not due to mutation alone (Ford, 1964, 1965). In other words, polymorphism is the variation in a local breeding population that exhibits distinct or sharp Mendelian segregation.

The term polymorphism as defined above excludes certain types of variation. It excludes purely phenotypic variation (because this is non-genetic); it excludes geographical variation (which does not exist in one population); it excludes polygenic variation (which does not segregate into discontinuous classes); and finally it excludes genetic variation due to new or recurrent mutation.

Polymorphic variation can be classified in several ways. A useful distinction is drawn between genetic polymorphism and chromosomal variation. The former is segregational variation in respect to homologous alleles of the same gene locus; the latter is polymorphism for chromosome types, such as sex chromosomes, or rearrangements, such as inversions.

Another distinction is that between transient and balanced polymorphism. In the case of transient polymorphism the diversity is temporary; it occurs while one form is in the process of replacing another under the controlling influence of natural selection. In balanced polymorphism the polymorphic types are more or less permanent components of the population, being preserved by selection in favor of diversity (Ford, 1964, 1965).

The various forms of polymorphism—genetic, chromosomal, transient, and balanced—are all common and widespread in the living world. Polymorphism is a virtually universal feature of populations in sexual organisms. An example involving the blood types in man will be described in the next section.

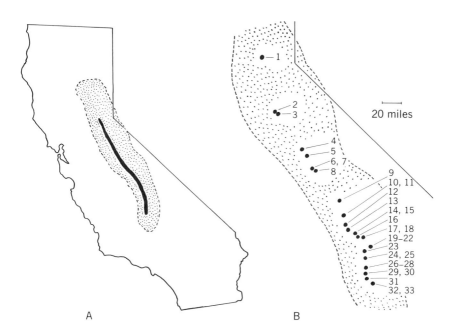

FIGURE 2.2
Geographical distribution of *Sequoiadendron giganteum*. (A) Map of California, showing distribution area of *Sequoiadendron* (black) in the Sierra Nevada (stippled). (B) Central and southern Sierra Nevada enlarged. Populations of *Sequoiadendron* are indicated by dots and numbered as in Table 2.1.

TABLE 2.1

Populations of giant sequoia *(Sequoiadendron giganteum)* on the west slope of the Sierra Nevada, California. (Sudworth, 1908; Jepson, 1909.)

Grove or forest	Size of area, acres	Number of trees
NORTHERN AREA		
1. North Grove	—	6
2. Calaveras Grove	51	101
3. Stanislaus Forest	1,000	1,380
4. Tuolumne Grove	10	40
5. Merced Grove	20	33
6. Mariposa Grove, A	—	365
7. Mariposa Grove, B	—	182
8. Fresno Forest	2,500	1,500
SOUTHERN AREA		
9. Dinky Grove	50	170
10. Converse Basin Forest	5,000	12,000
11. Boulder Creek Forest	3,200	6,450
12. General Grant Grove	2,500	250
13. Redwood Mountain Forest	3,000	15,000
14. North Kaweah Forest	500	800
15. Swanee River Grove	20	129
16. Giant Forest	8,000	20,000
17. Redwood Meadow Grove, A	50	200
18. Redwood Meadow Grove, B	10	80
19. East Kaweah Forest, A	1,500	3,000
20. East Kaweah Forest, B	20	80
21. Mule Gulch Grove	25	70
22. Homer's Peak Forest	5,500	1,500
23. South Kaweah Grove	160	300
24. North Tule River Forest	3,600	3,500
25. Middle Tule River Forest	15,000	5,000
26. Pixley Grove	850	500
27. Fleitz Forest	4,000	1,500
28. Putnam Mill Forest	4,000	900
29. Kessing Forest	2,800	700
30. Indian Reservation Grove	1,500	350
31. Deer Creek Grove	300	100
32. Freeman Valley Forest	1,000	400
33. Kern River Grove	700	200

The phenomenon of polymorphism leads us to the concept of the gene pool, which in turn gives us another way of looking at the local breeding population. Consider a population that is polymorphic for a gene A and contains the alleles A_1, A_2, and A_3. This population will form the diploid genotypes A_1A_1, A_1A_2, A_2A_2, etc., and these genotypes will be the observed units in any sampling of individuals, but clearly the underlying genetic polymorphism is a fundamental feature of the population in question. The population can be said to have a gene pool consisting of A_1, A_2, and A_3 alleles. Furthermore, these alleles will occur at characteristic frequencies in the gene pool; let us assume that their frequencies are 60%, 30%, and 10%, respectively. Thus the population can be described quantitatively in terms of the types and frequencies of genes in its gene pool.

It should be noted here that the gene pool concept is broader than the polymorphism concept. The gene pool of a population is made up of all of its genes. Thus the gene pool of our hypothetical population might be polymorphic for gene A, harbor a rare mutant allele of another gene B, and be monomorphic for other genes C and D.

The local breeding population can now be defined, or at least characterized, as a group of individuals sharing in a common gene pool (Dobzhansky, 1950). The individuals constituting the population in any given generation are the various genotypic products of the gametes drawn from its gene pool in the preceding generation.

Polymorphism in Human Blood Groups

Human individuals vary in their reactions to blood transfusions. Some transfusions result in agglutination or clumping of the red blood cells, whereas other transfusions do not. The agglutination reaction depends on the immunological relationships between the antigens in the red cells and the antibodies in the serum of the two bloods.

On the basis of the type of antigen present, four blood groups are recognized (A, B, AB, and O). Individual humans fall into one or the other of these four phenotypic classes. A person belonging to blood group A can donate blood to another person of type A without causing agglutination. Similarly, type B blood can be transfused into type B blood without agglutination. But transfusion of A into B, or B into A, results in heavy agglutination. It is not necessary to describe the reactions resulting from all combinations of blood types here (see Stern, 1960; Race and Sanger, 1962). In general, agglutination results from transfusions between members of different blood groups.

The blood types are determined by a series of three alleles: I^A, I^B, and I^O.

The allele I^O is recessive to I^A and I^B (and is sometimes written i). The alleles I^A and I^B are codominant. The six diploid genotypes of the three alleles give rise to the four phenotypic blood groups of the ABO series, as indicated in Table 2.2. Actually there are different but immunologically similar isoalleles of I^A (I^{A1}, I^{A2}, etc.), and therefore more than six possible genotypes, but we can ignore these fine differences in I^A for our present discussion.

TABLE 2.2
Genetics of ABO blood groups.
(Stern, 1960.)

Genotype			Phenotype (blood group)
$I^A I^A$	and	$I^A I^O$	Type A
$I^B I^B$	and	$I^B I^O$	Type B
$I^A I^B$			Type AB
$I^O I^O$			Type O

Human populations are generally polymorphic for the ABO blood groups. The frequencies of the blood-group phenotypes and of the underlying gene alleles are known for hundreds of local populations throughout the world. The allele frequencies in three populations are listed below (data from Mourant, 1954):

Uppsala, Sweden	$0.319\ I^A + 0.079\ I^B + 0.603\ I^O$
Punjab, India	$0.181\ I^A + 0.259\ I^B + 0.560\ I^O$
Navajo Indians, New Mexico	$0.133\ I^A + 0.000\ I^B + 0.867\ I^O$

These examples illustrate the point that human populations are alike in being polymorphic for the ABO series, but differ in their allele frequencies. Each local population has a gene pool with its own characteristic composition of I alleles.

The local populations are parts of larger, regional, racial groupings. Related local populations in the same region tend to have slightly different gene pools. Thus the frequency of I^A is 31.9% in the Uppsala, Sweden, population, as noted above, and is 28.4% in Falun, Sweden. Conversely, consistent differences in allele frequencies appear between geographical races.

Almost all native human populations in Western Europe have a high frequency of I^A and a low frequency (under 10%) of I^B. Central Asia shows a high frequency (20–30%) of I^B. Among American Indians, I^O is high in frequency, whereas I^B is either low or absent (Mourant, 1954). The polymorphic balance shifts in passing from one geographical area to another. We will return to the geographical distribution of the ABO blood groups in Chapter 13.

It is very interesting that parallel polymorphic variation in the ABO groups is found in the great apes. The chimpanzee is known to possess A and O blood types. The orangutan and gibbon have A, B, and AB types (Mourant, 1954; Wiener and Moor-Jankowski, 1971). The ABO polymorphism is thus more ancient than the human species itself, and is shared with man's closest relatives in the primate order.

There are several other systems of blood groups in man: the Rh system, MN system, and others. Human populations are polymorphic for these systems too (Race and Sanger, 1962). The polymorphic variation in the Rh and other systems seems to be independent of that in the ABO series.

Enzyme Polymorphism

The relatively new technique of gel electrophoresis makes it possible to detect polymorphisms in enzymes and some other proteins that could not be detected by ordinary genetic methods. The electrophoretic method makes use of the fact that different enzymes differ in mobility in an electric field. A tissue extract is put in a gel and exposed to an electric potential to set up an electric field. The different types of enzymes in the gel then become physically separated, and, when the gel is stained, show up as distinct bands. Fine differences between organisms in their gene-controlled enzymes can be revealed in this way.

Application of the electrophoretic technique to samples of natural populations has produced surprising results. Unexpectedly high levels of polymorphism have been found in populations of various species of organisms. The proportions of the enzyme loci assayed that turn out to be polymorphic are listed below (from summaries of Gottlieb, 1971, and Lewontin, 1973):

	Number of loci assayed	Percent polymorphic
Homo sapiens	71	28
Mus musculus (house mouse)	40, 41	20–30
Peromyscus polionotus (white-footed mouse)	32	23
Drosophila melanogaster	19	42
Drosophila pseudoobscura	24	43
Drosophila persimilis	24	25
Drosophila willistoni	20, 28	81–86
Limulus polyphemus (horseshoe crab)	25	25

Polymorphism is much more extensive, and involves more elements of the genotype, than was suspected on the basis of the classical approaches (Hubby and Lewontin, 1966; Lewontin and Hubby, 1966; Lewontin, 1974).

The Population Concept

The concept of genetically variable populations as reproductive units is by no means obvious. This concept did not exist in biology during the eighteenth and early nineteenth centuries, not does it exist in some branches of biology even today. According to Mayr (1972) it was introduced into biology by Darwin in 1859. The population concept was one of the elements in the Darwinian revolution in scientific thought.

The population concept stands in contrast to essentialism. Essentialism is the view that the observed things in the world are the expressions of an underlying essence. The things appear in varying forms, but the essence is immutable. The members of a class of objects, including the individuals in a population, are the variable expressions of the same essence.

Essentialism in one version or another was the traditional philosophy in Europe. Platonic philosophy, Christian theology, and philosophical idealism represented different versions of essentialism. Essentialism naturally dominated the thinking in the early history of biology. Here it took the form that Mayr (1957a, 1957b, 1972) has called typological thinking. This is the view that individual organisms are the imperfect and hence variable manifestations of the archetype of the species to which they belong.

Typological thinking is an obstacle to understanding evolution, which requires population thinking instead, since evolution is a change in the genetic composition of populations. A great but subtle accomplishment of the Darwinian revolution, according to Mayr (1972), was the replacement of typological thinking by population thinking in biology. The introduction of the population concept removed an old and strong obstacle to gaining an understanding of natural selection in particular and evolution in general.

3

The Statics of Populations

The Hardy-Weinberg Law

The gene pool of a local population normally contains various polymorphic genes in addition to the monomorphic ones. Furthermore, the polymorphic genes will exhibit certain definite allele frequencies in any given generation. Thus for a gene A present in two allelic forms, A and a, the gene pool in one generation might consist of 70% A alleles and 30% a alleles. The question arises: What are the expected allele frequencies in the next generation?

In a diploid population these alleles will be carried in the homozygous and heterozygous genotypes AA, aa, and Aa. These genotypes will also be found in certain proportions in any given generation. They are the parents of the genotypes in the next generation. The related question then arises: What are the expected proportions of genotypes in the second and later generations?

The expected allele frequencies and genotype frequencies are given by the Hardy-Weinberg law. This law is operative under the following conditions. The population is assumed to be large, so that errors of sampling do not affect the frequencies significantly from generation to generation. The individuals in the population are assumed to contribute equal numbers of functioning gametes; in other words, the various genotypes are equally successful in reproduction. Finally, random mating is assumed to prevail in the population.

Random mating, or panmixia, can be defined either in terms of individuals or in terms of gametes, with equal results. Considered from the standpoint of individuals, random mating occurs when individuals with different genetic constitutions mate without regard to their genotype. For example, an *AA* female might mate with *AA, Aa,* or *aa* males without any preference for one type of male.

The condition can be defined somewhat more precisely in terms of the array of gametes in the gamete pool. Random mating in this sense means that any given female gamete has a chance of being fertilized by any type of male gamete, and that this chance is directly proportional to the frequency of that type of male gamete in the gamete pool. In short, gametes carrying different alleles combine in pairs in proportion to their respective frequencies in the gamete pool. The individuals constituting the population in any generation then represent the products of different pairs of gametes drawn at random from the gamete pool of the preceding generation.

The Hardy-Weinberg law makes two statements about a population that conforms to the foregoing conditions: (1) The allele frequencies will tend to remain constant from generation to generation. (2) The genotypes will reach an equilibrium frequency in one generation of random mating and will remain at that frequency thereafter. These two generalizations will be illustrated in the following sections.

Allele Frequencies

We will illustrate the principle of constancy in allele frequencies by means of a numerical example. Assume that a diploid population polymorphic for *A* contains the following proportions of genotypes in the starting generation: 60% *AA,* 20% *Aa,* and 20% *aa.* We will trace the *A* alleles step by step through two generations.

(1) Allele frequencies in first generation. Since the genotype frequencies are given as

$$0.60\ AA + 0.20\ Aa + 0.20\ aa$$

the allele frequencies (q) in the same generation must be

$$q\,A = \frac{0.60 + 0.60 + 0.20}{2} = 0.70$$

$$q\,a = \frac{0.20 + 0.20 + 0.20}{2} = 0.30$$

(2) Gamete pool of first generation. The individuals are assumed to be equally fecund. Therefore the diploid individuals will produce haploid gametes in the proportions 70% A and 30% a. The allele frequencies in the gamete pool are the same as those in the original gene pool.

(3) Random mating. The gametes are drawn at random to form the zygotes of the second generation. The paired combinations are

♀ gametes		♂ gametes
0.70 A	×	0.70 A
0.70 A	×	0.30 a
0.30 a	×	0.70 A
0.30 a	×	0.30 a

(4) Zygotic frequencies in second generation. The products of the above system of random mating are

$$0.49\ AA$$
$$0.21 + 0.21 = 0.42\ Aa$$
$$0.09\ aa$$

The zygotes are assumed to be equally viable. Therefore these figures give the expected equilibrium frequencies of genotypes in the second generation.

It may be noted that the population was not in equilibrium as regards genotype frequencies in the first generation, but reached that equilibrium condition in just one generation of random mating.

(5) Allele frequencies in second generation. The gene pool of the second generation can be seen to contain the two alleles in the following frequencies:

$$q\,A = \frac{0.49 + 0.49 + 0.42}{2} = 0.70$$

$$q\,a = \frac{0.42 + 0.09 + 0.09}{2} = 0.30$$

Thus the allele frequencies are the same now as they were in the first generation.

The Hardy-Weinberg Formula

The Hardy-Weinberg formula provides a short, direct method of calculating the expected frequencies of genotypes in a random-mating population from the known allele frequencies in the gene pool.

Consider again a gene pool containing two alleles of a gene A. Let p represent the frequency of allele A and q the frequency of allele a (where $p + q = 1$). The random combinations of A and a gametes will then produce zygotes in the proportions given by the expansion of the binomial square $(p + q)^2$. The equilibrium frequency of genotypes, in other words, will be

$$p^2 \; AA$$

$$2 \, pq \; Aa$$

$$q^2 \; aa$$

We can apply this formula to our previous example (where q is used in a slightly different sense), and arrive at the same results by a simple operation. That is, if $p = 0.70$, the equilibrium frequency of AA is $p^2 = 0.49$.

Where the polymorphism involves a series of multiple alleles, the Hardy-Weinberg formula takes the form of a polynomial square. For three alleles $(A_1, A_2,$ and $A_3)$, whose frequencies are p, q, and r, so that $p + q + r = 1$, the equilibrium frequency of genotypes is given by the trinomial square $(p + q + r)^2$.

Effects of Inbreeding

Genotype frequencies quickly reach and then remain at an equilibrium condition under random mating in a large population. Departures from random mating bring about deviations from the Hardy-Weinberg equilibrium of genotypes. Let us consider another simple hypothetical example.

Assume that a population in generation 0 consists entirely of Aa heterozygotes. Reproduction is by self-fertilization. In generation 1 the population will contain 25% AA, 50% Aa, and 25% aa. The heterozygous class continues to decay at a regular rate in subsequent generations. After seven generations of selfing (i.e., in inbred generation 7), the population will contain almost 50% AA and almost 50% aa.

The formula giving the proportion of homozygotes derived from a heterozygous ancestor, for a single gene (A), after m generations of self-fertilization, is

$$\frac{2^m - 1}{2^m}$$

As applied to the case at hand, with seven generations of selfing, this formula gives the following expected genotype frequencies in inbred geeneration 7:

$$127/128 \ (AA + aa)$$
$$1/128 \ Aa$$

Inbreeding thus alters the genotype frequencies in the direction of a preponderance of homozygotes. But another aspect should be noted: the allele frequencies are not affected by the inbreeding. In our hypothetical example the allele frequencies were 50% *A* and 50% *a* in generation 0; and they remain the same in inbred generation 1 and again in generation 7.

Conclusions

In a large, polymorphic, panmictic population composed of equally viable and equally fertile individuals, the various homozygous and heterozygous genotypes will quickly reach certain equilibrium frequencies, which are dependent upon the existing allele frequencies. The genotype frequencies, having reached the Hardy-Weinberg equilibrium, will then remain constant during all succeeding generations of random mating.

Allele frequencies tend to remain constant from generation to generation in a large polymorphic population composed of equally viable and equally fertile individuals. This constancy is not dependent upon random mating. The aspect of the Hardy-Weinberg law dealing with allele frequencies is therefore more general than that dealing with genotype frequencies, and it is also more basic for evolutionary theory.

It is useful to examine the special conditions required for the operation of the Hardy-Weinberg law. These conditions exclude any factors affecting gene frequency other than the process of gene reproduction itself. Insofar as the level of variation in the gene pool is determined by gene reproduction per se, that level remains steady and unchanging through successive generations.

This is the statics of populations. Change in the level of variation of populations—the dynamic side of the question—requires the action of specific factors or forces. These are the forces of microevolution, to be considered next.

4

The Dynamics of Populations

Microevolution Defined

The level of variation in a large gene pool does not change by itself, as noted in the preceding chapter. Yet it is a fact of observation that local populations do undergo changes in allele frequencies during successive generations. The gene pool does change with time.

We can define microevolution as a systematic change in the frequencies of homologous alleles, chromosome segments, or chromosomes in a local population; that is, microevolution is any increase or decrease in the frequency of a variant form in the gene pool that continues to occur over a sequence of generations.

Examples of microevolutionary changes in natural populations are legion. It will suffice for our present purpose to describe one good example in *Drosophila pseudoobscura*.

Microevolutionary Changes in *Drosophila pseudoobscura*

Drosophila pseudoobscura is characterized by variation in the gene arrangement in a segment of chromosome III. The segment in question differs by inversions in different strains of the fly. The various inversion types are designated by names and letter symbols: Standard (*ST*), Arrowhead (*AR*), Chiricahua (*CH*), Tree Line (*TL*), Pikes Peak (*PP*), etc. There are 16 such inversion types in the species. These can be identified by their characteristic banding patterns in the salivary-gland chromosomes of the larvae.

Most natural populations of *Drosophila pseudoobscura* in western North America are polymorphic with respect to chromosome III. The polymorphic populations contain two or more inversion types, and consist of their various homozygous and heterozygous combinations (e.g., *ST/ST*, *CH/CH*, *ST/CH*). In any given population the inversion types tend to occur in certain average annual frequencies.

The common inversion types in *D. pseudoobscura* populations in the Sierra Nevada of California are *ST*, *AR*, and *CH*. Four other inversions occur at low frequencies in this area (*TL, OL, SC, PP*). Our story centers on one of these rare types (*PP*) in a particular local population (Mather). The Mather population, in the Yosemite region of California, has been studied intensively by Dobzhansky and his co-workers over a period of many years (see Dobzhansky, 1956, 1958, 1971).

The Pikes Peak inversion type was present but exceedingly rare in Sierran populations of *Drosophila* in and before 1945, and was unknown in the Mather population at that time. In 1946 *PP* appeared in the Mather population at the very low frequency of 0.3%. During the course of the next decade (1947–1957) *PP* rose rapidly in frequency to highs of 10 and 12%, as shown in Table 4.1. It fluctuated at somewhat lower but moderate frequencies during the following years (1959–1965). Since then it has dropped to a fairly low level (Table 4.1).

The rise in frequency of *PP* in the Mather population was accompanied by a decline of the formerly common *CH* to unprecedented lows of 2–6%. In recent years *CH* has come back up to its earlier level (Table 4.1). Corresponding adjustments took place in the proportions of *ST* and *AR* during the same period (Dobzhansky, 1971).

The spectacular increase in frequency of the *PP* inversion type was not confined to the Mather population alone. Parallel increases in *PP* were observed in other separate populations of *D. pseudoobscura* in scattered localities throughout California and Arizona. Thus in the San Jacinto Mountains in southern California, *PP* appeared for the first time in 1952, rose to peaks of 10 and 11% in the next few years, and declined again in the 1960s (Dobzhansky, 1971). The trends involving *PP* were general over a wide geographical area.

The causes of the observed microevolutionary changes are unknown. Enough is known to rule out some possible factors. Polymorphic laboratory populations of *Drosophila pseudoobscura* reared under favorable conditions conform to

TABLE 4.1
Frequency of two inversion types (*PP* and *CH*) in a population of *Drosophila pseudoobscura* at Mather, California, over a 28-year period. (Dobzhansky, 1971; Anderson et al., 1975.)

Year	Frequency of *PP*, %	Frequency of *CH*, %	Number of chromosomes assayed
1945	0.0	17	308
1946	0.3	17	336
1947	0.7	20	425
1950	3	17	812
1951	5	11	856
1954	12	13	1312
1957	10	4	316
1959	4	11	298
1961	6	3	350
1962	9	2	450
1963	7	6	446
1965	6	11	534
1969	2	3	312
1971	3	12	390
1972	6	17	576

Hardy-Weinberg expectations as regards the proportions of the inversion types and the corresponding diploid genotypes. Experimental evidence confirms theory in ruling out any internally directed changes in this case. The possible effects of chance are ruled out by the large size of the populations themselves and of the population samples studied. Mutation pressure is ruled out by the fact that parallel changes have taken place in widely separated local populations. Natural selection is almost certainly responsible for the observed changes. However, the selective factor involved remains unidentified despite extensive efforts to determine it experimentally (Dobzhansky, 1956, 1958, 1971).

The Primary Evolutionary Forces

The factors that bring about changes in gene frequencies or chromosome frequencies can be designated as the primary evolutionary forces. Four such forces are known. They are: mutation, gene flow, natural selection, and genetic drift.

Assume that a population gene pool consists predominantly of A_1 alleles and secondarily of A_2 alleles. The original gene frequencies can be altered by each one of the foregoing evolutionary forces, as follows.

The A_1 allele may mutate to A_2, occasionally or repeatedly, thereby increasing the frequency of the latter. Individuals or gametes carrying A_2 may migrate into the population from some other population in which A_2 is more common. Such migration or gene flow will also alter the pre-existing gene frequencies in the recipient population.

The carriers of A_1 and of A_2 may differ in phenotypic characteristics affecting their ability to survive and reproduce and contribute offspring to the next generation. If individuals carrying A_2 are superior to those carrying A_1 in these respects, A_2 will gradually increase in frequency. This is the force of natural selection, which is the most important of the factors controlling gene frequencies in natural populations.

Finally, the frequencies of A_1 and A_2 may shift significantly by chance alone in small populations. Genetic drift refers to the random component in gene frequency changes.

Each of the evolutionary forces varies in the intensity of action. The intensity can be quantified. Mutation rate (u) can vary from 0 to 1, where 0 is complete gene stability and 1 is complete instability of the gene. Similarly, the rate of gene flow can range from $m = 0$ (no migration) to $m = 1$ (complete swamping).

The selection coefficient (s) measures the average increase per generation of one allele relative to other, competing alleles. This coefficient can vary from 0 to 1, where $s = 0$ means no selection and $s = 1$ means complete gene replacement in a single generation.

The possibility of the action of genetic drift is expressed by a relationship between population size (N) and the other variables. Drift can be effective in controlling gene frequencies when N is low relative to s, m, and u.

Interactions Between Evolutionary Forces

The first two forces mentioned above, mutation and gene flow, produce variability. The second two forces, selection and drift, sort out this variability. The variation-producing forces start the microevolutionary process, and the variation-sorting forces go on to establish the variant types in new frequencies. Evolutionary change within populations can be viewed as a resultant of the opposing forces that produce and sort out genetic variations.

One of the old theories of evolution, the theory of orthogenesis, which still has some adherents today, postulates that evolutionary changes are directed mainly by the mutation process. There are two fatal arguments against this view.

First, mutational changes occur more or less at random with respect to the adaptive requirements of the organism, and could not alone bring about the adaptive characteristics actually seen in organisms. The additional factor of natural selection is necessary to take the process "from the chemical level of mutation to the biological level of adaptation" (Darlington, 1939, p. 127).

In the second place, selection happens to come into action after mutation in the time sequence of microevolution. No matter how strong or how oriented the mutation pressure may be in any population, its effects for good or bad are always subject to censure by natural selection. Selection always has the last word.

It is equally true that selection could not act without the mutation process to create a supply of new genetic variations. Microevolution is due not to the operation of any single force, but to the interaction between two or three or four forces.

The quantification of the primary evolutionary forces opens the way for a quantification of their interactions. The latter is a relatively simple matter in theory, and is sometimes feasible in synthetic experimental populations, but is extremely difficult to carry out in natural populations because of the numerous uncontrolled factors involved. Nevertheless, some approaches to the problem have been made in real populations.

Interaction between mutation and selection is exemplified by hemophilia in man. The bleeder disease, hemophilia, involving failure of clotting of blood, is usually fatal at an early age. Several types of hemophilia are known; we will restrict this discussion to the classical or A-type hemophilia. Hemophilia A is due to a recessive allele (hh) of the sex-linked or X-linked gene Hh. Heterozygous female carriers (Hh/hh) produce some sons of the constitution h/O who exhibit the disease (Stern, 1973).

The hemophiliac males mostly, though not invariably, die before reaching reproductive age, and therefore the hh allele has a low selective value in the semilethal range. The hh alleles are continuously being eliminated from human populations by selection. Yet they persist in the population at a low frequency. Their persistence is attributed to recurrent mutation from Hh to hh, the estimated mutation rate being $u = 0.00002-0.00004$ (Stern, 1973).

The persistent low frequency of hh in human populations thus represents an equilibrium between recurrent mutation and strong counterselection. This at least is a first approximation to the problem. There is some evidence that heterozygous females have a higher selective value than normal homozygous females for one of the other types of hemophilia (Rosin, Moor-Jankowski, and Schneeberger, 1958). If a parallel situation exists with respect to hemophilia A, it would tilt the selection-mutation equilibrium in the direction of a higher equilibrium frequency of hh in the population than would be expected without this complicating factor.

The production of variability by a combination of mutation and gene flow is illustrated by the land snail, *Cepaea nemoralis*, in Western Europe. The snail colonies are generally polymorphic for the presence or absence of brown bands on the shells. The phenotypic differences are controlled by a gene B present in two allelic forms; the dominant allele B determines bandless shells, and the recessive allele b, banded shells. The B gene mutates from bandless to banded and vice versa at an estimated rate of $u = 0.0001$–0.0005 (Lamotte, 1951).

The population structure of the European land snail differs in different areas. These differences affect the rate of gene flow. In the district of Aquitaine in France the colonies lie fairly close together and are frequently bridged by migrant individuals. Gene flow here is estimated to occur at the rate of $m = 0.003$–0.004. Gene flow is a more important source of population variability than gene mutation in Aquitaine. In the province of Brittany in France, by contrast, the colonies are widely spaced and only rarely connected by migrant individuals. Mutation is relatively more important and gene flow less so here than in Aquitaine (Lamotte, 1951).

This is by no means the whole story. The banding pattern in the snails is also controlled by selection and probably by drift, as will be discussed in a later chapter.

The action of genetic drift depends entirely on the interrelationship between the factors N, s, u, and m. For instance, drift can be an effective force controlling gene frequencies when $N < 1/2s$, if u and m are negligible. Conversely, moderate mutation rates and migration rates can nullify the action of drift. A most important interaction 'in nature is that involving the combination of drift and selection. These relationships will be described in more detail in Chapter 13.

The Population Genetics Approach to Evolution

The early Mendelian geneticists had considered the behavior of genes in family pedigrees. But the P, F_1, F_2, and F_3 of formal genetics are an artificial abstraction. In nature a mutant allele does not simply become united with the normal allele in an F_1 individual and become passed on to one-half of its gametes. Under natural conditions, outside the genetics laboratory or breeding plot, the mutation that develops in an individual organism does not remain within a family pedigree, but enters into the gene pool of a population. And there its fate, if it is either deleterious or beneficial, will be to change in frequency.

It follows that the genetics of populations cannot be studied properly without taking cognizance of the effects of natural selection on hereditary variations. The originally separate lines of thought of Mendelism and Darwinism must be merged.

This merger was made in the 1930s and led directly to the modern theory of evolution, or to the so-called synthetic theory of evolution. This theory began as a population-genetical approach to evolution. The titles of three classic works of the early period reflect the nature of the synthesis: *The Genetical Theory of Natural Selection* (Fisher, 1930); *Evolution in Mendelian Populations* (Wright, 1931); *Genetics and the Origin of Species* (Dobzhansky, 1937).

The population-genetical approach led directly to the solution of problems of microevolution. Furthermore, it turned out to be possible and useful to extend the same populational approach to other fields of biology, such as minor systematics and paleontology, thus elucidating the problems of speciation and macroevolution, and thereby broadening the base of modern evolutionary theory.

Collateral Readings

1. BASIC NINETEENTH-CENTURY WORKS ON ORGANIC EVOLUTION

> Darwin, C. 1859, 1872. *On the Origin of Species.* Ed. 1, 1859; ed. 6, 1872. John Murray, London. Reprints of ed. 1 by Watts & Co., London, 1950, and Harvard University Press, Cambridge, Mass., 1964. Various reprints available of ed. 6.
> Wallace, A. R. 1889. *Darwinism.* Macmillan, London and New York.

2. HISTORICALLY IMPORTANT WORKS THAT LAID THE FOUNDATIONS OF MODERN EVOLU-TIONARY THEORY (SYNTHETIC THEORY OF EVOLUTION)

> A. FROM THE STANDPOINT OF THEORETICAL POPULATION GENETICS

>> Fisher, R.A. 1930. *The Genetical Theory of Natural Selection.* Clarendon Press, Oxford. Second revised edition, Dover Publications, New York, 1958.
>> Haldane, J. B. S. 1932. *The Causes of Evolution.* Longmans, Green & Co., London. Reprinted by Cornell University Press, Ithaca, N.Y., 1966.
>> Wright, S. 1931. Evolution in Mendelian populations. *Genetics* **16**: 97–159. Reprinted in *Systems of Mating,* by S. Wright. Iowa State University Press, Ames, 1958.

> B. FROM THE STANDPOINT OF EXPERIMENTAL POPULATION GENETICS

>> Dobzhansky, Th. 1937, 1951. *Genetics and the Origin of Species.* Ed. 1, 1937; ed. 3, 1951. Columbia University Press, New York.

> C. EXTENDING THE THEORY TO ANIMAL SYSTEMATICS, PALEONTOLOGY, AND BOTANY

>> Mayr, E. 1942. *Systematics and the Origin of Species.* Columbia University Press, New York. Reprinted by Dover Publications, New York, 1964.
>> Simpson, G. G. 1944. *Tempo and Mode in Evolution.* Columbia University Press, New York. Reprinted by Hafner Publishing Co., New York, 1965.
>> Stebbins, G. L. 1950. *Variation and Evolution in Plants.* Columbia University Press, New York.

3. MODERN GENERAL WORKS ON EVOLUTIONARY THEORY

> Dobzhansky, Th. 1970. *Genetics of the Evolutionary Process.* Columbia University Press, New York.

Grant, V. 1963. *The Origin of Adaptations.* Columbia University Press, New York. Paperback edition, Columbia University Press, 1971.

Mayr, E. 1963. *Animal Species and Evolution.* Harvard University Press, Cambridge, Mass.

Mayr, E. 1970. *Populations, Species, and Evolution.* Harvard University Press, Cambridge, Mass.

Simpson, G. G. 1953. *The Major Features of Evolution.* Columbia University Press, New York.

Simpson, G. G. 1967. *The Meaning of Evolution.* Ed. 2. Yale University Press, New Haven, Conn.

White, M. J. D. 1973. *Animal Cytology and Evolution.* Ed. 3. Cambridge University Press, Cambridge and London.

5

Mutation

Introduction

A mutation is a sudden hereditary change in a phenotypic character caused by an abrupt structural and functional alteration in the genetic material. The genetic material is organized in a hierarchy of structural-functional units ranging from molecular sites within a gene to whole chromosomes and genomes. Correspondingly, there are different types of mutations from gene mutations to genome mutations.

Furthermore, sudden hereditary changes in phenotype can be caused by genetic processes other than structural changes in the genes or chromosomes. There are spurious as well as true mutations. The phenotypic changes do not in themselves reveal the genetic process involved. It is very difficult to distinguish between the various types of true and spurious mutations on the basis of direct observational evidence.

For our present purpose it is possible to disregard some of the complexities in the broad subject of mutations. The emphasis in this chapter is on some general features of gene mutations that are of primary evolutionary importance.

Gene Mutations

A gene mutation is an alteration that occurs in the nucleotide sequence within the limits of a gene and changes the mode of gene action. In other words, it is a molecular change in the gene that brings about an altered phenotypic effect. Suppose that a gene contains a triplet CTT at some point, which specifies the amino acid glutamic acid in a polypeptide chain. The triplet CTT could change to GTT by the substitution of a single nucleotide. The altered triplet now specifies glutamine instead of glutamic acid in the polypeptide product of gene action. The original and mutant protein molecules differ themselves and may well determine other second-order phenotypic differences.

The stability of genes through successive cell generations and individual generations, and hence the conservative aspect of heredity, are due to the exactness of the copying process during gene replication. But the copying process is not perfect. Copying errors occur occasionally. Gene mutations can be regarded as such errors of copying.

The new mutant allele then replicates itself faithfully until the next mutational change occurs. The net result of gene mutation is thus the existence of a pair or series of homologous alleles. Conversely, the presence of allelic variation in any gene goes back ultimately to the process of gene mutation.

Any gene in the genotype is presumably subject to mutation. At any rate, mutations do observably occur in genes controlling a wide diversity of characters. In *Drosophila melanogaster,* for example, there are wing mutants with slightly shriveled wings, greatly reduced wings, or no wings; eye-color mutants with white or vermilion eyes; bristle mutants of various sorts, and so on. Biochemical mutations affecting steps in metabolic pathways are well known in microorganisms and present though less easily studied in higher organisms. An array of leaf mutants in the wild tomato (*Lycopersicon pimpinellifolium*) is shown in Figure 5.1.

In terms of the magnitude of their phenotypic effects, gene mutations form a spectrum ranging from those with slight effects to those producing some major change in phenotype. The extreme types are called minor mutations and macromutations, respectively. The conspicuous but non-drastic mutations shown in Figure 5.1 are typical of the mid-region of the spectrum. Minor mutations are exemplified by mutant types in *Drosophila melanogaster* with statistically slight deviations from normal viability or normal bristle number. An example of a macromutation is the mutant *tetraptera* in *D. melanogaster* with four instead of two wings. It represents a departure from the two-winged condition characteristic of the family Drosophilidae and order Diptera.

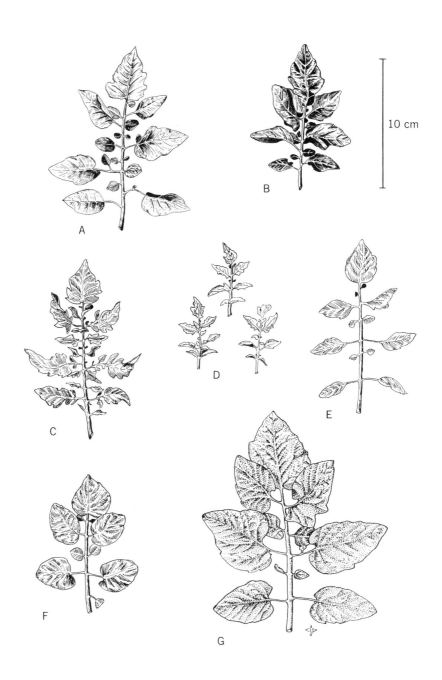

FIGURE 5.1
Mutant leaf forms of the wild tomato (*Lycopersicon pimpinellifolium*). (A) Normal form.
(B) Accumbens. (C) Bipinnata. (D) Diminuta. (E) Carinata. (F) Dwarf. (G) Bullosa.
(Redrawn and rearranged from Stubbe, 1960.)

In diploid animals and plants, a high proportion of new mutations are recessive, the wild-type genes being dominant. An important consequence of the recessive condition of many mutant alleles is that they are not exposed immediately and directly to the action of selection, but may be kept in storage for generations in the diploid population.

Mutation Rates

Spontaneous mutation rates for various genes in several organisms are presented in Table 5.1. Several points are worth emphasizing. First, there is a fairly wide range as to mutability between different genes in the same species, as indicated by the data given for man, *Drosophila,* and corn.

TABLE 5.1
Estimates of spontaneous mutation rates in several organisms.
(From various sources.)

Organism and gene	Average number of new mutants per 1 million gametes
Mouse *(Mus musculus)*	
Piebald	17
Pinkeye	8.5
Man *(Homo sapiens)*	
Dwarfness (achondroplasia)	40–120
Hemophilia A	20–40
Hemophilia B	5–10
Huntington's chorea	<1
Fruit fly *(Drosophila melanogaster)*	
Yellow body	120
Ebony body	20
Various lethals	10
Corn *(Zea mays)*	
Plant color	490
Color inhibitor	110
Sugary endosperm	2
Shrunken seeds	1
Waxy starch	<1
Bacteria	
Penicillin resistance in *Staphylococcus*	0.1
Phage resistance in *Escherichia coli*	0.03
Streptomycin resistance in *Escherichia coli*	0.0004

Second, the mutation rates in bacteria are much lower than those in higher organisms. The few bacterial genes listed in the table are representative of a larger body of data. The mutation rates for bacteria in the table lie in the range $u = 10^{-7}-10^{-10}$. The mutation rates for genes in corn are $u = 10^{-4}-10^{-6}$. *Drosophila,* man, and mouse have rates in the same range as corn. Bacterial genes seem to be more stable, in general, than genes in higher, multicellular organisms.

There are reasons for believing that at least some of the estimates of mutation rates in higher organisms are too high. Several possible sources of error are involved. One is the difficulty of distinguishing between true intra-genic mutations and rare recombinations of very closely linked genic units. The two events can lead to the same observable result, namely, an abrupt, true-breeding phenotypic change. Any large collection of mutations in a diploid organism, as observed at the phenotypic level, is likely to contain a fraction of undetected rare recombinations in addition to the true gene mutations, thus distorting the estimated mutation rate upward.

Another source of error is the undetected effect of regulatory genes on the gene expression of the supposedly mutable gene under consideration. The gene A may mutate from dominant to recessive ($A \rightarrow a$), and the homozygous recessive type aa may arise, and then apparently back-mutate to A. What is actually observed is an A phenotype in the inbred progeny of the usually true-breeding mutant type aa. Reverse mutation ($a \rightarrow A$) is one possibility. Another possibility is that aa is unchanged but its phenotypic expression is suddenly suppressed by an inhibitor gene at a different locus, and perhaps by a newly arisen mutation of the inhibitor gene. This process can easily lead to an overestimate of the mutation rate of A.

Even if the true gene mutation rates in higher organisms are lower by, say, one or two orders of magnitude than present estimates indicate, there would still be an adequate supply of mutational variations in populations. A moderate-sized population with an output of 10 to 100 million gametes would produce at least some new mutations each generation in an average gene.

Genotypic Control

A gene is known in *Drosophila melanogaster* that produces high mutation rates in other genes in the complement. This mutator gene is designated *Hi.* Flies homozygous for *Hi* have mutation rates 10 times higher than the normal rates; heterozygotes for *Hi* have 2–7 times the normal mutation rates. The gene *Hi* induces both visible and lethal mutations in many genes. It also causes inversions, a type of chromosome mutation (Ives, 1950; Hinton, Ives, and Evans, 1952).

We have described gene mutations previously as chance errors of copying during gene reproduction, and undoubtedly this is a correct view as far as it goes, but the evidence for mutator genes indicates that there is another aspect. The production of new mutational variation, which is important for the long-term success of the species in evolution, may not be left to chance entirely, but may be promoted by mutator genes. The mutation rate of a species may be, in part, a genotypically controlled component of its overall genetic system.

The closely related, tropical American species, *Drosophila willistoni* and *D. prosaltans*, have different mutation rates. The rates were measured for the class of lethal mutations on chromosome II and chromosome III. The mutation rates are as follows:

	Chromosome II	Chromosome III
D. willistoni	2.2×10^{-5}	3.0×10^{-5}
D. prosaltans	1.1×10^{-5}	2.1×10^{-5}

Drosophila willistoni, with the higher mutation rate, is common and widespread, occurring in many ecological niches, whereas *D. prosaltans* is rare and ecologically restricted. It is plausibly suggested that the high mutation rate of *D. willistoni,* by producing a supply of new variations, is a factor in its ecological versatility and hence in its abundance (Dobzhansky, Spassky, and Spassky, 1952).

Adaptive Value

Most new mutant types have a lower viability than the normal, or wild type. The reduction in viability ranges from a slight subvital condition to semilethal and lethal. A series of mutations in the X chromosome of *Drosophila melanogaster* were scored for viability. Ninety percent of the X-chromosome mutants had lower viability than normal flies, and 10% of the mutants were above normal, or supervital. The 90% fraction ranged from slightly subvital (45%) through intermediate degrees of reduced viability to semilethal (6%) and lethal (14%) (Timofeeff-Ressovsky, 1940).

More generally, the adaptive value of new mutants is usually reduced. Adaptive value includes fertility and the functional usefulness of morphological characters as well as physiological viability. Many mutant types are infertile whether or not they are normally viable. Morphological macromutations usually have impaired functional efficiency. Nearly 99% of a large sample of induced mutations in barley (*Hordeum sativum*) had a lowered adaptive value (Gustafsson, 1951).

These observations can readily be explained. The genes in the normal or wild-type genotype have all passed through many generations of natural selection; they have been screened for maximum adaptive value. Any changes in such genes would be expected to be, mainly, changes for the worse, in the same way that random tinkering with a clock is more apt to "impair its functional efficiency" than to result in improvements.

Gene mutations are frequently said to be "random" changes in the genes. The term random requires qualification in this context. Mutational changes may not, in fact, be random at the molecular level. Certain alterations in the nucleotide order in a DNA chain may occur more frequently than others. The so-called randomness of the mutation process has reference not to the molecular arrangement, but to the adaptive properties of the mutant genes. Mutations are random in the sense that they are not oriented in the direction of any present or future state of adaptation of the organism.

Nevertheless, a small fraction of the total array of gene mutations in genetically well-studied organisms prove to be superior to the standard type in one way or another. Between 0.1 and 0.2% of a sample of mutants in barley showed an increased yield in the standard or parental-type environment (Gustafsson, 1951).

A mutant type that is inferior in the standard environment may be adaptively superior in a different environment. The mutant *eversae* in *Drosophila funebris* has a reduced viability, 98% of that of wild-type flies, at 15°C, but superior viability (104%) at 24°C (Dobzhansky, 1951, p. 84). Six mutant types in the snapdragon *(Antirrhinum majus)* were all inferior to the parental strain in a normal greenhouse environment, but the mutants showed superior growth as compared with the parental strain in various abnormal greenhouse environments (Brücher, 1943; Gustafsson, 1951).

It is important to know that *some* new mutations are adaptively superior to the wild type, in either the standard or a new environment, for this indicates that the mutation process can be the starting point of evolutionary advances. The complementary fact, the high frequency of deleterious mutations, is also important. This latter fact rules out the possibility, still voiced in some quarters, that mutation pressure can be an orienting force in evolution.

Minor Mutations vs. Macromutations in Evolution

Most evolutionary geneticists have stressed the importance of minor mutations in evolution. A minority viewpoint, expressed by Goldschmidt (1940, 1952, 1953, 1955) and others, holds that macromutations are of primary importance in evolution. Controversies between these opposing viewpoints have developed in the past, but are unnecessary, since the two views are not mutually exclusive but on the contrary are complementary. Both minor mutations and macromutations play a role in evolution.

Minor mutations have certain distinct advantages as raw materials for evolutionary change. Each minor mutant produces only a slight phenotypic effect, for better or for worse. A slightly superior minor mutant allele can therefore be fitted into the pre-existing genotype without bringing about any drastic disharmonies. By employing a series of minor mutations at different loci, it is possible to built up to some adaptively quantitative effect without disrupting the functional efficiency of the organism during the interim stages.

Occasionally, a single mutation affecting a key structure or function can open up new possibilities for its possessor. Resistance to specific toxins is known to be caused by single-gene mutations in organisms as diverse as bacteria and mammals. The mutant toxin-resistant strain might be able to invade a toxic environment barred to the parental susceptible type. A mutant wingless fly might be favored in a windy island habitat where the normal winged form could not survive.

The single macromutation does not necessarily have to act alone. Its action can be regulated by a series of modifier genes with individually small effects. In such a case we have a combination of macromutations and minor mutations in evolution.

Direct evidence bearing on the question of the types of gene changes that are important in evolution is provided by genetic studies of interracial hybrids. The genetic evidence indicates that diverse gene systems play their respective parts in the differentiation of related races. Interracial character differences in plants and animals are commonly controlled by multiple gene systems, confirming the hypothesis of the evolutionary importance of minor mutations. But single-gene character differences are also found in sets of divergent natural populations. This situation is uncommon, however. A common type of gene system differentiating races in higher organisms is the combination of a major gene and several modifier genes. This type of gene system suggests an evolutionary change based on a conspicuous mutation associated with various minor mutations.

Chances of Survival of a New Mutation

Any particular mutant allele is rare when it first arises. It faces the problem of survival in the parental population. The Hardy-Weinberg law tells us that the new mutant allele will not increase in frequency by the process of gene reproduction per se. On the contrary, the mutant allele has a high probability of becoming extinct in the population by chance alone.

Fisher (1930, 1958) calculated the probability of survival of a new mutant allele appearing in a single individual in a large population. The probabilities are given in Table 5.2. We see that the chances of survival of the unique mutation decline markedly with the passage of generations. If the mutant allele is selectively neutral, it has a probability of 94% of becoming extinct by generation 31 and 98% by generation 127. If it has a small selective advantage (of 1%), its chances of becoming extinct are slightly lower (93% and 97% in generations 31 and 127) but still high.

TABLE 5.2
Probability of extinction of a mutation appearing in a single individual. (Fisher, 1930, 1958.)

Number of generations	Probability of extinction if:	
	Selectively neutral	1% selective advantage
1	0.368	0.364
3	0.626	0.620
7	0.790	0.782
15	0.887	0.878
31	0.941	0.931
63	0.970	0.959
127	0.985	0.973

Collateral Readings

Lindsley, D. L., and E. H. Grell. 1968. *Genetic Variations of* Drosophila melanogaster. Carnegie Institution of Washington, Washington, D.C., Publ. 627.
Neuffer, M. G., L. Jones, and M. S. Zuber. 1968. *The Mutants of Maize.* Crop Science Society of America, Madison, Wis.

6

Gene Flow

Introduction

A population may acquire a new allele (say, A_2 of gene A) either by mutation occurring in some individual of the same population, or by immigration of a carrier of this allele from another population. The latter process is gene flow.

Migration is a factor affecting population variation only if the foreign individuals are different from the population they enter. An immigrant carrying an A_1 allele into a population with an A_1 gene pool has no effect on that population's variability; an immigrant carrying an A_2 allele into the same recipient population does change its genetic composition.

The carrier of the new allele A_2 must have obtained it from some prior event of mutation. In the history of the species, A_2 presumably diverged by mutation from the ancestral form of the gene, A. Gene flow as a source of variation thus depends on the previous existence of mutations. We can regard gene flow as a delayed effect of the mutation process.

Nevertheless, the direct and immediate source of some variations in a population at any given moment may be the immigration of carriers of different alleles from other populations. Particular mutant alleles might never appear in the population, but might flow in from neighboring populations. The *ultimate* source of the variation in the recipient population is then gene mutation, occurring elsewhere at some previous time; but the *effective* cause of the variation, here and now, is gene flow.

Rate of Immigration

The rate of gene flow (m) varies quantitatively from 0 to 1, where $m = 0$ signifies no immigrant alleles, and $m = 1$ signifies swamping by immigrants in a single generation. In general,

$$m = \frac{\text{number of immigrants per generation}}{\text{total}}$$

In the case of one immigrant allele for every 1000 gametes in each generation, $m = 0.001$.

Let the frequency of an allele among the natives be symbolized by q_o and among the immigrants by q_m. Then the allele frequency (q) in the mixed population is (following Falconer, 1960)

$$q = m(q_m - q_o) + q_o$$

The change in gene frequency per generation (Δq) due to gene flow is (Falconer, 1960)

$$\Delta q = m(q_m - q_o)$$

For example, if the frequency of an allele among the natives is $q_o = 0.3$, and among the immigrants it is $q_m = 0.5$, and the migration rate is $m = 0.001$, then the change in allele frequency resulting from one generation of immigration is

$$\Delta q = 0.0002$$

The rate of change in gene frequency in a population due to gene flow thus depends on two factors: (1) the difference in allele frequency between the natives and the immigrants, and (2) the rate of immigration (Falconer, 1960).

Dispersal Patterns

The vehicles of gene flow are individual organisms in motile higher animals, free-swimming larvae in sessile marine animals, pollen and seeds in higher plants, and spores in lower plants and fungi. Dispersal is a normal part of the life history. All organisms, including sessile types, appear to have some free-living and motile stage in the life cycle when dispersal takes place.

Conversely, the vast majority of the dispersal units, including the individual organisms in motile animals, tend to remain close to their place of origin or home territory. Some degree of sedentariness is also normal. The two aspects— dispersal and sedentariness— work together to determine the spatial distribution of individuals and their genes from generation to generation.

The two complementary aspects are clearly shown in a study of the dispersal range of pine pollen. Pollen of Coulter pine *(Pinus coulteri)* was tagged with radioactive phosphorus and, after wind dispersal, was traced with a Geiger-Müller counter and by radioautographs. One-liter lots of the pollen were re-leased from inverted jars suspended at a height of 12 feet. Pollen release took place slowly on a fair day with gentle breezes. Pollen traps were placed at regular intervals along various radii from the point of release. The amount of pollen in the various traps could be estimated from the intensity of the radioactivity (Colwell, 1951).

The results are shown graphically in Figure 6.1. The main bulk of the pollen was dispersed downwind to a distance of 10–30 feet. Beyond this zone of maxi-mum concentration, the amount of dispersed pollen fell off rapidly. Only small amounts of pollen reached distances of 150 feet or more from the source point. The implication of these results is that a female cone in a pine forest will be swamped with pollen from neighboring trees, and will receive some pollen, but not very much, from trees standing a few hundred feet away (Colwell, 1951).

A local population of the Rusty lizard *(Sceloporus olivaceus)* in Texas was studied intensively over a six-year period. Most individuals of both sexes remain within 200–300 feet of the point of hatching. Some individuals, however, range more widely and up to 1650 feet from their place of origin. The tendency to dis-perse widely is somewhat more common in males than in females (Blair, 1960). The dispersal pattern of the Rusty lizard is shown graphically in Figure 6.2.

It has long been suspected that the difference between wide-ranging indi-viduals and more sedentary individuals in motile animals has a genotypic component. This suspicion has been confirmed recently in field voles *(Microtus pennsylvanicus* and *M. ochrogaster).* Protein genes used as genetic markers indicated that the actively dispersing voles differ genotypically from the non-dispersing voles. Also, the tendency to disperse is associated with a greater degree of aggressiveness (Krebs et al., 1973).

A study of dispersal designed to reveal rates of gene dispersion was carried out with *Drosophila pseudoobscura.* Genetically marked flies were released in nature and recaptured later at various distances from the point of release. As usual, most of the flies remained close to the starting point, but some dispersed far and wide. The results originally indicated that, ten months or several generations after release, 95% of the progeny of the released flies would be

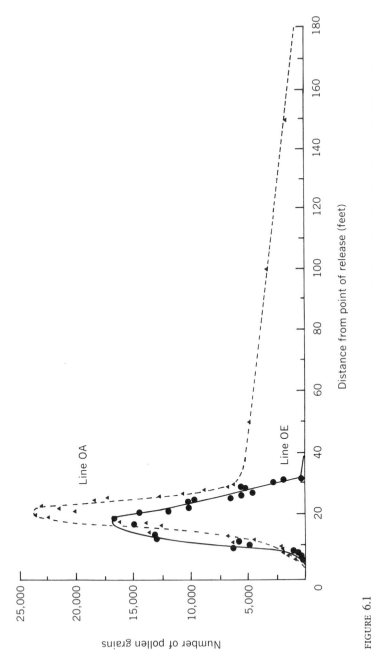

FIGURE 6.1
Amount of pine pollen transported by breezes to various distances from point of release. Line OA, directly downwind from pollen source. Line OE, radius 45° from line OA. (Colwell, 1951.)

found within a circular area with a radius of 1.76 km from the point of release (Dobzhansky and Wright, 1947). These studies have been repeated and re-analyzed recently, with results indicating that the dispersal rate of the flies is higher than originally estimated (Crumpacker and Williams, 1973; Dobzhansky and Powell, 1974; see also Johnston and Heed, 1976).

The pattern of much sedentariness combined with some long-range dispersal has been found to be widespread in nature. The examples described above are typical; other examples occur in mice, birds, honeybees, butterflies, insect-pollinated plants, bird-pollinated plants, and wind-borne seeds.

The population-genetical implication of this dispersal pattern is obvious. Interbreeding in an outcrossing organism will occur mostly between close neighbors. In the course of generations this will lead to a considerable amount of inbreeding despite an outcrossing breeding system. The close inbreeding is

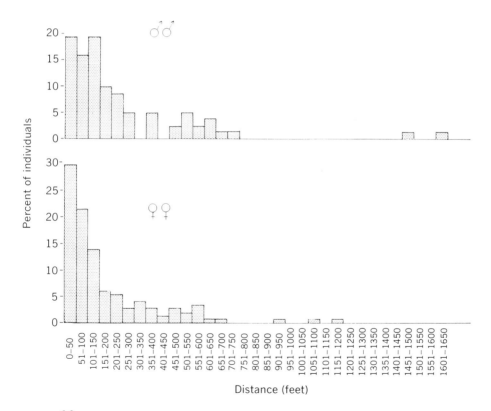

FIGURE 6.2

Percentages of male and female Rusty lizards dispersing various distances from point of hatching to place of sexual maturity. (Redrawn from Blair, 1960.)

supplemented, however, by some wide outcrosses involving immigrant individuals or gametes. The wide outcrosses provide for a small but significant amount of gene flow over larger distances in the population. In short, there is a combination of much inbreeding and some wide outcrossing in many populations.

Panmictic Units and Neighborhoods

The information on dispersal patterns gives us a deeper insight into population structure. The actual mating groups in a population of cross-fertilizing organisms are smaller, and usually much smaller, than the population itself. Certain subpopulational units need to be recognized.

The panmictic unit is the group within which random mating occurs (Wright, 1943a). This is an actual unit in those populations where random mating does occur locally. The number of individuals in such a panmictic unit is often symbolized by N or N_e, but since N is also used in a more general sense and N_e in a different sense, we will adopt the symbol N_n for the panmictic unit here to avoid any possible confusion. The size of N_n is affected by population density and migration rate. It should be noted that a certain proportion of inbreeding occurs in a panmictic unit, this proportion being $1/N_n$.

A subpopulational unit of more general applicability is the neighborhood (Wright, 1946). Some authors treat the neighborhood as though it were synonymous with the panmictic unit, and the two units are indeed the same under some conditions. However, Wright's (1946) original formulation of the neighborhood concept, which we follow here, recognizes a subtle difference between panmictic units and neighborhoods. The neighborhood is the same as the panmictic unit where panmixia occurs locally. But in many actual situations the breeding system brings about strong departures from local panmixia. In such cases the neighborhood corresponds to a derivative of a purely theoretical panmictic unit; or, to put the matter in other words, a standard is set up by a theoretical panmictic unit from which to derive the size of the neighborhood. Neighborhood size will also be denoted by N_n here.

The concept of the neighborhood starts with two assumptions. First, the chance of gamete union in a population falls off with distance between the parental individuals; and second, a proportion of inbreeding $1/N_n$ occurs in a panmictic unit (as noted above). As a result of this mating probability distribution, a large continuous population has a certain amount of local inbreeding per generation. The size of a panmictic unit with the same amount of inbreeding per generation gives the neighborhood size. That is, the factor $1/N_n$ is used to derive neighborhood size in cases where random mating does not occur locally.

Neighborhood area in a continuous population is the size of the space occupied, on the average, by N_n individuals. An important factor affecting this quantity is the standard deviation of the dispersal distances in the population (σ). It works out that a neighborhood, for a two-dimensional spatial distribution of individuals, has a circular area with a radius of 2 σ. Such a circle includes 86.5% of the parents of individuals at the center of the area.

If the population has a linear configuration, the neighborhood is a strip 3.5 σ long. A strip of this size includes about 92.4% of the parents of individuals within the strip. (For a mathematical expression of the above, see Wright, 1946; Dr. R. H. Flake has helped me to formulate the above verbal expression.)

A large continuous population can be visualized as a series of overlapping neighborhoods. Migration rate is an obvious factor affecting the size of the neighborhoods. As m increases, so does N_n. With high values of m, the large continuous population approaches (though does not necessarily attain) wide-scale panmixia. With low values of m, on the other hand, the large continuous population tends to break up into inbreeding groups, that is, into small neighborhoods.

On theoretical grounds it turns out that a great deal of local differentiation can develop in a large continuous population if the size of the random breeding group is small. Assume a continuous, outcrossing, polymorphic population, with no selection. If $N_n = 1000$ or more, little regional racial differentiation is expected to develop (apart from the effects of selection). But if $N_n = 100$, regional races can develop; and if $N_n = 10$, local races may develop without selection (Wright, 1943a).

Estimates of Neighborhood Size

Attempts have been made to estimate the size of neighborhoods in several organisms. Owing to technical difficulties this is no easy task. Several estimates are given in Table 6.1.

Relatively small neighborhoods are indicated for most species listed. The Rusty lizard *(Sceloporus olivaceus)* has a neighborhood size of $N_n = 225–270$, probably closer to 225, and a neighborhood area the size of a circle with a radius of 213 meters. The phlox *(Phlox pilosa)* has a neighborhood size of 75–282 plants, and a neighborhood area 4.4–5.2 meters in diameter. Both of these cases represent considerable departures from wide-scale panmixia.

More extreme departures from wide-scale panmixia are found in *Lithospermum caroliniense* (Boraginaceae), where the neighborhood area is about 4 meters in diameter and the effective size is about 4 plants. Similarly, in some wild populations of the house mouse *(Mus musculus)*, the breeding groups are very small, isolated subpopulations of about 4 individuals (DeFries and

TABLE 6.1
Estimates of neighborhood size in several organisms.

Organism and place	N_n	Reference
Drosophila pseudoobscura, California	500–1000	Dobzhansky and Wright, 1943
	Less	Wallace, 1966
	More	Crumpacker and Williams, 1973; Dobzhansky and Powell, 1974
Drosophila pseudoobscura, Colorado	3240–6480	Crumpacker and Williams, 1973
Cepaea nemoralis, England	Mostly 190–2850, sometimes up to 12,000	Greenwood, 1974, 1976
Sceloporus olivaceus (Rusty lizard), Texas	225–270	Blair, 1960; Kerster, 1964
Linanthus parryae (Polemoniaceae), California	14–27	Wright, 1943*b*
	More	Epling, Lewis, and Ball, 1960
Phlox pilosa (Polemoniaceae), Illinois	75–282	Levin and Kerster, 1968
Liatris aspera (Compositae), Illinois	30–191	Levin and Kerster, 1969
Lithospermum caroliniense (Boraginaceae), Illinois	ca. 4	Kerster and Levin, 1968

McClearn, 1972; see also Selander, 1970). With such small breeding groups the situation is favorable for local racial differentiation, as a result of inbreeding alone, and without the action of selection.

In the land snail *(Cepaea nemoralis)* and *Drosophila,* on the other hand, neighborhoods are often fairly large.

Collateral Readings

Andrewartha, H. G., and L. C. Birch. 1954. *The Distribution and Abundance of Animals.* University of Chicago Press, Chicago.

Bateman, A. J. 1950. Is gene dispersion normal? *Heredity* **4:** 353–363.

Blair, W. F. 1960. *The Rusty Lizard: A Population Study.* University of Texas Press, Austin.

Brussard, P. F., and P. R. Ehrlich. 1970. The population structure of *Erebia epipsodea* (Lepidoptera: Satyrinae). *Ecology* **51:** 119–129.

Ehrlich, P. R. 1961. Intrinsic barriers to dispersal in checkerspot butterfly. *Science* **134:** 108–109.

Ehrlich, P. R., R. R. White, M. C. Singer, S. W. McKechnie, and L. E. Gilbert. 1975. Checkerspot butterflies: a historical perspective. *Science* **188:** 221–228.

Greenwood, J. J. D. 1974. Effective population numbers in the snail *Cepaea nemoralis*. *Evolution* **28**: 513–526.

Levin, D. A., and H. W. Kerster. 1968. Local gene dispersal in *Phlox*. *Evolution* **22**: 130–139.

Levin, D. A., and H. W. Kerster. 1971. Neighborhood structure in plants under diverse reproductive methods. *Amer. Nat.* **105**: 345–354.

Levin, D. A., and H. W. Kerster. 1974. Gene flow in seed plants. *Evol. Biol.* **7**: 139–220.

7

Recombination

Introduction

Mutation and gene flow can produce variability in a population with respect to single genes. When allelic variation arises in two or more genes, as a result of these primary processes, the stage is set for the action of a second-order process, recombination. Recombination can combine the novel alleles, which at first are likely to be carried by different individuals, in a single genotype. Furthermore, recombination can multiply the number of different genotypes in the population. It converts a small initial stock of multiple-gene variation into a much greater amount of genotypic variation.

The Process

Assume that new mutations arise in two independent genes, A and B, in a diploid sexual population. Assume further that the mutant alleles (a and b) are originally carried by different individuals with the genotypes $AaBB$ and $AABb$, respectively. The process of recombination can now proceed through the following series of steps. (1) Crossing of the carriers of the different mutant alleles: $AaBB \times AABb$. (2) Production of the double heterozygote $AaBb$ in F_1 (along with other types). (3) Independent assortment of A and B at gamete formation to produce four classes of gametes: *AB, Ab, aB,* and *ab.* (4) Production of nine classes of genotypes in F_2: *AABB,* . . . , *aabb.*

Most of the nine genotypes are new. The population at the beginning of the process contained three genotypes (*AABB, AaBB,* and *AABb*); two generations later it contains nine genotypes, including such new recombination types as *aaBb* and *aabb.*

Complete independence of the genes *A* and *B* is not a necessary assumption as far as recombination is concerned. The genes *A* and *B* can recombine whether they are borne on different chromosomes or at separate loci on the same chromosome. Linkage lowers the frequency of the recombination types but does not prevent their formation unless the genes involved are very tightly linked.

Amount of Genotypic Variability

Let the number of separate genes present in two allelic forms increase on an arithmetic scale (2, 3, . . . , *n*). The number of diploid genotypes then increases exponentially (3^2, 3^3, . . . , 3^n). In general, the number of possible diploid genotypes (*g*) is $g = 3^n$.

We saw above that two separate genes (*A* and *B*) present in two allelic forms each can produce nine genotypes, which is $3^2 = g$. It is well known in Mendelian genetics that a tri-hybrid cross, involving three genes (*A, B,* and *C*), can yield 27 genotypes ($g = 3^3$).

Linkage disturbs the frequencies but not the total possible numbers of recombination types. If the separate genes are unlinked, their double or multiple heterozygote will produce the various recombination types in characteristic frequencies. If the genes are linked but separable by crossing-over, the recombinants will still be produced, but in lower frequencies proportionate to the strength of the linkage.

Polymorphic genes are commonly represented by multiple alleles in natural populations. In such cases, therefore, we do not raise 3, but rather some larger number, to the *n*th power in order to determine the number of genotypes. The general formula for the possible number of diploid genotypes (*g*), involving *n* (number of separate genes) and *r* (number of alleles of each gene), is

$$g = \left[\frac{r(r+1)}{2}\right]^n$$

Consider the application of this formula to the case of just two separate genes with various numbers of alleles. The results are shown graphically in Figure 7.1.

Individual variability due to recombination is seen to rise rapidly with an arithmetic increase in the number of alleles at the two loci.

Consider next the genotypic variability possible for multiple alleles at more than two loci. Some examples are listed in Table 7.1. We see that a half million genotypes can be produced by recombination between 5 genes containing 10 alleles each. Going beyond the table, if a series of 10 or more alleles exists at each of 6 separable loci, the number of diploid recombination types is in the billions.

Most natural populations of higher animals and plants that have been studied genetically do contain polymorphisms for different genes. The foregoing

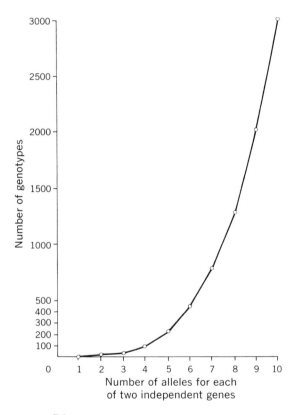

FIGURE 7.1
Increase in genotypic variation due to recombination with increase in number of alleles at each of two separate loci. (From V. Grant, *The Origin of Adaptations,* copyright 1963, Columbia University Press, New York; reproduced by permission.)

TABLE 7.1
The number of diploid genotypes that can be produced by recombination among various numbers of separate genes, each of which possesses various numbers of alleles. (Grant, 1963.)

Number of alleles of each gene	Number of genes				
	2	3	4	5	n
2	9	27	81	243	3^n
3	36	216	1,296	7,776	
4	100	1,000	10,000	100,000	
5	225	3,375	50,625	759,375	
6	441	9,261	194,481	4,084,101	
7	784	21,952	614,656	17,210,368	
8	1,296	46,656	1,679,616	60,466,176	
9	2,025	91,125	4,100,625	184,528,125	
10	3,025	166,375	9,150,625	503,284,375	
r	$\left[\dfrac{r(r+1)}{2}\right]^2$	$\left[\dfrac{r(r+1)}{2}\right]^3$	$\left[\dfrac{r(r+1)}{2}\right]^4$	$\left[\dfrac{r(r+1)}{2}\right]^5$	$\left[\dfrac{r(r+1)}{2}\right]^n$

SOURCE: V. Grant, *The Origin of Adaptations,* copyright 1963, Columbia University Press, New York; by permission.

numerical examples are not based on unrealistic assumptions; they are if anything too conservative.

Recombination is clearly a mechanism for generating tremendous amounts of individual genotypic variation. Given a moderate amount of polymorphism in just a few separable genes, this genic variation can be built up by recombination into astronomical numbers of genotypes. With only a moderate amount of genic variation, the number of possible recombination types can easily exceed the total number of individuals in the species. Recombination is the reason why no two individuals developing from different zygotes are exactly alike genotypically in sexually reproducing organisms.

Recombination and Mutation

Linked genes are recombined following crossing-over. If the linked genes occupy closely neighboring loci, crossing-over will occur only rarely. A recombination type will then appear in the progeny as a rare event. The recombinant behaves like a mutant.

The similarity between rare recombination types and mutations can be illustrated by the following model. Two closely linked genes, *A* and *B,* control similar processes and can substitute for one another. In other words, a normal phenotype is produced by the two dominant alleles (*AB*) or by single dominant alleles (A or B). The double heterozygote $\dfrac{Ab}{aB}$ has a normal phenotype and usually breeds true. But crossing-over occurs rarely between A and B and yields some ab gametes. These produce $aabb$ zygotes, which exhibit a "mutant" phenotype and breed true for the deviant character. If the crossing-over between A and B occurs at a rate comparable to mutation rates, it is not possible in ordinary practice to distinguish the rare recombinant from a mutant.

A number of cases conforming to the above general model have been revealed by refined genetic analysis in *Drosophila melanogaster, Zea mays,* and fungi. Crossover values as low as 0.00026, and possibly even lower, have been found in *Drosophila.* The compound gene A in corn (*Zea mays*), governing the color of the kernels and other plant parts, consists of two adjacent subgenes, which recombine rarely to yield types that superficially resemble mutants (see Grant, 1975, Ch. 4, for review).

The implication of these findings is that any sample of unanalyzed mutant types is likely to include a fraction of rare recombination types in addition to true gene mutations. Existing estimates of spontaneous gene mutation rates, particularly in higher animals and plants, may therefore be too high, as was noted in Chapter 5. True gene mutations may well be rarer events in higher organisms than present estimates indicate.

The Role of Recombination

Complex phenotypic characters are determined not by single genes, but by gene combinations. Such gene combinations are necessarily composed of alleles that work harmoniously together. Recombination is the mechanism that assembles the gene combinations.

Recombination is important in all organisms. Some means of exchanging genetic material exists in all kingdoms of organisms, including the bacteria and viruses. Sexual reproduction, which is the chief means of bringing about recombination in eukaryotic organisms, has parasexual counterparts accomplishing the same net result in prokaryotes.

Although recombination is found in all major groups, its relative importance varies widely between groups. In higher animals there is the highest premium on recombination, which is promoted by obligatory sexual reproduction, high chromosome numbers, and other features of the genetic system. Bacteria and viruses, at the other extreme, manage to live successfully with a bare minimum of recombination. Higher plants are intermediate between these extremes in respect to the amount of recombinational variation normally generated.

These broad differences are correlated with the complexity of the organisms. Bacteria are relatively simple organisms with a relatively simple genotype. Important life functions, such as the ability to synthesize some vital metabolite, may be determined by single genes; the origin of new simple functions of this sort can often be brought about by gene mutation in combination with natural selection.

Single genes have relatively less individual importance in the enormously complex genotype of a higher animal. Here the adaptively valuable phenotypic characters are mostly determined by gene combinations, and usually by very complex gene combinations. Mechanisms for producing gene recombinations are therefore essential.

The evolution of new complex characters begins with the origin of multiple-gene variation and ends with the fixation of a new adaptive gene combination in a population. Recombination is an important midpoint, but only a midpoint, in this process.

Mutations in two or more genes appear first and may be dispersed subsequently by gene flow. The mutant alleles, if recessive, may be stored unexpressed for many generations in the diploid condition. The diploid state is a storehouse for recessive mutational variation, and sexual reproduction is the key to this storehouse, exposing the homozygous recessive segregates and the various recombination types. Certain new allele combinations may prove to be adaptively superior to the ancestral genotype in either the ancestral or some new environment. Recombination has assembled them. Their establishment in populations now depends on other factors yet to be considered.

In summary, although mutation is the ultimate source of genetic variation, recombination is the chief effective source of variability in sexual organisms. Recombination produces most of the individual genotypic variations in the breeding population that are destined to be worked over by the forces of natural selection and genetic drift.

Collateral Readings

Carson, H. L. 1957. The species as a field for gene recombination. In *The Species Problem*, ed. by E. Mayr. American Association for the Advancement of Science, Washington, D.C.

Darlington, C. D. 1958. *The Evolution of Genetic Systems*. Ed. 2. Basic Books, New York.

Grant, V. 1975. *Genetics of Flowering Plants*. Columbia University Press, New York. Chapters 1, 4, 7, 23, 24.

8

Basic Theory of Selection

Introduction

Natural selection includes a diverse array of processes. On the one hand we have the original Darwinian concept of selection, and on the other, the modern genetical theory of selection. Within the limits of the latter, moreover, we have to recognize different modes of selection. Furthermore, the selective processes operate at different levels of biological organization. The simplest form of natural selection is that envisioned in the single-gene model.

The Single-Gene Model

A large population is assumed to be variable for a gene A. The population contains an ancestral or wild-type allele A in high frequency, and a new mutant allele a in low frequency. If the carriers of a contribute more progeny to the next generation than do the carriers of A, and if this differential reproduction of the two alleles continues systematically generation after generation, then a will gradually rise in frequency in the population and A will decline in frequency.

This is the simplest form of natural selection. It suggests a definition applicable to the single-gene model. Natural selection in this sense is the differential and non-random reproduction of different alternative alleles in a population.

Natural selection takes place when the carriers of one allele (*a*) are more successful in reproduction than the carriers of an alternative allele (*A*), consistently and systematically during successive generations. The differential reproduction of the alternative alleles is non-random.

The selective advantage of *a* over *A* does not have to be great in order to lead to a change in allele frequency. The favored allele *a* may have only a slight advantage over other competing alleles, and *a* will still increase in frequency.

The Selection Coefficient

The selective advantage of one allele over the alternative allele (or alleles) can be expressed as a percentage or as a coefficient, the selection coefficient (*s*); *s* can have a range of values from 0 to 1.

The quantitative value of the selection coefficient is derived from the relative rates of reproduction of the alternative alleles. Assume that *a* is the favored allele and *A* the unfavored allele in a large population. For every 100 *a* alleles passed on to the next generation in this population, some number of *A* alleles (from 100 to 0) will also be passed on. The selection coefficient is a function of this ratio. The formula for *s* is

$$s = 1 - \frac{\text{reproductive rate of unfavored allele}}{\text{reproductive rate of favored allele}}$$

Consider the following numerical examples:

(1) The relative rate of reproduction of *a* and *A* per generation is 100 *a* : 99 *A*. Hence $s = 1 - 99/100 = 0.01$. Or, *a* can be said to have a 1% selective advantage.

(2) The contributions of the alternative alleles to the next generation are in the ratio 1000 *a* : 999 *A*. Then $s = 0.001$, and the selective advantage of *a* is 0.1%.

(3) The ratio is 100 *a* : 50 *A* per generation; $s = 0.5$.

(4) The extreme case is 100 *a* : 0 *A*. Here $s = 1$. This means complete gene substitution in a single generation.

(5) The opposite extreme is 100 *a* : 100 *A*. In this case $s = 0$. No selection is taking place.

It is important to note that the selective advantage of *a* is not an all-or-none advantage, except in case 4. In every other possible case the selective advantage of the favored type is a statistical difference in reproductive rates. In any given generation in the breeding population, some individual carriers of *a* may fail

to reproduce, and some individual carriers of *A* may be more successful reproductively than the average *a* types. But what counts, as regards changes in the
gene pool, is the net reproductive contribution of all *a* carriers and all *A* carriers
combined in the whole large population.

There is a random component in reproduction in the population. This random
component may affect the relative rate of production of *A* and *a* either locally
or in particular generations. But these random differences in reproduction of
A and *a* are not selection. Selection occurs only if there is also a non-random
component in the differential reproduction of the alternative alleles.

This is why the definition of selection given earlier specifies both "differential
and non-random"; it is also why we have specified a large size in the population.

Rate and Extent of Change

The changes in allele frequency due to selection can run full course to complete
replacement. The new allele *a* may be present initially at a low frequency as a
result of mutation or gene flow. But so long as *a* has a selective advantage over the
pre-existing allele *A,* even a slight advantage, *a* will continue to increase in frequency, and *A* will decline. The ultimate result, if enough time is available, may
be the complete substitution of *a* for *A* in the population.

It is evident that these changes will take place faster with a high selection
coefficient and slower with a low coefficient. The rate of change in allele frequency is directly proportional to the value of *s*.

However, for any given value of *s,* the rate of change in allele frequency is not
constant at all allele frequencies. If the population is diploid, and the favored
allele is dominant, change in allele frequency is rapid at low and intermediate
frequencies, but falls off at high frequencies. For instance, a dominant allele
with a 1% selective advantage ($s = 0.01$) can increase in frequency from 0.01%
to 98% in about 6000 generations; but it requires over 5000 generations to increase from 98% to 99%, and about 1,000,000 generations to increase from 99%
to 100%. If the favored allele is recessive, on the other hand, change in frequency
is very slow at low frequencies, but picks up at intermediate frequencies.

It has been assumed in the preceding discussion that the favored allele will
progress all the way to complete fixation in the population. This is a simplifying
assumption, but not necessarily a realistic one, for several reasons.

The selection coefficient may not remain constant throughout the long course
of selection, but could undergo fluctuations and reversals. A given gene might
then have a selective advantage at one stage in the history of the population and

a selective disadvantage at another stage. Furthermore, during the course of time, other alleles are likely to be introduced into the population by mutation and gene flow. In this connection, the assumption of just two competing alleles in the population is unrealistic in itself.

Finally, if the population is diploid, the two or more alleles are incorporated into the respective homozygotes and heterozygotes, e.g., *A* and *a* in *AA, aa,* and *Aa;* the real discriminating effect of selection will then be not among the alleles, but among their genotypes. The genotype *aa* might well be superior to *AA,* but the heterozygote *Aa* might be superior to *aa,* and then the population will remain permanently polymorphic for *A* and *a.* In this case *a* will never reach complete fixation as a result of selection. We will discuss this situation in more detail in Chapter 11. Meanwhile let us note that population geneticists are more and more coming to the conclusion that many alleles controlled by selection do tend to level off at varying frequencies lower than 100%.

Fitness

What is the basis for these changes in allele frequency? Why does *a* undergo systematic increases in frequency relative to *A?* In the first place, the basis of selective advantage must lie in some phenotypic character difference controlled by the *A* gene. In the second place, the phenotypic character involved must have some importance in survival and reproduction. The carriers of *a* must be superior to the carriers of *A* in some aspect of life affecting reproductive success.

The question of the selective advantage or disadvantage of any given allele thus opens up large questions of gene action, gene-controlled development, gene-gene interaction, and genotype-environment interaction. We will explore some of these questions in the next chapter. For the present we will simply assume that the *a* types have some superiority over the *A* carriers in respect to viability or fecundity.

This brings us to the concept of fitness in population genetics (also known as adaptive value or selective value). The fitness of a genotype is the average number of progeny left by that genotype relative to the average number of progeny of other, competing genotypes. Fitness is thus a purely quantitative and operational concept. It is a quantitative measure of reproductive success.

Fitness (symbolized by w) is a function of the selection coefficient ($w = 1 - s$). If the ratio of the reproductive rates of *A* and *a* is 99 A : 100 a, then $s = 0.01$, as already noted, and we can now add that $w_A = 0.99$. In short, $w_A = 99\ A\ /\ 100\ a = 0.99$.

High fitness, as the term is used in population genetics, does not necessarily carry the connotation of "survival of the fittest"; the synonymous term, adaptive value, does not necessarily mean that the genotype with a high value is especially well adapted to its environment. This may be the case. But, alternatively, the high fitness or high adaptive value might result solely from high fecundity. High fitness refers to genetically determined success in reproduction; it does not refer to the specific reasons for that success, which can vary widely from case to case.

The Darwinian Concept of Selection

Population genetics and modern evolutionary theory equate natural selection with differential reproduction of alternative forms of genes, genotypes, or other reproducible units. The original Darwinian concept was somewhat different. Darwin and his followers emphasized the life-and-death value of some variable characteristics of a species in "the struggle for existence."

Darwin's thesis in *The Origin of Species* (1859, 1872) can be summed up as follows. The individuals of a species compete for the means of life. These individuals differ with respect to minor variations in their characteristics and the variations are often hereditary. Some variant forms are better adapted for survival in the struggle for existence than others. Consequently the former will reproduce preferentially and pass their favorable characters on to future generations.

Darwin used the giraffe to illustrate his thesis. He postulated, plausibly, that the individuals in an ancestral population of giraffes differed slightly in the length of the neck and forelegs. The taller and the shorter animals would have entered into competition for leaves in times of food scarcity in their native savanna. In such times the taller individuals would have been able to reach leaves high in the trees that were not accessible to the shorter individuals. The latter therefore perished and their short-necked and short-legged characters disappeared with them. Conversely, the long neck and forelegs of the modern giraffe are a result of the preferential survival and reproduction, generation after generation, of the taller individuals (Darwin, 1872, Ch. 7).

In Darwin's own words (1859, Ch. 4):

> Can it . . . be thought improbable [that] variations useful in some way to each being in the great and complex battle of life should sometimes occur in the course of thousands of generations? If such do occur, can we doubt (remembering that many more individuals are born than can possibly survive) that individuals having any advantage, however slight, over others, would have the best chance of surviving and of procreating their kind? On the other hand, we

may feel sure that any variation in the least degree injurious would be rigidly destroyed. This preservation of favourable variations and the rejection of injurious variations I call Natural Selection.

Components of Fitness

Differential reproduction is the end result of numerous components, which can be listed as follows:

> Differential mortality in early and juvenile stages
> Differential mortality in adult stage
> Differential viability
> Differential mating drive and mating success
> Differential fertility
> Differential fecundity

Darwin's theory of natural selection emphasized the first two or three components. His later theory of sexual selection (Darwin, 1871) added the fourth. The population-genetics theory of selection, on the other hand, explicitly defines natural selection in terms of the end result, differential reproduction; in practice it often places the emphasis on the last three components, the reproductive processes involved themselves. The difference between the two schools of thought reflects differences in professional orientation. Darwin and his followers were field naturalists, whereas the foundations of modern selection theory were laid by laboratory men and statisticians.

In retrospect, it might have been desirable to have assigned different technical terms to the different sets of component processes, as Darwin himself did in the case of sexual selection. But in fact we now have one term applied to a very wide variety of processes. We can live with the actual terminological situation, since there is no inherent contradiction between the Darwinian and the population-genetics concepts of selection, but in living with it we must keep the composite nature of the present selection concept in mind, and be prepared to distinguish the separate subprocesses when necessary.

We should distinguish between two basic forms of selection. In the first case the selective advantage of type a over type A resides in a better adaptedness to a critical environmental situation. Selection then improves the adaptations of the organisms constituting the population, as envisioned by Darwin. In the second case the a types leave more progeny than the A types because of greater sex drive or fecundity, whether they are superior in other secular aspects of life or not. This form of selection does not necessarily lead to improvements in adaptation. On the contrary, it can result in a spreading of characteristics that turn out to be injurious to the population or species as a whole.

It follows as a corollary that we also need to recognize different kinds of fitness. High fitness in population genetics refers to a relatively large production of progeny, as noted earlier, and in itself it does not specify the cause of the high fitness. It is useful to recognize two broad categories of fitness: (1) adaptedness, or the degree of adaptation of the individual or population to its environmental conditions, and the ability to leave more progeny for this reason; and (2) reproductive success per se.

Collateral Readings

Darwin, C. 1859, 1872. *On the Origin of Species*. Eds. 1 and 6. John Murray, London. Various modern reprints available.

Dobzhansky, Th. 1970. *Genetics of the Evolutionary Process*. Columbia University Press, New York. Chapter 7.

Grant, V. 1963. *The Origin of Adaptations*. Columbia University Press, New York. Chapter 9.

Lerner, I. M. 1958. *The Genetic Basis of Selection*. John Wiley & Sons, New York.

Mayr, E. 1970. *Populations, Species, and Evolution*. Harvard University Press, Cambridge, Mass. Chapter 8.

9

Gene Expression in Relation to Selection

Introduction

The question can be raised: Does natural selection act on genes or on phenotypes? Selection acts directly on phenotypes and only indirectly on the underlying genes. Between the direct action of selection and the resulting change in gene frequency, therefore, lies the whole complex train of events by which gene action is translated into phenotypic characteristics.

Consequently, in order to understand the workings of natural selection, we must take the phenotypic expression of genes into consideration. One set of questions centers on the kind of phenotypic expression in relation to the environmental background; another on the degree of phenotypic modifiability; and still another on the multifarious modes of action and interaction of genes.

The Relativity of the Selective Value

Let us first consider the implications of a very obvious relationship. A given genotype determines one and the same phenotypic character expression in a variety of environments. The phenotype in question has a certain selective advantage in some environments; in other environments it might have a different selective advantage, or no advantage, or even a selective disadvantage. Under such conditions selection will favor the genotype determining that phenotype in some environments, but not to the same degree, or not at all, in other environments.

In short, the selective value of a particular allele or genotype is not one of its intrinsic properties. It is instead a function of the phenotype-environment interrelationship.

A simple example is provided by shell color in the European land snail, *Cepaea nemoralis*. The snail lives in a variety of habitats, from closed dark beech woods to open sunny meadows. It is polymorphic for shell color, with brown, pink, and yellow forms. A single polymorphic gene determines these shell-color differences. The brown allele produces brown shells, and the yellow allele produces yellow shells, in the full range of environments inhabited by the snail.

The snails are subject to predation by thrushes and other birds that rely on their visual senses. The brown shells provide concealing coloration against the bird predators in beech woods, and the yellow snails are concealingly colored in meadows. As a result of the visual selective component in bird predation, the brown shell color is prevalent in snail populations in woodlands, and the yellow color is prevalent in meadow populations. The selective values of the brown and yellow alleles are thus correlated with the type of habitat, and become reversed with change of habitat (for further details, see Cain and Sheppard, 1954; Sheppard, 1959; Lamotte, 1959; Jones, 1973).

Drosophila pseudoobscura in western North America is polymorphic for inversions on chromosome III. The types of inversions are assigned names and code letters: Standard (ST), Chiricahua (CH), Arrowhead (AR), etc., as noted in Chapter 4. It has been found that certain inversion heterozygotes have a selective advantage over the corresponding homozygotes. But this adaptive superiority of the heterozygous genotypes is manifested only under particular environmental conditions. Thus ST/CH has a higher selective value than either ST/ST or CH/CH at warm temperatures (21–25°C), but not at cool temperatures (16°C). The high selective value of ST/CH is manifested only when the flies are reared in crowded population cages. At a given warm temperature (21°C), ST/CH flies are superior when feeding on one kind of yeast *(Kloeckera)*, but not on another kind of yeast *(Zygosaccharomyces)* (see Dobzhansky, 1970, Ch. 5, for review).

Phenotypic Modifiability

The phenotype is the product of genotype-environment interaction during the course of development. A particular phenotypic characteristic is a product of environmental influences during development, as well as of gene action.

Furthermore, the role of the environmental component in the formation of a phenotypic character varies among different characters and among different organisms. In man, for example, the environmental component is slight for the blood types, considerable for body weight, and very large in the sphere of mental and behavioral traits.

If the process of selection is working on a phenotypic character that is determined predominantly by the genotypic component in development, with only a slight environmental influence, the effect of selection on the composition of the gene pool will be immediate and relatively direct. But selection will be much less effective — its effects will be more delayed — in the case of phenotypic characters that are molded by the environment to a large extent. The capacity of a genotype to respond by appropriate phenotypic modifications to a wide range of environmental conditions tends to hamper the effectiveness of environmental selection.

Pleiotropy

A gene usually has different and often unrelated phenotypic effects, or in other words, the gene affects more than one phenotypic character. This condition is known as pleiotropy. Pleiotropy complicates the action of selection.

Suppose that a particular allele has two pleiotropic effects, one of which is favorable and the other unfavorable. The selective value of the allele for all practical purposes will then be the net value when its selective advantages are weighed against its disadvantages. At the point when the selective disadvantages outweigh the advantages, selection ceases to favor the allele and begins to discriminate against it.

Plant and animal breeders confront this problem in artificial selection constantly. Some economically useful characters, such as very high yield, simply cannot be fixed in a population of domesticated plants or animals if they are developmentally correlated with strongly disadvantageous characters, such as reduced fertility. Pleiotropic effects undoubtedly hold back the effects of natural selection in many natural populations in a similar way.

Expressivity Modifiers

The phenotypic expression of a particular gene or gene combination is affected by other genes of the complement, known as modifiers. In this and the next section we will discuss the roles of two classes of modifier genes: expressivity modifiers and dominance modifiers.

Expressivity is the degree of phenotypic expression of a gene or gene combination. A series of individuals with the same constitution for a given gene, when grown or reared in a standard environment, may exhibit different degrees of phenotypic expression of that gene. Thus different individual *Drosophila* flies, and lines of flies, may have different numbers of body bristles, although they have the same constitution for the main bristle-controlling gene and have been reared in the same environment. The gene is said to have variable expressivity.

The variable expressivity may be caused by expressivity modifiers. Plus modifiers enhance the phenotypic expression of the major gene and minus modifiers suppress it. The individual organisms, though genotypically uniform with respect to the major gene, are genetically different in their sets of expressivity modifiers. Furthermore, the action of the modifier genes may be affected by environmental factors, thus introducing another complicating factor. The variable expressivity will then be a result of complex interactions between the plus and minus modifiers and the environmental factors. This in turn complicates the action of selection.

Let us assume that a population is monomorphic for a major gene or gene combination determining a phenotypic character, but contains latent variability in the system of modifier genes. As a result of minus expressivity modifiers, the phenotypic character is not expressed in one environment (E_1), in which it would be adaptively valuable, and is only weakly expressed in another environment (E_2), in which it has a positive selective advantage. Of course selection will be ineffectual in environment E_1. But selection can be expected to build up systems of modifiers that enhance the phenotypic expression in environment E_2. The new set of modifiers may then have effects that carry over to environment E_1. The character comes to expression and can be stabilized by selection in E_1 as well as in E_2.

Waddington (1953, 1956) carried out interesting experiments with *Drosophila melanogaster* that seem to conform to the above model. He made use of the known fact that certain abnormal phenotypes (crossveinless wings, the bithorax condition of the thorax) can be induced in some adult flies by treating populations of eggs or pupae with environmental shocks (high temperatures, ether). The induced phenotypic alterations are not ordinarily hereditary, but they eventually became so in Waddington's experiments.

Waddington applied the environmental shock—the heat treatment or ether treatment—to starting populations of flies in the egg or pupal stages; he obtained the appropriate phenotypic responses in some of the adult flies, and he selected for the abnormal phenotypes. This process was repeated each generation for 24 to 29 generations. It was, in effect, artificial selection for an abnormal phenotypic trait in an abnormal environment in which that trait is expressed.

The selection was effective. The descendent generations of flies at the end of the experiment contained a significantly higher frequency of individuals that exhibited the phenotypic response to the environmental shock than did the ancestral generations at the beginning of the experiment. Furthermore, some of the derived flies exhibited the abnormal phenotype not only in the abnormal environment in which the selection was carried out, but also in the normal environment, for which they were not selected. The advanced generations contain some flies that develop abnormal phenotypes without the stimulus of the abnormal environment.

These results can be, and have been, interpreted in different ways. A plausible interpretation is that the selection process resulted in the building up of sets of modifiers that enhanced the expressivity of the altered wing or thorax character in the abnormal environment. The new, strong modifiers had their effects on phenotypic expression, however, not only in the abnormal environment but also, as a by-product, in the normal environment. (For the original accounts of the experiments see Waddington, 1953, 1956; for differing interpretations see Waddington, 1957; Stern, 1958, 1959; Bateman, 1959; for a review see Grant, 1963, pp. 206–211.)

The model involving expressivity modifiers presented above has implications for evolutionary theory. The model indicates that selection for a set of modifier genes in one environment may produce a genotype that gives rise to unpredicted phenotypic expressions in other environments. New potentialities for phenotypic expression can sometimes be created inadvertently by selection for expressivity modifiers. In this way a population being selected for one new environment could acquire genotypically controlled phenotypes that, as a side effect, "preadapt" it for still other new environments.

Dominance Modifiers

The relationship of dominance and recessiveness in an allele pair in a diploid may stem from the relative strengths of action of the two alleles themselves, but this is not the whole story, for dominance and recessiveness are also brought about by the action of other genes, known as dominance modifiers. A particular heterozygous allele pair may produce a dominant phenotype when in one genetic background and an intermediate phenotype in another background. One genetic background has a set of strong dominance modifiers, modifier genes that enhance the phenotypic expression of the dominant allele, while the other genetic background has a weak or otherwise different set of modifiers.

It will be recalled that the majority of new mutations are both deleterious and recessive. The usually deleterious effects of mutations are an inevitable consequence of the randomness of the mutation process in an organism that has previously gone through many generations of selection. But the recessiveness is not an intrinsic property of the mutation process. Why then should most new mutations be recessive?

Fisher (1930, 1958) suggested that natural selection in diploid organisms has built up systems of modifiers that strengthen and stabilize the action of the normal, wild-type alleles. The dominance modifiers would be advantageous insofar as they protect the diploid organism from the immediate ill effects of most new mutations. Dominance modifiers enable the diploid population to store mutations, deleterious or otherwise, in the recessive state, and to expose them to selection slowly in a proportionately small number of homozygous recessive segregates.

Fisher's theory of the origin of dominance is consistent with much evidence concerning both dominance and mutation, and provides a unifying explanation of diverse observed facts, but has not been confirmed as regards the phylogenetic aspects of the problem. We cannot say for any diploid species that dominance has developed in the manner suggested by Fisher. The theory remains attractive but unconfirmed and open to question. The theory has been criticized by some authors (Crosby, 1963; Ewens, 1965), and defended, successfully in my opinion, by others (Mayo, 1966; Sheppard and Ford, 1966).

The Genotype as a Unit of Selection

The single-gene model of selection is obviously an oversimplification. The single-gene model is biologically realistic in some actual cases—for example, in a competition among clones of bacteria differing with respect to a single gene. But as applied to most situations in eukaryotic organisms this model has to be regarded as an abstraction useful for clarifying the more complex, real processes.

The phenotypic characters exposed to the action of selection in higher organisms are, with rare exceptions, not single-gene characters, but instead are products of gene combinations. Selection is not given the opportunity of discriminating between the alleles of a single gene, but must operate on the alternative forms of a gene system composed of many components.

The primary form of selection takes place not among genes themselves, or even gene systems, but among individuals in a population. And the individual organism, particularly in the more advanced forms of life, is a complex machine composed of many organs with different functional roles. The diverse organs and functions must be coordinated and harmonized. A change in one character may well be advantageous in relation to its own particular function, but have disadvantageous side effects on other functions of the organism. Selection will then favor or reject the new character on the basis of its net advantage or disadvantage to the individual organism as a whole.

The end result in cases where selection has opposing tendencies is often a compromise. Life is full of such compromises. The colorful plumage of birds of paradise serves a useful role in courtship, but also advertises the presence of the birds to predators. The conflict is resolved by a compromise. The brilliant plumage is confined to the male birds, which are expendable, while the females remain plain-colored.

Collateral Readings

Waddington, C. H. 1957. *The Strategy of the Genes.* George Allen & Unwin, London.

10

Examples of Selection

Types of Evidence

The efficacy of natural selection—its ability to bring about changes as predicted by selection theory—has been tested and confirmed in countless studies over the years. The studies have been carried out with diverse species of animals, plants, and microorganisms. The studies can be grouped, according to the approach, into three classes: selection experiments in the laboratory or breeding plot; observations in uncontrolled natural populations; and the tracing of historical changes due to artificial selection in domesticated plants and animals. A few examples representing each of these three approaches will be presented in this chapter.

The Illinois Corn-Selection Experiment

A long-term selection experiment with corn *(Zea mays)* has been carried on at the Illinois Agricultural Experiment Station since 1896. Artificial selection was carried out for several characters in different lines derived from a variable foundation population. The characters are protein content of kernels, oil content of kernels, and height of ear. The results have been reported by Winter (1929), Woodworth, Leng, and Jugenheimer (1952), Bonnett (1954), and Leng (1960).

Protein content of the grains was subject to selection in two parallel lines: a high line (for high protein content) and a low line, derived from the same foundation stock in 1896 with 10.9% protein. In each subsequent generation (corn is an annual plant producing one generation per year), a large number of corn ears were harvested and some of their kernels assayed for protein content. The fraction of the ears with the highest protein was then assigned to the high line and used as seeds for the next generation. Similarly, the fraction of ears with the lowest protein was used as seeds for each new generation in the low line. This fraction was 24 out of 120 ears in each line in the early years of the experiment, and 12 out of 60 ears in the later years. The selection experiment reached generation 60 in 1959.

Response to the selection was immediate and long-continued. In the high line, protein content rose from 10.9% in generation 0 to 13.8% (in generation 5) and 19.4% (in generation 50). In the low line, protein content went down from 10.9% (generation 0) to 9.6% (generation 5) and 4.9% (generation 50). Curves showing the progress of selection in the two lines up to generation 60 are presented in Figure 10.1. Response to selection for oil content of the kernels was even more marked than that for protein (Woodworth, Leng, and Jugenheimer, 1952; Leng, 1960).

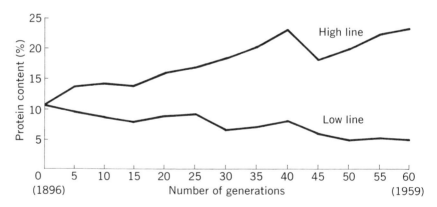

FIGURE 10.1
Response to selection for high or low protein content of grains in corn (*Zea mays*) during a 60-generation period. (Drawn from data of Woodworth, Leng, and Jugenheimer, 1952, and Leng, 1960.)

The selection for protein or oil content led to unexpected correlated changes in various morphological features. Thus the derivative high-protein line had small ears with translucent flinty kernels, and the low-protein line had large ears with long starchy kernels (Woodworth, Leng, and Jugenheimer, 1952).

Selection for height of ears above ground started with a foundation population grown in 1903 in which the average height varied from 43 to 56 inches in different individuals. The plants with the ears closest to the ground in each generation were chosen as parents of the next generation in the low line. A parallel high line was continued by a similar procedure. The two lines were continued for 25 generations until 1928.

Here again the response to selection was marked. In generation 24 (in 1927) the average height of ears was 8.1 inches in the low line and 120.5 inches in the high line. As before, unpredicted but correlated changes took place in the highly selected lines. The low-ear strain came into flowering early in the season, whereas the high-ear strain was 10 to 14 days later in flowering (Bonnett, 1954).

Zea mays is an outcrossing and highly heterozygous plant whose populations contain large reserves of genetic variability. The gradual and long-continued responses to selection in the Illinois experiment are consistent with the idea that the characters selected are determined by many genes that were polymorphic in the foundation populations. It is probable that new recombinational variation was released during the long course of the experiment, permitting the selection to continue to produce upward or downward changes, and it is quite possible that new mutational variation could have arisen too. The selection brought about correlated changes in other characters, thus demonstrating the complex and integrated nature of the genotype.

Viability in *Drosophila*

Dobzhansky and Spassky (1947) designed an experiment with *Drosophila* so that the changes in the populations would be guided by natural selection rather than by the conscious will of the experimenter. It is thus complementary to the Illinois corn-selection experiment, in which artificial selection was the guiding factor.

The foundation populations were strains of *Drosophila pseudoobscura* with reduced viability. Their viability was expressed quantitatively as a percent of normal viability. Seven related strains were used as foundation populations for seven replicate cultures. Each replicate was subdivided into four subcultures, which were treated differently. The treatments are designated briefly as follows: (1) irradiated and crowded; (2) crowded but not irradiated; (3) irradiated but not crowded; (4) not crowded and not irradiated.

The lines given treatments 1 and 2 were made homozygous in certain chromosomes to facilitate the expression of new mutations; the lines under treatments 3 and 4 were maintained in heterozygous condition for the same chromosomes.

Treatments 3 and 4 serve as controls for treatments 1 and 2, respectively. All lines were maintained for 50 generations in the laboratory. Viability and other characteristics were tested every few generations and at the end of the experiment.

Let us first consider the changes in one line of the seven given treatment 2. The viability of this line at the start of the experiment was 29% (of normal viability). Males and females were placed in a bottle containing food material and allowed to lay unlimited numbers of eggs, which soon resulted in the container's becoming overpopulated. A random sample of the offspring was then transferred to a new culture bottle and allowed to breed again. The line was continued in this fashion for 50 generations.

It was assumed that the crowded conditions in the bottles and the strong competition for food would lead to an increase in the frequency of any mutant types possessing greater vigor than their siblings. This expectation was realized. The viability of the strain rose from 29% to 90% in the course of the experiment.

Most of the other replicate lines of treatment 2 also underwent improvements in viability. One line showed viability increases from 60% to nearly 100%, another from 30% to over 80%. Viability rose in five of the seven lines under treatment 2.

The seven lines given treatment 1 were handled generally in the same way as the foregoing series, except that the male parents of each new generation were treated with X rays before they were introduced into the fresh culture bottle. The purpose of the X-ray treatment was to increase the number of new mutations. A gain in viability was exhibited by six of the seven irradiated lines. Some of the viability gains were quite large, as from an initial 29% to a final 103% in one line, and from 65% to 115% in another line.

Altogether 11 of the 14 lines maintained under the conditions of intense competition for food in crowded culture bottles underwent marked improvements in viability during 50 generations. Similar improvements did not occur in the control lines (treatments 3 and 4). The microevolutionary changes in the first two series of lines (treatments 1 and 2) are associated with a partially homozygous constitution facilitating the expression of new genetic variations, and with differential mortality in relation to food getting. In short, natural selection was at work and produced observable effects in the crowded lines but not in the controls.

Melanism in the Peppered Moth

The peppered moth *(Biston betularia)* occurs widely in England. It flies by night but spends the day at rest on the trunk or branches of trees, where it is vulnerable to the attacks of robins, thrushes, and other insectivorous birds.

These birds hunt by sight. Two polymorphic forms of the moth concern us here: the so-called typical form, which is speckled gray, and a melanic form with black wings and body, known as carbonaria. The gray moths blend well into a background of lichen-covered bark, and bird predation is reduced by this concealing coloration. The melanic form, on the other hand, is conspicuous against a background of gray bark but is concealingly colored on soot-covered bark (Figure 10.2) (Ford, 1964, Ch. 14; Kettlewell, 1956, 1973).

The typical and carbonaria forms differ with respect to a single gene, *C*. The carbonaria form is the result of a rare dominant mutation in this gene. The gray form is *cc* and the melanic type, *Cc* or *CC*.

The background environment of the peppered moth was changed drastically by the Industrial Revolution in England. One side effect of industrialism was an outpouring of smoke and soot over the formerly unpolluted rural English countryside. The lichens disappeared and the tree trunks and branches turned black with soot. These changes affected profoundly the composition of *Biston* populations in the industrialized areas.

The peppered moth populations consisted almost exclusively of the typical gray form in pre-industrial England, and this situation still exists in unpolluted parts of England. The melanic carbonaria form was first noticed as a rare mutant in 1848 at Manchester. From 1848 to 1898 the carbonaria form rapidly increased in frequency in industrial districts; it became the common type, while the typical gray form became rare. The frequency of the carbonaria allele *C* is estimated to have increased from 1% to 99% during the 50 generations from 1848 to 1898. A very high selective advantage of the carbonaria allele is required to account for the observed change in gene frequency (Ford, 1964; Kettlewell, 1956, 1973).

The selective factor is visually oriented predation by insectivorous birds. The typical form is protected on lichen-covered bark and the carbonaria form on soot-covered bark (Figure 10.2). Release and recapture studies show that the survival rate of the typical form is twice as high as that of the carbonaria form in an unpolluted area; conversely, the survival rate of carbonaria exceeds that of the typical form by a factor of about two near industrial Birmingham (Kettlewell, 1956). The historical change in substrate brought about by industrialization thus reversed the relative selective values of the *c* and *C* alleles.

The *C* gene has pleiotropic effects on physiological vigor that may lead to secondary selective values in relation to other factors. The carbonaria type is more viable than the typical form in stressful environments in the laboratory and probably also in nature. This favorable pleiotropic effect of the carbonaria allele is overruled by strong counterselection in the form of bird predation in unpolluted areas. In sooty areas, however, the moth populations can exploit the physiological advantages of the carbonaria allele.

Parallel changes from a gray or brown to a melanic form took place in several

FIGURE 10.2
The speckled-gray and the melanic forms of the peppered moth (*Biston betularia*) perched on a lichen-covered trunk and on a soot-covered trunk. In the left-hand picture the gray moth is barely visible below and to the right of the melanic moth. (Photographs courtesy of Dr. H. B. D. Kettlewell.)

other species of moths in England during the same period; a parallel evolution of melanism in moths occurred again on a somewhat later time schedule in industrial central Europe and the northeastern United States (Owen, 1961; Ford, 1964; Kettlewell, 1973).

Shell Color in Land Snails

The European land snail *(Cepaea nemoralis)* is common in a variety of habitats — woods, meadows, hedgerows — in England and western Europe. The populations are usually polymorphic for the color of the shell, which may be brown, pink, or yellow (Figure 10.3). Shell color is determined by a series of multiple alleles, in which brown is top dominant and yellow the bottom recessive (Cain and Sheppard, 1954; Sheppard, 1959; Lamotte, 1959; Jones, 1973).

Shell color plays the central role in the concealing coloration of the snails. Thrushes and other birds hunt the snails by sight, and take a higher proportion of contrastingly colored snails than of cryptically colored ones. What constitutes contrasting or cryptic coloration depends, of course, entirely on the background.

One of the habitats of *Cepaea nemoralis* is dense beech woods. These woods have a carpet of red-brown leaf litter, which remains about the same throughout the year. Here the brown and pink snails are protectively colored, and here these color forms are common. Yellow snails are rare in beech woods but common in green meadows (Cain and Sheppard, 1954; Sheppard, 1959).

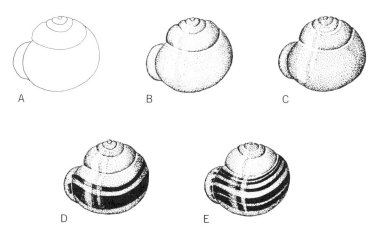

FIGURE 10.3
Polymorphic types of the European land snail (*Cepaea nemoralis*). (A) Yellow unbanded. (B) Pink. (C) Brown. (D) Two-banded. (E) Five-banded.

Still another habitat is mixed deciduous woods. The character of the background changes with the seasons in these woods. In early spring the floor is brown with leaf litter, but in summer the floor turns green. And here the selective values of the different shell types also change with the seasons. In early spring the brown and pink snails have an advantage in relation to bird predation, whereas in summer the yellow snails have the advantage. Populations in deciduous woods undergo cyclical changes in the relative frequencies of the different polymorphic forms throughout the year (Cain and Sheppard, 1954; Sheppard, 1959).

Concealment is not the only function of shell color. There is some evidence indicating differences in physiological hardiness between the various color forms. The yellow form seems to possess better tolerance for temperature extremes, for both heat and cold, than the brown and pink forms. The greater temperature tolerance of the yellow type would of course be advantageous in open meadow habitats and in warm-summer areas. The yellow form does attain a high frequency in meadow habitats, as already mentioned, and also increases in frequency on a geographical transect from northern to southern Europe (Lamotte, 1959; Jones, 1973).

Another feature of the shells is the banding pattern. The shells may be banded or unbanded; if banded, they have from 1 to 5 bands (Figure 10.3). The presence or absence of bands is determined by a gene *(B)* closely linked to the shell-color gene. At this *B* locus, unbanded is dominant over banded. A separate locus controls the number of bands. Recombination produces the various possible combinations of shell color and banding pattern (Cain and Sheppard, 1954; Sheppard, 1959).

The role of natural selection is less clear in the case of banding pattern than in that of shell color. In other animals dark bands around the body are known to contribute to concealing coloration in semi-shady habitats with broken patterns of light and shadow. A similar function probably exists in *Cepaea nemoralis*. In some habitats, at least, the banded snails are at an advantage in relation to bird predation. It is also probable that some of the variation in banding pattern is controlled by genetic drift; we will return to this aspect of the problem in a later chapter.

Resistance to Toxins

Penicillin, streptomycin, and other antibiotics, when first introduced into medical use in the 1940s and 1950s, were effective in small doses in controlling disease bacteria. Shortly after the medical use of these antibiotics became widespread, however, their effectiveness in controlling bacterial infections began to decline, and higher doses had to be employed to achieve the desired results. There are antibiotic-resistant and antibiotic-susceptible strains of bacteria. The resistant types arise by spontaneous mutations at a given low rate. The application of antibiotics in light or moderate doses then sets in motion a selection process in favor of the resistant strains.

Such microevolutionary changes have been reported in laboratory experiments. An example is a selection experiment carried out with a strain (no. 209P) of *Micrococcus pyrogenes aureus (Staphylococcus aureus)*, a pathogenic bacterium that causes suppuration in wounds and food poisoning (McVeigh and Hobdy, 1952).

The original population of this strain was susceptible to various antibiotics and its growth was blocked by minute concentrations of these agents. Isolates from the original population were subcultured and grown in a succession of growth media containing increasing concentrations of penicillin and other specific antibiotics. Resistance to these antibiotics developed in the various lines. The greatest increase in resistance occurred during the first 10 to 20 subcultures; increase thereafter was slower. The increase in resistance to the various antibiotics was as follows:

Chloromycetin	193-fold
Aureomycin	210-fold
Sodium penicillin	187,000-fold
Streptomycin	250,000-fold

The derived resistant types exhibit other characteristics. They generally grow more slowly than the original, susceptible strain. The penicillin-resistant type had lost much of its pathogenicity and its ability to grow anaerobically. The susceptible form is thus superior to the resistant types in the ordinary antibiotic-free environment. Removal of antibiotics from the environment of the bacteria leads to selection in the reverse direction in favor of the susceptible form.

Parallel microevolutionary changes have been observed in several species of insect pests in relation to the widespread application of insecticides. The development of DDT-resistant strains of the common housefly *(Musca domestica)* is one well-known example. The evolution of cyanide-resistant strains of scale insects *(Aonidiella aurantii* and other species) in California citrus orchards is another.

Various toxins occur in the natural world, and the process of evolutionary adjustment to them takes place in nature as well as in agriculture and medicine. For example, the senita cactus *(Lophocereus schottii)* of Mexico contains alkaloids that are toxic to most species of *Drosophila*. But *Drosophila pachea* is resistant to this alkaloid and breeds in the rotting stems of the senita cactus. In its unique breeding sites, *D. pachea* is free from competition from other species of *Drosophila* (Kircher and Heed, 1970).

Domestication

Scores of species of plants and animals have been transformed greatly by a combination of artificial and natural selection in the course of domestication. The histories of domesticated plants and animals, where known, provide numerous good examples of the efficacy of selection. Darwin, in *The Origin of Species*, cited the differences between the wild rock pigeon and the domestic pigeon, including such specialized breeds as the tumbler and pouter, as an example of the power of artificial selection to bring about profound evolutionary changes. Other cases come readily to mind: the dog, cattle, tomato, squash, etc. In the following section we will consider one such case, maize.

Evolution of Corn

Indian corn or maize *(Zea mays)* is a member of the tribe Maydeae of the grass family (Gramineae). It is strikingly different from all other members of its tribe and family in a number of characters, particularly in the distribution of the sex organs and in the structure of the corn ear or cob. The corn cob is a very complex and highly specialized plant part; nothing like it is found anywhere else in the grass family. Nevertheless, modern corn with its unique cob and other features has evolved from wild grass ancestors in tropical and subtropical America.

The evidence has been gained from comparative morphological, taxonomic, genetic, cytogenetic, ethnobotanical, archeological, and palynological studies. Leading workers in the gathering of this evidence have been Mangelsdorf, Galinat, Harlan, and others. Some unresolved disagreements exist among the various workers. More details will be found in the following interesting and significant papers: Mangelsdorf (1958, 1959); Mangelsdorf, MacNeish, and Galinat (1964); Mangelsdorf, Dick, and Camara-Hernandez (1967); Anderson and Brown (1952); Galinat (1970, 1971*a*, 1971*b*); De Wet, Harlan, and Grant (1971); De Wet and Harlan (1972); Wilkes (1967, 1972). A definitive monograph is that of Mangelsdorf (1974).

In *Zea mays* the staminate and the pistillate flowers (spikelets) are segregated into different inflorescences, the tassels and ears, respectively; and corn is thus monoecious, an unusual arrangement in the Gramineae. The pollen-bearing tassel develops at the top of the corn stalk. The grain-bearing ears develop in the axils of leaves in the mid-region of the stalk (Figure 10.4). The corn ear is a spike consisting of a stout central axis bearing many rows of pistillate spikelets and, later, as many rows of mature kernels. The whole ear is enveloped by modified leaf sheaths (the husk). The grains are firmly attached to the cob and

A   B

FIGURE 10.4
Form of the shoot in two varieties of corn *(Zea mays)*. (A) Dent corn. (B) Popcorn.

are naked, that is, not covered by bracts (glumes or chaff) as in other grasses (Figure 10.5).

In other grasses the grains are protected individually by the glumes or chaff, and separate individually from their inflorescence, so that they can function as units of seed dispersal. The opposite character combination in the corn ear — the firmly attached naked grains and the husk surrounding the whole ear — represents a loss of a mechanism for seed dispersal. But these same characters are useful in agriculture in that they make it possible to harvest the corn efficiently and with minimum loss of kernels. A feature of the more highly developed varieties of modern corn is the large size of the individual grains and of the whole ear (Figure 10.5A). In some primitive varieties of *Zea mays,* as in its wild relatives, the grains are tiny and the ears are small (Figure 10.5B). The grains are hard and flinty as well as small in primitive corn, but are floury or sugary and thus easier to grind or chew in advanced corn.

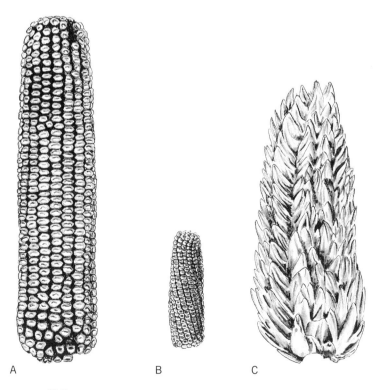

A   B   C

FIGURE 10.5
The corn cob in three varieties of *Zea mays.* (A) Dent corn. (B) Argentine popcorn. (C) Pod corn. (Redrawn from Mangelsdorf, 1958.)

The increased yield of modern corn requires a large leafy shoot to produce the stored food materials. This requirement is met; modern corn is a towering plant for an annual herb, as large as some bamboos, and is much taller than the primitive varieties of corn (Figure 10.4). These and other features of the advanced varieties of corn are useful to man and have been bred into corn during many generations of artificial selection.

Corn has been cultivated by various agricultural groups of American Indians for at least 7000 years. Centers of corn culture developed in eastern North America, the Southwest, Mexico, Central America, and the Andean region. Some primitive characters are found in varieties of corn still being grown in these centers; others are found in long-extinct varieties preserved in archeological deposits.

Thus popcorn is relatively small in stature, with several stalks from the base of the plant, and has small ears bearing tiny hard kernels (Figure 10.5B). Pod corn exhibits another primitive character, namely, the envelopment of the kernels by chaff, as in other grasses (Figure 10.5C).

An important fossil corn cob is that from Bat Cave, New Mexico, a cave inhabited by an agricultural people who cultivated a primitive type of corn. The Bat Cave corn ear was small, smaller than an ear of popcorn. The kernels were tiny, as in popcorn, were borne on long pedicels from a relatively slender central axis, and were enclosed by chaffy bracts, as in pod corn (Figure 10.6). The tip of the spike may have borne staminate spikelets. The oldest specimens of Bat Cave corn are dated by radiocarbon and other methods as 3500–4300 years old. Older fossil corn ears, from Tehuacan Valley, Mexico, about 7000 years old, also possessed the combined features of popcorn and pod corn, and were bisexual with a staminate tip.

A still older fossil deposit, from Mexico City, contains pollen grains like those of modern corn. This deposit is estimated to be about 80,000 years old. Although identification on pollen grains alone, without other plant parts, is necessarily uncertain, the Mexico City fossil pollen is very probably from some ancestral wild form of corn.

Mangelsdorf crossed modern popcorn with pod corn to produce a synthetic pod-popcorn containing the primitive characteristics of each parental type. The synthetic pod-popcorn has a slender, bisexual, grass-like inflorescence with female flowers in the lower part and male flowers above. The grains are enveloped by chaffy bracts and separate readily from the inflorescence when ripe. Synthetic pod-popcorn thus has a grass-like inflorescence. It exemplifies a possible early stage in corn evolution (Figure 10.7B).

Two groups of wild grasses in the American tropics and subtropics are closely related to *Zea mays* and have been considered as possible wild ancestors of

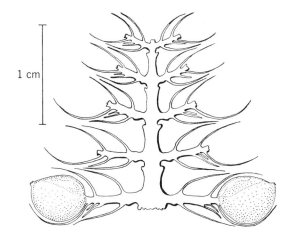

FIGURE 10.6
Longitudinal section of the basal part of a fossil ear of
Bat Cave corn. The tip is missing. (Mangelsdorf, 1958.)

corn. The first group is *Zea mexicana* (teosinte), and the second, the small genus
Tripsacum (gamagrass) (Figure 10.7C, A). Of the two groups, *Zea mexicana*
is the more closely related to *Zea mays*.

Teosinte *(Z. mexicana)* is a widespread annual plant in Mexico and Central
America that frequently occurs as a weed in or near corn fields. *Zea mays* and
Z. mexicana can be crossed artificially, and hybridize naturally in Mexico, and
the F_1 hybrids are fertile. *Zea mexicana* has separate pistillate and staminate
inflorescences, like *Z. mays,* but in the former the pistillate inflorescence is a
branched system of slender spikes surrounded by a husk. In *Z. mexicana* the
pistillate spikelets on a spike are protected individually by special envelopes
(cupules); the cupules are present but reduced and serve a strengthening rather
than protective function in *Z. mays*. In *Z. mexicana* the axis of the spike dis-
articulates, so that the grains separate freely from one another. Teosinte does
not have a cob (Figure 10.7C).

The several species of *Tripsacum,* including *T. dactyloides,* are widely
distributed in the warm parts of the New World. *Tripsacum* differs from *Zea
mays* and *Z. mexicana* in being perennial and having bisexual inflorescences
(Figure 10.7A). The inflorescence of *Tripsacum* is not too different from that
of synthetic pod-popcorn, mentioned earlier. *Tripsacum* can be crossed successfully with *Zea mays,* but not with *Z. mexicana*, and the F_1 hybrids, when
obtained, are more or less sterile.

The candidates for the ancestral stock of *Zea mays* are thus: (1) *Tripsacum,*

(2) some extinct wild form of pod-popcorn (Mangelsdorf, 1958, 1974), or (3) *Zea mexicana* (Galinat, 1971*a*; De Wet and Harlan, 1972). The third hypothesis, postulating an origin of corn from teosinte *(Z. mexicana),* is favored by some students but lacks supporting fossil evidence. With regard to the second hypothesis, it should be noted that the Bat Cave and Tehuacan fossil corn were

FIGURE 10.7
Grain-bearing inflorescences of three relatives of modern corn with primitive characteristics. (A) *Tripsacum dactyloides.* Upper spikelets staminate and lower spikelets pistillate. (B) Synthetic pod-popcorn. Upper spikelets staminate and lower spikelets pistillate. (C) *Zea mexicana.* Branched series of pistillate spikes. (Redrawn from Mangelsdorf, 1958, and Wilkes, 1967.)

types of pod-popcorn. Although differences of opinion continue to exist among students of corn evolution, the weight of evidence seems to favor hypothesis 2.

The morphological differences between *Zea mays* and its wild grass ancestor, whatever the exact identity of that ancestor was, are very great. Nevertheless, these differences can be accounted for as a result of interaction between genetic variability and natural and artificial selection. Modern races of corn contain large stores of genetic variation. Some of this variation is mutational in origin; some stems from spontaneous hybridization between the races of corn and between corn and its relatives, teosinte and *Tripsacum* (Mangelsdorf, 1974).

Some large changes in particular morphological features in *Zea mays* are known to be produced by single genes and simple gene systems. Pod corn, with its chaffy bracts, for example, is determined by a mutant allele (*Tu,* tunicate) of the gene *Tu* on chromosome 4. The difference between sessile and stalked pistillate spikelets is another example of a genetically simple but phylogenetically significant character difference in corn. The difference between flinty and relatively soft kernels is a single-gene difference (Mangelsdorf, 1958; Galinat, 1971*a*). Some of the morphological changes in the evolution of corn could then be due to the artificial selection of a relatively small number of major genes and their modifier genes. Other morphological and physiological changes in corn evolution are due to the selection of more complex systems of multiple factors.

Conclusions

We have examined cases where selection of mutations in single genes has brought about relatively simple evolutionary changes (black peppered moths, toxin-resistant bacteria). Other cases involve selection for multifactorial characters (protein content of grain in corn). But in the evolution of corn itself from its wild ancestors we are confronted with changes of greater magnitude, changes in a complex of characters determined by a combination of genes. A black peppered moth is still a peppered moth, but modern corn represents an essentially new type of plant in the grass family.

Critics of selection theory have long argued that selection is a negative force; it can eliminate the unfit, but, in combination with blind mutations, it cannot be expected to produce anything new (e.g., Koestler and Smythies, 1969).

This criticism misses the point that character complexes are determined not by simple genes, but by gene combinations. Sexual reproduction is a mechanism for assembling the combinations of alleles that determine new character complexes, as we have seen earlier. And selection is a mechanism that can then establish the new allele combinations in populations if they are favorable. Seen in this light, selection is a mechanism for bringing about highly improbable events, as pointed out by Fisher (1930, 1958) and Huxley (1943, pp. 474–475).

A corn cob is an improbable phenomenon in the grass family. What are the chances of a corn cob's developing *without* the mechanism of selection?

Collateral Readings

Darwin, C. 1875. *The Variation of Animals and Plants Under Domestication.* Ed. 2, 2 vols. John Murray, London.

Ford, E. B. 1964. *Ecological Genetics.* Ed. 1. Methuen, London.

Heiser, C. B. 1973. *Seed to Civilization.* W. H. Freeman and Co., San Francisco.

Mangelsdorf, P. C. 1974. *Corn: Its Origin, Evolution, and Improvement.* Harvard University Press, Cambridge, Mass.

Sheppard, P. M. 1959. *Natural Selection and Heredity.* Philosophical Library, New York.

11

Modes of Selection

Directional Selection

The selection process appears in various forms and at various levels of biological organization. The main modes and levels will be discussed in this chapter.

The mode of selection discussed in the preceding chapters is known as directional (or progressive) selection. Directional selection brings about a progressive or unidirectional change in the genetic composition of a population (Figure 11.1). It occurs when a population is becoming adapted to a new environment, or, conversely, when the environment is changing progressively and the population is keeping pace with it. A good example of the first situation is the rise in frequency of the melanic form of *Biston betularia* and other moths when confronted with a sudden change in environment, namely, industrial soot (Chapter 10). Environmental changes that occur as gradual trends, such as many climatic changes, illustrate the second situation that evokes directional selection.

Stabilizing Selection

Stabilizing selection can be contrasted with directional selection. If a population is well adapted to a given environment that remains stable, the main action of selection is to eliminate such ill-adapted and peripheral variants as arise by

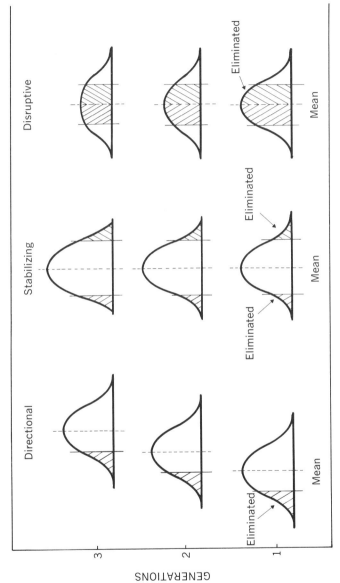

FIGURE 11.1
The effects of three modes of selection on the genetic variation of a population.

mutation, gene flow, segregation, and recombination. A range of genotypes of proven adaptedness is thus preserved while the ill-adapted types are weeded out. This is stabilizing selection.

The effect of stabilizing selection is shown graphically in Figure 11.1. The genetic variation in a population is assumed to have a normal distribution, with numerous individuals near the mean and a few individuals at the extremes for any given variable and measurable character. Under stabilizing selection the peripheral variants on both extremes of the normal curve are eliminated, generation after generation. The preferential reproduction of the individuals possessing characteristics around the mean for the population results in the preservation of a constant modal condition through time. This is contrasted with directional selection, in which the elimination of the genetic variation is one-sided and the mean consequently shifts during a succession of generations (Figure 11.1, left).

Prout (1962) carried out an experiment on stabilizing selection in *Drosophila melanogaster*. The variable character used was time of development of the flies. In the line subjected to stabilizing selection, the individuals close to the mean for development time were selected in each generation. After a number of generations of such stabilizing selection, the variance of development time decreased in this line relative to that in the starting generation and in unselected control lines.

Stabilizing selection is ubiquitous, though unspectacular, and is the most common mode of selection in nature. We have previously referred to the striking example of directional selection in *Biston betularia,* but note the complementary role of stabilizing selection in the moth populations. Prior to industrialization, the rare melanic mutants were weeded out of the populations by stabilizing selection, a selective process that probably went on for many centuries; and in the altered industrial environment occupied by the new, predominantly melanic populations, stabilizing selection is again at work, weeding out the occasional gray types.

Disruptive Selection

In disruptive selection (also known as diversifying selection), the extreme types in a polymorphic population are selected for, and the intermediate types are selected against. The result is to preserve and accentuate the polymorphism (Figure 11.1, right).

Consider again Prout's (1962) selection experiment on development time in *Drosophila melanogaster.* Disruptive selection was practiced in one line by mating the earliest flies with the latest flies to emerge in each generation. The result, after a number of generations, was an increase in the variance of development time as compared with the unselected control lines.

A series of experiments on disruptive selection in *Drosophila melanogaster* has been carried out by Thoday and his co-workers. One of these experiments (Thoday and Boam, 1959), to be described here, involves selection for number of body bristles (chaeta number) under interesting experimental conditions.

Selection was carried out concurrently for high bristle number in a set of high lines and for low bristle number in a set of low lines. This disruptive selection tended to subdivide the original population into different "high" and "low" subpopulations. But the high and low lines were constantly inter-crossed so that gene flow between them was always present and was working in opposition to the disruptive selection. The experiment was continued in this manner to generation 36.

The high and low lines did not diverge significantly during the early generations. They began to diverge in generation 14 and the divergence became large after generation 30. The divergent selection curves for one high and one low line are shown in Figure 11.2. It should be noted that the high and low lines here represent a different situation from that of the high and low lines in the corn-selection experiment described earlier (Figure 10.1); the high and low lines

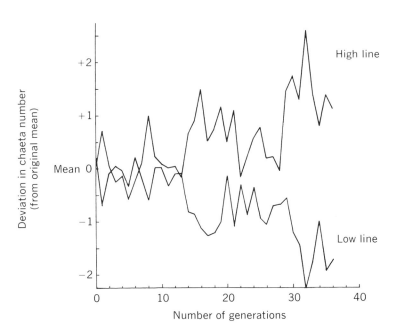

FIGURE 11.2
Response to disruptive selection for bristle number in *Drosophila melanogaster* under conditions of maximum gene flow. Further explanation in text. (Redrawn from Thoday and Boam, 1959.)

were kept isolated in the corn experiment but were cross-mated in the present experiment. This experiment shows, therefore, that disruptive selection can prevail over gene flow.

Experiments of Streams and Pimentel (1961) with *Drosophila melanogaster* are interesting in quantifying the interaction between disruptive selection and gene flow. Four levels of gene flow were used (6%, 20%, 50%, and 0% in the control lines) and two intensities of disruptive selection (moderate and strong). Strong disruptive selection produced its effects in spite of 20% gene flow but was ineffective with 50% gene flow. Moderate disruptive selection, however, was overcome by even slight (6%) gene flow.

For further details concerning experiments on disruptive selection in *Drosophila* the reader is referred to the following papers: Thoday and Boam (1959); Millicent and Thoday (1961); Thoday and Gibson (1970); Thoday (1972); Streams and Pimentel (1961); Soans, Pimentel, and Soans (1974).

Disruptive Selection in Nature

The efficacy of disruptive selection is now well established under experimental conditions. The next question is to assess the role of this force in natural populations. Here we are on less certain ground; we must speak now in terms of probabilities rather than certainties.

One situation in nature in which disruptive selection probably comes into play is that where well-differentiated polymorphic types have a definite selective advantage over poorly differentiated polymorphic types. This situation is realized in the case of sexual dimorphism. Females and males with well-differentiated secondary sexual characters have greater success in mating and reproduction than various types of intermediates (intersexes, homosexuals, etc.).

A second possible situation arises where a polymorphic population occupies a heterogeneous habitat. The polymorphic types are assumed to be specialized for different niches or subniches in the habitat and to live mainly in their respective special niches. Thus polymorphic type *A* might be adaptively superior in subniche *A'* but inferior in subniche *B'*, whereas morph *B* would flourish in subniche *B'* but not in *A'*. This alleged correspondence between polymorphic types and the array of subniches in a heterogeneous habitat could be brought about by disruptive selection. And conversely, disruptive selection in this situation would enable a polymorphic population to colonize a wider range of subniches in its habitat (Levene, 1953). This hypothesis is plausible enough but is not yet supported by very much hard evidence.

An example that apparently conforms to the above model occurs in the sulphur butterfly, *Colias eurytheme.* The females of this North American species are polymorphic for wing color; there are orange-winged females and white females. This polymorphism is controlled by a single gene.

In midsummer at several localities in California the white and the orange forms reach their peaks of activity at different times of day. The white form is more active in the early morning and late afternoon, and the orange form in midday, suggesting that the two polymorphic types have different temperature and humidity preferences. There is also a seasonal change in the frequency of the two color forms in California populations, the white form rising in frequency as temperatures decline in the fall (Hovanitz, 1953; Remington, 1954). Physiological studies show that wing pigmentation plays an important role in temperature regulation in *Colias eurytheme* (Watt, 1969). The color polymorphism in this species thus appears to broaden the temperature range and perhaps the length of the season in which the populations can remain active.

A third situation in which disruptive selection operates in nature is found in plants under certain special conditions. The plant population is sedentary but outcrossing and stands astride two ecologically different zones—for instance, two soil types or two topographic zones. It is not uncommon to find a different set of adaptive characteristics developed in the two halves of the plant population. This differentiation is maintained in spite of interbreeding. The chief controlling factor involved is almost certainly disruptive selection.

A high-montane species of pine, *Pinus albicaulis,* occurs at and just above tree line in the Sierra Nevada of California. The populations on mountain slopes up to timberline consist of erect trees, the common growth form of the species. Above timberline the species is represented by a low, horizontal, elfinwood form. The arboreal and the elfinwood subpopulations are contiguous and are cross-pollinated by wind, as shown by the presence of some intermediate individuals, but maintain different growth forms in their respective subalpine and alpine ecological zones under the influence of divergent selective pressures. Parallel racial variation across the narrow timberline belt occurs in other associated high-montane species of conifers and willows (Clausen, 1965).

Another set of examples illustrating the same set of forces involves pasture grasses in lead-mining districts in Wales. In such districts the soil changes abruptly from ordinary pasture soils without significant quantities of lead to mine soils containing much lead. *Agrostis tenuis* and other species of grasses in the region form continuous populations cutting across the different adjacent soil types. The populations are held together genetically by wind cross-pollination. Nevertheless, minor racial divergences have developed between lead-intolerant subpopulations on the pasture soils and lead-tolerant subpopulations on the mine soils (Jain and Bradshaw, 1966; Antonovics, 1971).

Balancing Selection

Heterozygotes are often superior to the corresponding homozygous types in general vigor or in some specific component of viability, such as competitive ability or disease resistance. This superiority can occur in genotypes that are heterozygous for either a single gene or a block of genes.

When a heterozygote *Aa* has a selective advantage over one or both homozygous types, an effect of selection will be to preserve both alleles (*A* and *a*) in the population. An equilibrium frequency of *A* and *a* will develop in the gene pool, the exact level of which will depend upon the relative selective values of the alternative polymorphic types. But neither the *A* nor the *a* allele will go to extinction in the population insofar as the allele frequencies are controlled by selection. A state of balanced polymorphism will be maintained.

Selection in favor of heterozygotes is known as balancing selection (also heterozygote superiority or heterozygote advantage).

Under balancing selection, the alternative alleles or gene blocks do not go to complete fixation or extinction, as noted above, but instead the favored unit of selection is a heterozygous allele pair or gene combination. This has a bearing on the expected effects of directional selection in a balanced polymorphic system. Directional selection in combination with balancing selection is expected to bring about not a replacement of one gene or gene block by another in the gene pool, but a series of shifts from one heterozygous combination to another, e.g., $A_1A_2 \rightarrow A_2A_3 \rightarrow A_3A_4$ (Lerner, 1954, pp. 113–114).

Balanced polymorphism based on heterozygote superiority is a fairly widespread phenomenon. Among animals it is found, for example, in grasshoppers, chickens, mice, man, and various species of *Drosophila;* among plants it is found in corn *(Zea mays),* barley *(Hordeum sativum), Arabidopsis,* and *Oenothera.* The most thoroughly investigated case is that in *Drosophila pseudoobscura,* which will be described in the next section.

Heterozygote Superiority in *Drosophila pseudoobscura*

The wild North American species *Drosophila pseudoobscura* is remarkable for the variation in its chromosome III. Some 16 types of chromosome III, differing from one another with respect to inversions, are known in the species (as was mentioned in Chapter 4). The inversion types, which can be identified cytologically in the salivary gland chromosomes of the larvae, are designated by names (Standard, Arrowhead, Chiricahua, Timberline, Pikes Peak, etc.) or by letter symbols (*ST, AR, CH, TL, PP,* etc.). Most populations are polymorphic for some of the inversion types and produce the various possible homozygotes and heterozygotes.

The inverted chromosome segments differ in their genic contents as well as in their cytological features. A population polymorphic for *ST, CH,* and *AR* contains not only three types of chromosomes but also three different sets of genes on the inverted segments; and the *ST/CH, ST/AR,* and *CH/AR* geno-types are genic heterozygotes as well as inversion heterozygotes. Now inversions prevent effective gene recombination in inversion heterozygotes. Consequently the inversion polymorphism in *Drosophila pseudoobscura* is at the same time a polymorphism for blocks of genes that remain intact from generation to generation.

Dobzhansky and his school have explored the inversion polymorphisms in natural populations of *D. pseudoobscura* throughout the range of the species. The most extensive sampling has been carried out in certain populations in the Sierra Nevada and San Jacinto Mountains in California that have been studied over a period of many years. Much experimental work, using population cages in the laboratory, has also been carried out with strains of flies collected from the same natural populations in California.

The common inversion types in these California mountain populations are *ST, CH,* and *AR;* also present in low frequencies are *TL, PP,* and *SC.* The populations often exhibit an excess of inversion heterozygotes over expectation on the basis of a Hardy-Weinberg equilibrium. One determines the actual frequency of the various inversion types in the gamete pool, by appropriate methods of population sampling; and from this information one calculates, by the Hardy-Weinberg formula, the expected frequency of heterozygous geno-types. The latter is then compared with the known actual frequency of inversion heterozygotes. A statistically significant excess of heterozygotes is regularly found in the natural populations at certain seasons of the year.

Thus a population in the San Jacinto Mountains, polymorphic for five in-version types, was sampled in May, 1952, and found to have the following observed and expected frequencies of inversion heterozygotes and homozygotes (data of Epling, Mitchell, and Mattoni, 1953):

	Observed	Expected
ST/CH	0.232	0.168
ST/AR	0.232	0.168
All heterozygotes	0.830	0.724
ST/ST	0.134	0.184
All homozygotes	0.170	0.276

There is a consistent excess of heterozygotes and a corresponding deficiency of homozygotes as compared with the expected Hardy-Weinberg proportions.

The observed deviations from the Hardy-Weinberg equilibrium could be explained on the basis of selection for heterozygotes. One way to test this hypothesis is to compare the frequencies of inversion heterozygotes and homozygotes in an egg sample taken from the natural population with the frequencies in a sample of adult flies. The egg samples were found to contain the heterozygous and homozygous genotypes in Hardy-Weinberg proportions. The deficiency of the homozygous classes in the adult samples must therefore be due to a differential mortality during development between the egg and adult stages, operating in favor of the heterozygotes (Dobzhansky and Levene, 1948).

Parallel evidence for heterozygote superiority was obtained in artificial populations reared in population cages. The founders of the artificial populations were strains of flies from the San Jacinto Mountain natural population carrying *ST, CH,* and *AR* chromosomes. The population cages were maintained at warm temperatures and allowed to become overcrowded. In the egg stage the inversion homozygotes and heterozygotes were found to be present in Hardy-Weinberg proportions. But the same artificial populations exhibited a significant excess of inversion heterozygotes in the adult stage (Dobzhansky, 1947*a*).

Developmental and behavioral studies reveal that the superior fitness of the inversion heterozygotes can be resolved into a number of different components: greater pre-adult viability, development rate, longevity, fecundity, and mating speed (Moos, 1955; Dobzhansky, 1970, pp. 137–138).

It should be stressed again that the heterozygote superiority is manifested only under certain conditions of temperature, food medium, and population density; when these conditions are not fulfilled, in either the natural or artificial populations, the inversion heterozygotes cease to have a selective advantage over the homozygotes.

The important point is that the environmental conditions bringing heterozygote superiority to expression do exist periodically in natural populations. Balancing selection acts intermittently if not continuously in these populations. Its intermittent action is sufficient to maintain a permanent balanced polymorphism. The *AR* and *CH* chromosomes rise in frequency in the natural populations in the cool early summer (May and June), and the *ST* chromosomes increase in the hot summer and fall (July to September or October), but the changes never go all the way to complete fixation or extinction (Dobzhansky, 1943, 1947*b*, 1948*a*). The balancing selection thus plays a role in increasing the seasonal range of a population over what it would presumably be if it were monomorphic.

Sexual Selection

A pronounced sexual dimorphism exists in many groups of animals. The dimorphism usually consists of the development of sex-limited characters in the males—characters that are involved indirectly in reproduction. Such secondary sexual characters fall into two broad classes: characters of size, strength, and armature (e.g., the superior size of male sea lions, the antlers of male deer); and characters of ornamentation and display (e.g., colorful plumage of male ducks, gorgets of male hummingbirds, special songs, courtship displays).

In *The Descent of Man and Selection in Relation to Sex* (1871), Darwin reviewed the then known instances of secondary sexual characters in male animals, belonging to a wide range of animal groups, and advanced the theory of sexual selection to account for them.

Darwin conceived of sexual selection as a process supplementary to the more general and pervasive natural selection. The latter, in Darwin's terminology, was the process that built up the adaptive features of a species as a whole, including the secular adaptations common to the two sexes and the primary sex differences related directly to reproduction. When all of these general characteristics were accounted for, by natural selection in the original Darwinian sense of the term, a large class of special, male-limited characters was found to be left over and unaccounted for. Such secondary sexual characters as the antlers of male deer and the colorful plumage of male ducks are not adaptively beneficial to the species as a whole, nor are they necessary for reproduction. But these characters can be hypothesized to enhance the mating success of the male individuals that possess them. The theory of sexual selection was introduced at this point in order to explain the development of such special male characters.

Darwin describes the proposed process as follows (1871, Ch. 8): ". . . sexual selection . . . depends on the advantage which certain individuals have over others of the same sex and species solely in respect of reproduction." He goes on to explain (and I am paraphrasing his statement here while preserving his phraseology). When the two sexes follow exactly the same habits of life, yet the male has sensory or locomotive organs more highly developed than those of the female, the males have probably often acquired their characters not from being better fitted to survive in the struggle for existence, but from having gained an advantage over other males in reproduction. "Sexual selection must here have come into action."

The theory of sexual selection has had a tortuous history since 1871. It was controversial in Darwin's time. The subject later fell into a state of neglect. When revived in the modern period of evolution studies, the subject entered upon a new scene, one in which the same phenomena were being viewed from other alternative standpoints. Sexual selection has become mildly controversial again. It is difficult if not impossible to assess the role and importance of sexual selection with any degree of precision at the present juncture.

There is no particular theoretical problem in regard to sexual selection. Sexual selection is one of several modes of natural selection (and not an alternative process, as in the nineteenth-century classification of concepts). More particularly, it is a mode of selection, again one of several, wherein the selective discrimination operates between competing members of a subgroup of a population. In this case the subgroup is the male sex and the characters undergoing selection are those affecting success in mating.

A serious difficulty arises from the fact that secondary sexual characters constitute a very varied assemblage of features. They serve diverse functions in the life of the population. Their formation cannot be attributed to any one single mode of selection. In size differences between sexes we have to consider selection for ecological divergence (See Part V) as well as sexual selection. The weapons of males may serve to establish territories or to establish social dominance as well as to secure females. Male adornment, displays, songs, and scents may play roles as courtship stimuli and as species-specific recognition signals. In the latter case they could be due to selection for reproductive isolation (see Part V) as well as sexual selection. We have, then, a mixture of selective processes at work, and under these conditions it is difficult to sort out the effects of sexual selection alone.

One set of cases in which sexual selection can be concluded to play an important role is that of mammals and birds with a polygamous mating system. Under polygamy the strongest males collect and guard a harem of females, or a territory containing their females, and drive off their weaker brothers, who consequently possess few or no mates.

Many mammals and some birds are polygamous. Among mammals polygamy occurs in deer, cattle, sheep, most antelope, elephants, seals, sea lions, walruses, and baboons; among birds it occurs in chickens, pheasants, and peacocks. Secondary sexual characters are well developed in the males of these animals, in marked contrast to the condition in related, non-polygamous groups.

Thus in polygamous chickens, pheasants, and peacocks the cocks are notably larger, more pugnacious, and better decorated than the hens, while in monogamous partridge, grouse, and ptarmigan the differences between the sexes are relatively slight. In walruses and sea lions, the males are huge; in many ungulates they possess antlers or horns; and in baboons they are large and aggressive. By contrast, the sexes are nearly equal in size and strength in monogamous wolves and certain monogamous species of monkeys, in the members of the cat family that form maternal families, and in colonial but non-polygamous rodents (Darwin, 1871, Ch. 8).

The reproductive behavior of bighorn sheep *(Ovis canadensis)* has been carefully studied in the wild and in captivity by Geist (1971). The adult rams have large heavy horns and a thick skull, as is well known. The rams engage in butting contests during the rutting season, when they are competing for access to the ewes. The fights are often severe and may result in the wounding or death of the vanquished animals. The observed facts of rutting behavior in sheep and other ruminants are at variance with a commonly held opinion in animal behavior studies that male animals engage mainly in non-injurious bluff and other displays. Displays of symbols of rank in the dominance hierarchy of rams are present, to be sure, but are backed up by combat if necessary.

Some individual rams emerge from combat as dominant members of their social group. These dominant males then chase other males off from the ewes, but copulate freely with the ewes themselves. The mating system in bighorn sheep does not preclude some copulation between subordinate males and ewes, but it does ensure preferential breeding by the dominant males. The latter therefore apparently sire more lambs than the subordinate males. It follows that some horn and skull characters of male bighorn sheep, and their offensive and defensive skill in inter-male combat, have presumably been developed by sexual selection (see Geist, 1971).

The case for a sexual selective origin of male display characters in non-polygamous groups is less clear. This is not to say that sexual selection is not involved, in some instances at least. The development of male display characters in some non-polygamous groups may be a product of the joint action of ordinary individual selection and sexual selection. The initial stages of development of the characters could be fostered by directional selection, to promote courtship and mating; then sexual selection might carry the character development to greater extremes, on the basis of the mating advantage of the more highly decorated males (Mayr, 1972*b*).

Selection at Subindividual Levels

The modes of selection discussed so far all take place at the individual level of organization, and this is indeed the basic form of selection. But selective processes can also discriminate between reproducible biological units lower than the individual organism in the hierarchy of organization. Several types of subindividual selection will be introduced in this section.

In flowering plants the male gametes are carried by an independent unit, the male gametophyte, consisting of the pollen grain and pollen tube. A heterozygous plant produces genetically different classes of pollen, and, in some cases, the pollen segregates for genetic factors affecting the viability, germination ability, or growth rate of the pollen. Furthermore, pollen is normally produced and delivered to the stigma in excess of the numbers required for fertilization, so that a competition among pollen grains or tubes ensues.

The result of this competition, where the pollen is segregating for growth factors, is that some classes of male gametes are more effective than others in fertilization and hence in the production of embryos or endosperm in the seeds. Let some marker gene determining a morphologically visible trait be linked to the pollen growth factor. Then, instead of the expected Mendelian ratio for the marker gene, one observes an altered ratio. The morphological type determined by the marker allele borne by the superior class of pollen is present in excess, and the opposite morphological type is deficient, in the seed or seedling generation.

Altered segregation ratios due to selective discrimination against certain classes of pollen are well known in various species of flowering plants. A well-analyzed case in corn *(Zea mays)* concerns the linked genes *Su* (for type of endosperm) and *Ga* (controlling pollen tube growth). The *Su/su* heterozygote usually produces corn kernels segregating into starchy vs. sugary endosperm classes in Mendelian ratios. But let the sugary allele *(su)* be linked to the *ga* allele (for slow pollen tube growth) in a heterozygote *(Su Ga / su ga),* and let the heterozygote be used as a male parent, so that the linked segment *su-ga* is transmitted through the pollen. Then a marked deficiency of sugary types appears in the kernels in the next generation (Mangelsdorf and Jones, 1926).

Certain gene-controlled aberrations of meiosis in *Drosophila melanogaster* cause one chromosome of a homologous pair to become included in more functioning sperms than the other. The distorted chromosome segregation at meiosis leads to an increase in frequency of one chromosome type over that of its homolog in the gamete pool. This process is known as meiotic drive.

Corresponding changes in frequency take place in the genes borne on the two homologous chromosomes and in the phenotypes that they determine. In one case the sex ratio is altered in the direction of an excess of females; in another case a recessive lethal is caused to rise in frequency (Sandler and Novitski, 1957; Hiraizumi, Sandler, and Crow, 1960).

Meiotic drive is essentially a process of differential reproduction of homologous chromosomes. It therefore seems to represent a form of inter-chromosomal selection.

The single-gene model of selection was introduced in Chapter 8 as a useful abstraction, as a stepping-stone toward the understanding of selective processes in real populations. However, it is a logical necessity to suppose that single-gene selection was itself a real process at very primitive stages of evolution, when the prevailing units of biological organization were gene-like particles. A close approximation to gene selection is still found today in the differential reproduction of virus particles or bacterial cells differing in a single viral or bacterial gene.

Interdeme Selection

We turn next to the question of selection at the level of the local population, or interdeme selection. Interdeme selection is differential reproduction of different local populations.

A parenthetical note on terminology is needed. Interdeme selection is the general term. Population selection is a synonym. The term group selection is also sometimes used as a synonym of interdeme selection, but will be used here for a special case of interdeme selection in which the populations involved are social groups. Kin selection, as the term suggests, is selection on the basis of kinship; it represents a still more special case. Kin selection comes close to group selection where, as is often the case, the social group is composed of kin.

In this section we will discuss interdeme selection in a general way, and will go on to mention probable examples of group selection. Group selection is discussed again in Chapter 37.

Population biologists are far from being of one mind concerning the reality of interdeme selection. Williams (1966) and others argue that it is unnecessary to postulate a process of interdeme selection—that modifications of individual selection are capable of bringing about the effects attributed to population selection. Wright (1931, 1956, 1960) and other students, including myself, believe that interdeme selection is a real process in some situations in nature. Some of the opposing points of view are presented by Williams (1971) and Alexander (1974).

Wright (1931, 1960) compares two types of population systems—a large continuous population and a series of small semi-isolated colonies—with respect to the theoretical effectiveness of selection. The total size *(N)* is assumed to be the same in both population systems, and the organism is assumed to be outcrossing.

In the large continuous population, selection is relatively ineffective in raising the frequency of favorable but rare recessive mutations. Furthermore, any tendency of a favorable allele to rise in frequency in one part of the large population is opposed by interbreeding with neighboring subpopulations in which the allele is rare. In the same way, favorable new gene combinations that succeed in becoming assembled in one local segment of the population are broken apart and swamped out by interbreeding with neighboring segments.

These difficulties are largely obviated in the island-like population system. Here selection, or selection and drift jointly, can quickly and effectively raise the frequency of a rare favorable allele in one or more small colonies. Favorable new gene combinations can also be established readily in one or more small colonies. Isolation then protects the gene pools of these colonies from the swamping effects of migration from and interbreeding with other, less-favored colonies. The model up to this point invokes only individual selection or individual selection combined with drift, in certain colonies.

We will now suppose that the environment of the population system has changed in such a way as to lower the adaptedness of the old genotypes. The new favorable genes or gene combinations established in certain colonies have a high potential adaptive value for the population system as a whole in the new environment. The stage is now set for interdeme selection to come into play. The less-fit colonies dwindle and become extinct, while the better-adapted colonies expand and replace them throughout the range of the population system. The subdivided population system acquires a new set of adaptive characteristics as a result of individual selection within some colonies followed by the differential reproduction of different colonies.

There are some features in the organic world that are difficult to explain on the basis of individual selection, but that can readily be explained as products of interdeme selection. The first of these is sexual reproduction, which benefits the population in future generations, but not the parental types themselves. The second is altruistic behavior in social animals, which benefits the social group but often endangers the altruistic individual. The third is the adaptive characteristics of the sterile worker caste in social insects.

The mode of selection responsible for the evolution of sex has been discussed extensively over the years and is under renewed discussion currently. Some evolutionists, including myself, favor the idea that interdeme or interspecific selection has played an important role in the evolution of sex. The reason for holding this idea is that sexual reproduction benefits the population in future generations, by providing a supply of recombinational variations, but the parental genotypes are broken up in the recombination process. However, it has recently been shown by models that individual selection *can* favor sexual reproduction under certain assumed conditions (Williams, 1975; Felsenstein and Yokoyama, 1976). Whether the development of sex by individual selection is a likely process or a far-fetched one is another question. The problem remains open.

One problem with respect to altruistic behavior is to analyze the selective advantage of this type of behavior into individualistic and social components. Consider a socially useful behavioral trait in a social animal, such as defensive action against enemies. What fraction of this characteristic is due to the maximal reproductive success of the individual animals possessing the trait, and what fraction, if any, to the enhanced reproductive success of the social group? It is difficult to isolate the individual and its success or failure from the social group in a social animal, since what benefits the individual often benefits the group, and vice versa. My own particular hunch is that some fraction of altruistic behavior has resulted from interdeme selection.

Altruism is attributed to kin selection in current thinking. However, this does not mean that interdeme selection is necessarily ruled out, as is sometimes implied, since kin selection, as noted earlier, comes close to deme selection in many actual cases.

As an example of social insects, let us consider the honeybee *(Apis mellifica)*. The beehive or colony is a type of population consisting of females (queens), males (drones), and neuters (workers). The queens and drones are sexually fertile but economically useless.

The important functions of food getting, defense and maintenance of the colony, and rearing of the broods are carried out by the workers, which possess bodily and psychic adaptations for performing these functions. The chief secular adaptations by which the colony lives reside in these worker bees. Yet the workers, being sexual neuters, do not reproduce as individuals, and consequently have no opportunity to pass the genes determining their adaptive characteristics on to the next generation. That task is performed by the queens and drones. If, therefore, the queens and drones do not carry genes making for adept and efficient worker bees, the hive will not thrive, and may be eliminated by competition from other hives containing better-adapted workers. Such a process of replacement of one strain of beehives by another has actually been observed in recent years in various parts of the world (see Michener, 1975). The unit of selection in this case definitely seems to be the colony as a whole (see Darwin, 1872, Ch. 8; also Grant, 1963, p. 260).

Social bees represent one of the extremes of social integration in the animal world. And they exhibit the effects of interdeme selection in a pronounced fashion. The corollary is that we might expect less highly socialized animals, such as many mammals and birds, to be less affected by interdeme selection, yet to possess *some* characteristics produced by this mode of selection.

Selection at the Species Level

Several modes of selection operate at the species level of organization: (1) selection for ecological divergence; (2) selection for reproductive isolation; and (3) species replacement. These are merely mentioned here in order to round out our survey; they will be discussed in Part V on speciation phenomena.

Collateral Readings

1. DISRUPTIVE SELECTION

   Thoday, J. M. 1972. Disruptive selection. *Proc. Roy. Soc. London, B* **182:** 109–143.

2. BALANCING SELECTION

   Dobzhansky, Th. 1970. *Genetics of the Evolutionary Process.* Columbia University Press, New York. Chapters 5, 6.

3. SEXUAL SELECTION

   Campbell, B., ed. 1972. *Sexual Selection and the Descent of Man 1871–1971.* Aldine-Atherton, Chicago.

   Darwin, C. 1871. *The Descent of Man and Selection in Relation to Sex.* Ed. 1. John Murray, London. Various modern reprints available.

   Geist, V. 1971. *Mountain Sheep: A Study in Behavior and Evolution.* University of Chicago Press, Chicago.

   Maynard Smith, J. 1958. Sexual selection. In *A Century of Darwin,* ed. by S. A. Barnett. Heinemann, London.

4. LEVELS OF SELECTION

   Grant, V. 1963. *The Origin of Adaptations.* Columbia University Press, New York. Chapter 10.

   Lewontin, R. C. 1970. The units of selection. *Ann. Rev. Ecol. Syst.* **1:** 1–18.

   Wright, S. 1956. Modes of selection. *Amer. Nat.* **90:** 5–24.

5. SELECTION AT SUBINDIVIDUAL LEVELS

   Grant, V. 1963. *Op. cit.* Chapter 10, in part.

   Grant, V. 1975. *Genetics of Flowering Plants.* Columbia University Press, New York. Chapter 13, in part.

   Lewontin, R. C. 1970. *Op. cit.*

6. INTERDEME SELECTION

   Alexander, R. D. 1974. The evolution of social behavior. *Ann. Rev. Ecol. Syst.* **5:** 325–383.

   Grant, V. 1963. *Op. cit.* Chapter 10, in part.

   Lerner, I. M. 1954. *Genetic Homeostasis.* Oliver & Boyd, Edinburgh and London.

   Williams, G. C. 1966. *Adaptation and Natural Selection.* Princeton University Press, Princeton, N.J.

   Williams, G. C., ed. 1971. *Group Selection.* Aldine-Atherton, Chicago.

Wilson, E. O. 1975. *Sociobiology: The New Synthesis.* Harvard University Press, Cambridge, Mass. Chapter 5.

Wynne-Edwards, V. C. 1962. *Animal Dispersion in Relation to Social Behaviour.* Oliver & Boyd, Edinburgh and London.

7. KIN SELECTION

Emlen, J. M. 1973. *Ecology: An Evolutionary Approach.* Addison-Wesley, Reading, Mass. Chapter 3, in part.

Hamilton, W. D. 1964. The genetical evolution of social behavior. *J. Theor. Biol.* **7:** 1–52.

Wilson, E. O. 1975. *Op. cit.* Chapter 5.

12

Effects of
Individual Interactions

Competition

Interactions between individuals of one sort or another affect the process of selection. The most general of these individual interactions is competition for necessary raw materials, energy sources, or living space. Natural selection operates most effectively under conditions of competition; conversely, the selective pressures are relaxed when competition is absent from the scene.

This generalization is supported by scores of selection experiments in *Drosophila pseudoobscura*. The selective advantages of the various inversion homozygotes and heterozygotes (ST/ST, ST/CH, ST/AR, etc.) appear only when the population cages are crowded, not in thinly populated cages (see Chapter 9).

It is probably correct to say that selection *can* operate in the absence of competition, particularly when the selective forces are physical environmental factors, but that the effectiveness of the selection is much greater when competition exists.

Sukatchev (1928) compared the viabilities of three asexually reproducing strains of the dandelion *(Taraxacum officinale)* in experimental plots in Leningrad. The strains are designated as biotypes A, B, and C. They were grown in pure open stands (with plants spaced 18 cm apart), in pure dense stands (individuals 3 cm apart), and in mixed stands (A, B, and C). Natural mortality in the different plantings was measured at the end of two years.

We are concerned here with the two types of pure stands. The observed viabilities of the three strains, as represented by the percentage of individuals surviving through two years, was as follows:

Strain	Survival, open stand	Survival, dense stand
A	77%	27%
B	69%	49%
C	90%	24%

Strain C is superior in open stands. But under competition, in dense stands, B is superior to both C and A. Apparently the genotypically determined ability to weather physical environmental factors is different from the ability to succeed in a strong competition (Sukatchev, 1928).

Subsequent experiments along similar lines in both plants and insects have led to the concept of competitive ability as a special gene-controlled property of the organism. In barley *(Hordeum sativum)* and flour beetles *(Tribolium castaneum* and *T. confusum)*, different genotypes differ in their degree of competitive ability (Sakai and Gotoh, 1955; Lerner and Ho, 1961). The barley experiments show that competitive ability is a character separate from general vigor.

In the higher vertebrates, competitive ability often assumes the form of aggressive behavior. Aggressive behavior is usually associated with competition, is manifested most strongly and sometimes exclusively under conditions of crowding, and is a technique for coping with competition (Wilson, 1971).

Mixtures of Genotypes Under Competition

A given genotype often reacts very differently, as regards competitive ability, in a pure dense culture and in a mixed dense culture. In wheat *(Triticum vulgare)* and oats *(Avena sativa)*, two varieties grown together in competition give a larger number of plants at harvest and a higher total yield than either variety sown alone (Gustafsson, 1951). Chromosomally polymorphic experimental populations of *Drosophila pseudoobscura* likewise exceed monomorphic populations in total biomass of flies produced from a given amount of food (Beardmore, Dobzhansky, and Pavlovsky, 1960).

Furthermore, the genotype that succeeds well in competition with other individuals of the same genotype is not necessarily a superior competitor in a polymorphic mixture of genotypes. And conversely, a relatively poor competitor in a dense pure stand may become the dominant member in a mixed stand (Gustafsson, 1951).

This is shown by another part of Sukatchev's (1928) experiment with *Taraxacum*. He grew strains A, B, and C in pure dense stands and in a mixed dense stand in the experimental garden, and counted the number of flowering heads per individual, a good measure of plant vigor and fruitfulness. The results were as follows:

Strain	Heads/individual, pure dense stand	Heads/individual, mixed dense stand
A	20–35	1–8
B	34–43	12–20
C	8–11	16–23

There is a reversal in the relative competitive ability of genotype C in going from a pure culture to a polymorphic mixture.

Similar results have been observed repeatedly in single-variety vs. mixed plantings of barley, wheat, timothy *(Phleum pratense),* and other grasses (Gustafsson, 1951). And again, parallel results are obtained in experimental monomorphic and polymorphic populations of *Drosophila melanogaster* and *Tribolium castaneum* (Lewontin, 1955; Sokal and Karten, 1964).

Density-Dependent Factors

Some factors in the environment that operate as selective agents become more severe as population density increases. Those environmental factors whose severity of action on the population increases with population density are called density-dependent factors. They stand in contrast to density-independent factors, the strength of action of which is not correlated with population density.

In general, physical factors in the environment, such as temperature, moisture, floods, lava flows, etc., tend to be density-independent, while biotic factors tend to be density-dependent.

Infectious diseases provide a good example of a density-dependent factor. As the population becomes more dense, infections and mortality increase; and at some point in the rising population density, epidemics may develop. As population size decreases, on the other hand, the disease factor becomes less active, and, at the lower extreme of population dispersion, may become quite inactive.

Competition is a density-dependent factor. The action of herbivorous animals on a plant population, and of predators on a population of prey animals, are other such density-dependent factors.

Density-dependent factors tend to have a stabilizing effect on population number. When the population becomes very dense, the factor reduces its size; but when the population is thinned down, the factor plays a permissive role, allowing the numbers to build up again.

Density-dependent factors generally exercise their selective role at and above certain threshold population densities. Disease organisms set in motion a selection for disease resistance when the population reaches some level of density, and this selective process increases in intensity at higher densities. The selective value of a disease-resistant gene is not constant, in other words, but varies with population density. The variation in selective value of the gene may span a wide range, being virtually nil in a dispersed state of the population, acquiring a positive selective value at a threshold density, and having a high selective advantage at high population densities.

Frequency-Dependent Selection

A number of cases have been found in both insects and plants where the selective value of a gene or genotype varies with the frequency of that gene or genotype. This situation is known as frequency-dependent selection. The selective value usually varies with frequency in a negatively correlated way. That is to say, the gene or genotype usually has a higher selective advantage when at low frequency than it has at higher frequencies. The opposite situation of low-frequency selective disadvantage has also been reported.

Good examples of frequency-dependent selective values in nature are provided by Batesian mimics in various species of butterflies. The success of the mimicry as a means of protection against bird predators depends on the mimics' being relatively uncommon in comparison with the number of the models. The insectivorous bird learns to avoid the model species of butterfly, which it recognizes by sight, through experience with its noxious or poisonous qualities. The mimic species benefits indirectly from its resemblance to the model. If the Batesian mimic becomes abundant, however, it confuses the learned reaction of the bird predator to its own disadvantage. Therefore the selective advantage of the mimetic form is high when it is rare and drops when it becomes common.

In several species of *Drosophila*, including *D. melanogaster*, *D. pseudoobscura*, *D. paulistorum*, and *D. willistoni*, mutant males and wild-type males have been compared with respect to mating success in experimental populations. It has been found in these cases that the mating success of a type of male varies with the frequency of that type in the experimental population. The type of male that is rare, whether it is the mutant or normal type, has the advantage in mating (Ehrman and Spiess, 1969).

Other examples of frequency-dependent selective values in insects are an enzyme locus and an inversion type in *Drosophila melanogaster,* and the black mutant in *Tribolium castaneum* (Kojima and Yarbrough, 1967; Nassar, Muhs, and Cook, 1973; Sokal and Karten, 1964). Parallel examples in plants include a gene for seed-coat pattern in *Phaseolus lunatus* and variant corolla types in *Phlox drummondii* (Harding, Allard, and Smeltzer, 1966; Levin, 1972).

13

Genetic Drift

General Considerations

According to the Hardy-Weinberg law a selectively neutral allele will tend to remain at a constant frequency from generation to generation. But it will be recalled (from Chapter 3) that this law is applicable only to a very large population. The predicted constant allele frequency represents a statistical average for many trials; gene reproduction in a large population fulfills this condition of many trials. In any set of few trials, such as occurs in gene reproduction in a small population, one expects deviations from the average allele frequency due to chance alone.

The chance variations in allele and genotype frequencies from generation to generation in a small population are known as genetic drift. Genetic drift refers to the random component in the rate of gene reproduction. Drift leads to two characteristic phenomena in a small polymorphic population: first, to fluctuations in allele frequency from generation to generation; and eventually, to complete fixation or extinction of the allele.

We have assumed that the allele is selectively neutral. This assumption is simplifying but unnecessary. Let us assume alternatively that the allele has a slight selective advantage or disadvantage. The predicted change in allele frequency from generation to generation is again a matter of statistical averages and is again subject to chance deviations.

If, for example, a population is polymorphic for A and a, and a has a 0.1% selective advantage ($s = 0.001$), the predicted contributions of the two alleles to the gene pool in the next generation are in the ratio 1000 a : 999 A. This average proportion will be realized in a large population. But significant deviations from the ratio due to random factors are expected to occur in a small population.

Thus the action of selection does not in itself rule out the possibility of the action of drift. In fact, there is reason to believe that the most important evolutionary role of genetic drift lies in its joint action with selection.

Effect of Population Size

Whether allele frequencies in a population will or will not be affected significantly by drift depends on four factors: (1) the size of the population (N); (2) the selective value of the allele (s); (3) the mutation pressure (u); and (4) the amount of gene flow (m). These four factors interact. Their interrelationships have been investigated and quantified by Wright (1931).

Let us first consider population size. As noted previously, random fluctuations in allele frequency will be negligible and insignificant in a large but not in a small population. In the small population, by chance alone, an allele may change from a low to a high frequency, or reach fixation, in one or a few successive generations.

The theoretical distribution of allele frequencies in very small isolated populations has been shown by Wright (1931) to be represented by a U-shaped curve (Figure 13.1). Assume that a given allele (A_1) is selectively neutral ($s = 0$), and is a polymorph in a series of small isolated populations (of size N). This allele is expected to be approaching either extinction or fixation in most of the polymorphic populations, and to be present in medium frequencies in only a relatively few populations, as shown by the curve for N in Figure 13.1.

The distribution of allele frequencies in small populations (of size N) can be compared with that in a series of large populations (of size 4 N). Here most populations have the allele at medium frequencies (curve for 4 N in Figure 13.1).

The graph in Figure 13.1 can also be read to indicate the theoretical distribution of allele frequencies (abscissa) in a series of polymorphic genes (ordinate) in a single population. Most polymorphic genes are represented at extreme frequencies in a small population (curve N), but at medium frequencies in a large population (curve 4 N).

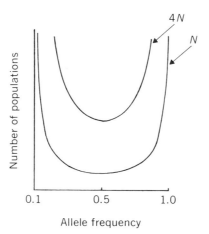

FIGURE 13.1
Frequency of a given allele in a series of
populations of two size classes. The
populations are either small (*N*) or large
(4 *N*); *s* = 0. (Rearranged from Wright, 1931.)

Effect of Selection

The polymorphic allele was assumed to be selectively neutral in the preceding discussion, and this assumption is reflected in the symmetry of the U-shaped curves. In a series of small populations the selectively neutral allele approaches either extinction or fixation in equal measure. If the allele has a small selective value in the small populations, however, the curve remains U-shaped but is skewed. The direction of the skewness depends on whether the allele is selectively advantageous or disadvantageous, and the degree of skewness on the magnitude of the selective value (Figure 13.2A). An allele with a small selective advantage tends to be present at either high or low frequencies, but more often at high frequencies, in a series of small isolated populations.

Here again sets of small populations differ from sets of large populations with respect to the expected behavior of genes. The gene frequencies are controlled more effectively by selection in large populations than in small ones at equivalent values of *s* (Figure 13.2).

Another way of verbalizing the curves in Figure 13.2 is to say that selection, theoretically, has a relatively small effect on gene frequencies in populations below a certain critical size, whereas chance alone is likely to be very effective in controlling gene frequencies under these same conditions (Wright, 1931). Weak selective pressures are likely to be overruled by drift in a small population.

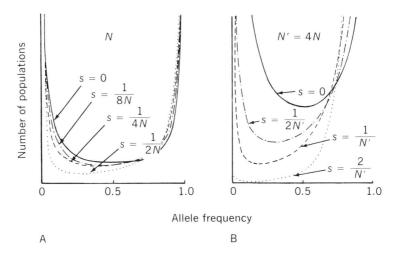

FIGURE 13.2
Expected distribution of allele frequencies for different values of s and two values
of N. (A) In a small population. (B) In a population 4 times larger, with
equivalent intensities of selection. (Wright, 1931.)

This brings us to the question of how small is "small" and how large is "large"
for the action of drift. The critical value of N at which drift becomes effective
is a function of s. The relations between N and s and drift are as follows (Wright,
1931):

Allele frequencies controlled by	When N is	When s is
Drift	$N \leq \dfrac{1}{2\,s}$	$s \leq \dfrac{1}{2\,N}$
Selection	$N \geq \dfrac{1}{4\,s}$	$s \geq \dfrac{1}{4\,N}$
Selection and drift	$N = \dfrac{1}{4\,s}$ to $\dfrac{1}{2\,s}$	$s = \dfrac{1}{4\,N}$ to $\dfrac{1}{2\,N}$

These relations can be visualized when plotted graphically on a linear scale
(Figure 13.3). We see that when N is low relative to s, drift prevails; that when
N is relatively large, selection prevails; and that a range of overlap exists where
drift and selection may act jointly.

These general relations can readily be translated into concrete figures. Suppose
an allele has a selective value of $s = 0.01$. Its frequency is controlled by drift

if $N \leqq 50$. But if the selective value of the allele is $s = 0.001$, its frequency
is controlled by drift in a population of $N \leqq 500$. Thus, in general, if the selective value of the allele is *fairly* low, it can be fixed or lost by drift in a quite small
population; but if the selective value is *very* low, drift can control its frequency
in a medium-sized population.

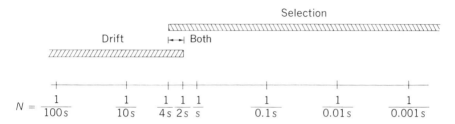

FIGURE 13.3
Realms of action of drift and selection corresponding to different proportions of N and s.
(From V. Grant, *The Origin of Adaptations,* copyright 1963, Columbia University Press, New
York; reproduced by permission.)

The range of overlap in which both drift and selection are operative also
varies with the magnitude of s. In the numerical examples given above, that
range is $N = 25-50$ for $s = 0.01$, and $N = 250-500$ for $s = 0.001$.

The possibility of joint action of selection and drift can, theoretically, be very
important in evolution. Wright (1931, 1949, 1960) points out that a favorable
gene can be established much more rapidly by selection and drift in an island-
like population system than by selection alone in a single large population of
the same overall size. The comparative probability of fixation of an initially
rare gene with a slight selective advantage in the two types of populations – the
single continuous and the subdivided populations, each containing a total of
10^6 individuals – has recently been calculated and found to be an order of magnitude greater in the subdivided population (Flake and Grant, 1974).

Effect of Gene Flow

Wright's (1931) equations describing the interaction between gene flow and
drift are as follows. Allele frequencies are controlled by drift when

$$N \leqq \frac{1}{2m} \quad \text{or} \quad m \leqq \frac{1}{2N}$$

Allele frequencies are controlled by gene flow when

$$N \geqq \frac{1}{4\,m} \quad \text{or} \quad m \geqq \frac{1}{4\,N}$$

and by both forces in the intermediate range,

$$N = \frac{1}{4\,m} \quad \text{to} \quad \frac{1}{2\,m}$$

For example, if $N = 100$, the critical value of m is $\frac{1}{2\,N} = 0.005$. Thus immigration of foreign alleles at rates greater than $m = 0.005$ can prevent drift from occurring in a population of 100 or less breeding individuals.

In general, small amounts of gene flow can prevent drift. A small population must be rather well isolated for drift to occur. But when this condition exists, when N and m are both low, drift can affect gene frequencies significantly.

The effect of mutation rate is described by equations parallel to those for gene flow. Allele frequencies are controlled by mutation pressure when $N \geqq \frac{1}{4\,u}$ and by drift when $N \leqq \frac{1}{2\,u}$. High mutation rates could prevent drift in a small population.

The Fixation of Gene Combinations

The joint action of selection and drift in small populations can promote the fixation not only of single genes, but also of gene combinations. The latter case can be very important in evolution.

Suppose that a large population contains two rare mutant alleles, a and b, of the independent genes A and B. Most individuals in this population have the diploid genotype $AABB$; in addition there are a few mutant-carrying individuals, $AaBB$ and $AABb$. Suppose further that the gene combination $aabb$ has a high adaptive value in relation to new environmental conditions.

In the large population, the sexual process will produce $aabb$ only rarely, and will then immediately break it up again. It is difficult for selection to "get a hold on it," and therefore its rise in frequency occurs at a slow rate.

But a small isolate of this population might have an intermediate or high frequency of the otherwise rare alleles, a and b, as a result of chance. Proportionally more $aabb$ zygotes would be produced and exposed to selection in each generation. Selection could then act effectively to raise the frequency of $aabb$

further. Thus the new gene combination can be established more rapidly by selection and drift in certain small colonies than by selection alone in the large population.

Experimental Evidence

Kerr and Wright (1954) tested the theory of drift in a series of very small populations of *Drosophila melanogaster* polymorphic for a body-bristle gene, forked (*f*). They started 96 replicate lines with four female and four male flies each. The initial frequency of the mutant allele forked was 0.5 in each line. The lines were continued by a random choice of the parental individuals (four females and four males) for the next generation. The 96 lines were maintained in this way to generation 16.

By generation 16, the wild-type allele had become fixed in 41 lines; forked was fixed in 29 lines; the remaining 26 lines were still polymorphic.

A parallel experiment made use of the eye mutant, Bar (*B*), in *Drosophila melanogaster*. Here 108 lines were started and continued as before. The Bar type is selectively disadvantageous. As of generation 10, the wild-type was fixed in 95 lines; Bar was fixed in 3 lines; 10 lines were still polymorphic (Wright and Kerr, 1954).

It is apparent that the polymorphic genes do drift into fixation in a high proportion of the small experimental populations. Sometimes they become fixed in spite of counterselection. More often they become fixed by the combination of drift and selection, as in the case of the 95 ultimately wild-type lines in the experiment with Bar.

The joint action of selection and drift has been demonstrated in other experimental studies. One set of experiments involves inversion types in laboratory populations of *Drosophila pseudoobscura* (Dobzhansky and Pavlovsky, 1957; Dobzhansky and Spassky, 1962). Another experiment carried out over a 17-year period with an annual herbaceous plant, *Gilia*, involves vigor and fertility in a series of related inbred lines (Grant, 1966*a*).

Drift in Natural Populations

In nature there are three common situations in which population size can be small enough to permit the effective action of drift, with or without selection. (1) The population system consists of a series of permanently small, isolated colonies. (2) The population is usually large, but is decimated periodically, and then builds up again from a few survivors. (3) A large population gives rise to isolated daughter colonies, the latter being founded by one or a few founder individuals. The new colonies thus pass through a bottleneck of small size in their early generations, although they may build up in size later (the so-called founder principle of Mayr, 1942, 1963).

If drift is playing an effective role (again with or without selection, but probably with), we would expect the variation pattern of the colonies to show the following characteristic features. First, the small colonies—the sister colonies in mode 1 and the daughter colonies in the early generations in modes 2 and 3—should be relatively uniform genetically. Second, there should be a marked colony-to-colony variation in genetically determined characters. This local racial differentiation is expected to be displayed most clearly in series of small sister colonies (mode 1), but should also be evident in some series of larger derived populations descended from small colonies (modes 2 and 3). And third, the intercolonial variation, when plotted on a distribution map, should be somewhat irregular and haphazard.

Variation patterns have been carefully studied in a number of plant groups with colonial population systems. In some of these groups the variation pattern conforms to the above expectations and thus suggests the effective action of drift.

Species of cypress trees *(Cupressus spp.)* in California exist as series of isolated groves; the groves are uniform, but each grove has its distinctive morphological characteristics (Wolf, 1948; Grant, 1958). The same type of variation pattern is found in herbaceous plants with colonial population systems, such as *Gilia achilleaefolia* in California, the *Erysimum candicum* group in the Aegean Islands, and the *Nigella arvensis* group in the Aegean Islands (Grant, 1958; Snogerup, 1967; Strid, 1970). The case for drift in these examples is strengthened by the fact that related species of *Juniperus, Gilia,* and *Nigella* in other areas form large continuous populations with a different type of variation pattern, namely, gradual intergradation along geographical transects.

Haphazard local variation is found again in some population systems of the European land snail, *Cepaea nemoralis,* for the presence or absence of bands on the shells (Figure 10.3). This shell character is determined by a single allele pair, with bandless dominant over banded, as noted in Chapter 10. The frequency of the bandless phenotype and bandless allele fluctuates widely from one colony to the next in areas of France where *Cepaea nemoralis* has a colonial population structure (Figure 13.4). But the bandless allele varies in frequency in a more gradual manner along geographical transects through large populations of the European land snail (Lamotte, 1951, 1959).

The pattern of irregular local racial variation occurs widely in plants, snails, butterflies, mice (Figure 13.5), and other groups in which it is correlated with a

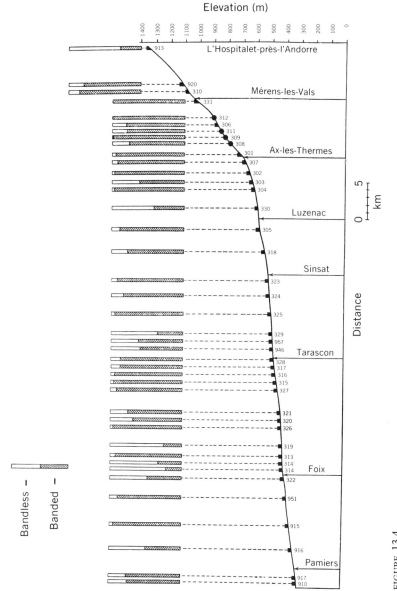

FIGURE 13.4
Proportion of banded and bandless shells in colonies of *Cepaea nemoralis* in the Ariège Valley of the Pyrenees, France. (Lamotte, 1951.)

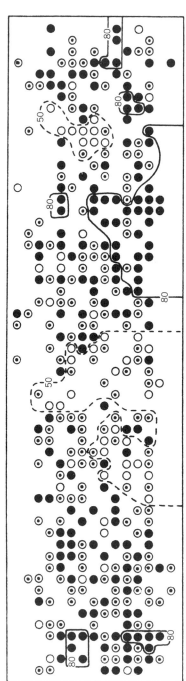

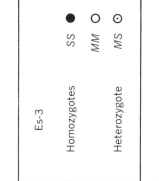

Es-3

Homozygotes   SS   ●

              MM   ○

Heterozygote   MS   ⊙

FIGURE 13.5

Fine-scale distribution of three genotypes for an enzyme locus (esterase-3) in mice (*Mus musculus*) in one side of a barn in Texas. The dots indicate the location of individual mice when trapped. Note clustering of like genotypes in small areas and local differentiation in allele frequencies. (Selander, 1970.)

colonial type of population system. The variation pattern and its correlation with population structure are strongly suggestive of the effective action of drift in such cases.

Blood-Group Alleles in Human Populations

Some of the strongest evidence for drift in natural populations is provided by the human species. Population size has been favorable for drift in many parts of the world throughout human history. Small isolated or semi-isolated populations consisting of 200–500 adult individuals were common in the food-gathering and hunting stages of culture. Small isolated farming and fishing communities exist in various parts of the world. Some religious sects form small isolated breeding populations because of religious beliefs prohibiting inter-marriage with outsiders.

The large body of data available on the ABO and other blood groups in human populations, both small and large, together with the simple genic basis of the blood types, makes the blood types a useful indicator of the genetic similarities or differences between populations. The polymorphism in the ABO blood groups was described briefly in Chapter 2.

An interesting case is that of the Polar Eskimos near Thule in northern Greenland. This small band of 271 or fewer individuals existed in complete isolation for generations. When first contacted by another band of Eskimos from northern Baffin Island, who spent several years trying to reach them, they believed they were the only people in the world.

The Polar Eskimos were found to differ markedly from the main populations of Eskimos in the frequencies of the blood-group alleles. The I^A allele varies from 27–40% in the larger Eskimo populations of Greenland (Table 13.1). Similar frequencies of I^A are found in the Eskimos of Baffin Island, Labrador, and Alaska. But the Polar Eskimos deviate from this norm with an allele frequency of I^A of 9% (Table 13.1). On the other hand, the Polar Eskimos have a particularly high frequency of I^O as compared with those of other Eskimo populations in Greenland and elsewhere (Laughlin, 1950).

Other small isolated human populations showing a marked local differentiation in respect to the ABO blood types are the aboriginal tribes of southern Australia, the Dunker religious sect in eastern North America, the Jewish community of Rome, and certain village communities of islands and mountains in Japan (Birdsell, 1950; Glass, Sacks, Jahn, and Hess, 1952; Dunn and Dunn, 1957; Nei and Imaizumi, 1966).

The Dunker religious sect was founded in Germany in the early 1700s and

TABLE 13.1
Frequency of the ABO blood-group alleles in Eskimo
populations in Greenland. (Laughlin, 1950.)

Region	Allele frequency, %		
	I^A	I^B	I^O
Nanortalik, Julianehaab District, southern Greenland	27	3	70
South of Nanortalik	35	5	60
Cape Farewell	33	3	64
Jakobshavn	29	5	66
Angmagssalik, eastern Greenland	40	11	49
Thule, northern Greenland	9	3	84

later immigrated to the eastern United States. The members of this sect marry largely among themselves and have thus remained reproductively isolated for generations from the surrounding German and American populations. Some Dunker communities are quite small; one in southern Pennsylvania consisted of about 90 adult individuals when studied in the early 1950s. It is significant that the Pennsylvania Dunkers deviate from the general German and American populations in blood types and other characteristics (Glass, Sacks, Jahn, and Hess, 1952).

Table 13.2 gives the frequencies of the *I* alleles in the Pennsylvania Dunkers and in the racially similar large populations of West Germany and the eastern United States. It is apparent that the general German and American populations are similar in gene frequency. But the Dunkers deviate from their German ancestors and modern American neighbors in this respect, the I^A allele having a significantly higher frequency and the I^B allele being close to extinction (Glass, Sacks, Jahn, and Hess, 1952).

TABLE 13.2
Frequency of the ABO blood-group alleles
in three related Caucasian populations.
(Glass, Sacks, Jahn, and Hess, 1952.)

Population	Allele frequency, %		
	I^A	I^B	I^O
West Germans	29	7	64
Eastern Americans	26	4	70
Dunkers in Pennsylvania	38	2	60

Other characters in the Pennsylvania Dunkers, such as type of ear lobe or hair, showed similar deviations from those in the ancestral or neighboring populations. In five different genes the allele frequencies found in this small breeding group diverged significantly from those typical of the surrounding populations (Glass, Sacks, Jahn, and Hess, 1952).

The I alleles are not selectively neutral. Some human populations show a positive correlation between the incidence of stomach ulcers and an $I^O I^O$ genotype, and between the incidence of stomach cancer and the genotypes $I^A I^A$ and $I^A I^O$. The indication that the I alleles have a selective value is sometimes used as an argument against drift. But this argument is based on the misconception that drift and selection are mutually exclusive forces, and is therefore not cogent.

Conclusions

The conclusion that gene frequencies are controlled to a significant extent by drift in small populations is demanded by the laws of probability and confirmed by experimental studies. The next question is the probable role of drift in natural populations. Does drift play a significant role in evolution?

This question has been the crux of a long-lasting controversy in modern evolution studies. The parties to the controversy can be categorized roughly as the believers and the non-believers. In my opinion, the non-believers have raised some weak objections and false issues. For example, it is often contended that, if selection can be shown to be operative in a population, drift is thereby ruled out as unnecessary (Fisher and Ford, 1947; Ford, 1955, 1964; Mayr, 1963, pp. 203ff.). This argument carries the implication that the issue is drift vs. selection, with selection always winning; by the same token, it diverts attention from the evolutionarily more important issue of selection-and-drift vs. selection alone.

Consider the I gene in small isolated human populations, for example. The non-believers argue that, since selection is involved, drift is gratuitous. But no one is contending that drift is the *only* factor controlling gene frequencies of I in small isolated populations; the contention is rather that drift is *a* factor in such situations.

Another argument often used against drift is that this factor has not been proven to be effective in natural populations. This argument misapplies the rules of evidence applicable to experimental populations to the vastly more complicated situation in nature. We cannot control the variables in natural populations, as we do in experiments, and therefore we cannot positively identify and quantify the various factors involved. But we can look for patterns in nature that are (or are not) consistent with theoretical and experimental findings.

The best evidence for the effectiveness of drift in microevolution is the pattern of haphazard local differentiation in series of permanently or intermittently small isolated colonies. This pattern has been found repeatedly in various groups of plants and animals with a colonial population system. The pattern may not prove, but it does strongly suggest, that drift plays a significant role in such population systems.

Collateral Readings

Cavalli-Sforza, L. L., and W. F. Bodmer. 1971. *The Genetics of Human Populations.* W. H. Freeman and Co., San Francisco. Chapter 8.

Ford, E. B. 1964, 1971. *Ecological Genetics.* Eds. 1 and 3. Ed. 1, Methuen, London. Ed. 3, Chapman & Hall, London. Chapters 2–4, 9.

Grant, V. 1963. *The Origin of Adaptations.* Columbia University Press, New York. Chapter 11.

Wright, S. 1931. Evolution in Mendelian populations. *Genetics* **16:** 97–159. Reprinted in *Systems of Mating,* by S. Wright. Iowa State University Press, Ames, 1958.

Wright, S. 1960. Physiological genetics, ecology of populations, and natural selection. In *Evolution After Darwin,* Vol. 1, ed. by S. Tax. University of Chicago Press, Chicago.

14

Cost of Selection

Genetic Load

Natural selection has both a positive and a negative aspect. It entails the preferential survival and reproduction of some genotypes and the preferential elimination and non-reproduction of others. In this chapter we are concerned with the second aspect.

The loss of a certain proportion of genotypes from a population as a result of the action of selection is obvious when the selective elimination takes the form of mortality in the Darwinian sense. Then there is a visible corpse. But selection also involves success or failure in the reproductive phase alone, as we have seen earlier (in Chapter 8). Individuals of a given genotype may be physically vigorous but still make a reduced contribution to the next generation because of low fecundity. In this case there is no visible corpse. But the genes for low fecundity decline in frequency and may eventually die out anyway.

The non-reproduction of an individual due to selection in its negative aspect, whether it comes as selective mortality or as failure of reproduction per se, is known as genetic death (or selective death). Genetic deaths lower the reproductive potential of the population.

The population would be stronger in numbers without its quota of genetic deaths. Therefore the latter represent a burden to the population, at least potentially. This burden is referred to as genetic load. Genetic load is defined (by Wallace, 1968) as $1 -$ (mean population fitness). It can also be thought of as the sum total of genetic deaths per generation.

Various types of genetic load can be recognized (Brues, 1969; Crow, 1970). The main types are the mutational load, segregational load, and substitutional load. These types of load are correlated with different types of selection.

The mutational load is an inevitable by-product of the mutation process. This process generates deleterious mutations, which must be weeded out by stabilizing selection.

A segregational load exists in populations that exploit the advantages of heterozygote superiority. The superior heterozygotes segregate the less fit homozygotes in each generation. The latter lower the average fitness of the population. They represent the cost of balancing selection.

Directional selection involves a substitutional load. The substitution of an old allele by a new, superior allele entails the genetic deaths of the carriers of the old allele. The total number of selective deaths involved in a complete gene substitution—the overall substitutional load—may be quite high, since the old allele, which is destined to be replaced, is normally present at a high frequency at the beginning of the substitution process.

Protein Polymorphisms

Electrophoretic assays of enzyme loci in a number of populations of *Drosophila*, vertebrates, and plants have revealed that rather high proportions of these loci are polymorphic. Thus 43% of 24 enzyme loci in *Drosophila pseudoobscura* are polymorphic, and the proportion is even higher in *D. willistoni* (see Chapter 2). It is generally assumed that the percentage of polymorphic loci detected by electrophoretic methods can be taken as representative of the genotype as a whole.

The occurrence of large numbers of polymorphic genes—some observed, others predicted by extrapolation from the observed ones—runs counter to expectations based on considerations of genetic load. Each balanced polymorphism in a population, since it entails the continual segregation of homozygotes of reduced fitness, produces a certain segregational load. With an increase in the number of independent balanced polymorphisms this segregational load goes up exponentially. At some point the population should be unable to bear the segregational load. Yet the highly polymorphic populations of *Drosophila* are flourishing.

One school of workers, led by Kimura, has attempted to solve the dilemma by postulating that the protein polymorphisms are not in fact maintained by selection. The alleles of the enzyme genes are supposed to be selectively neutral and to be drifting at various frequencies in populations (Kimura, 1968; King and Jukes, 1969). The problem indeed disappears on this supposition.

But is the supposition correct? The evidence in *Drosophila* is that it is not. Here the protein polymorphisms do not exhibit random fluctuations and random geographical variation, as would be expected if the alleles were selectively neutral, but instead show regular patterns of variation, which suggest that they are maintained by balancing selection (Ayala and Anderson, 1973; Ayala and Tracey, 1974; Ayala and Gilpin, 1974).

One could readily suspect that some polymorphic genes are selectively neutral or nearly so. But many others are controlled by balancing selection and hence must be contributing to a segregational load. The problem cannot be dismissed by simplistic assumptions of widespread selective neutrality. The correct solution remains to be discovered.

Perhaps we should reexamine the assumption that electrophoretically detectable enzyme loci are representative of the genotype as a whole. It may not be correct to estimate the overall rate of polymorphism by simple extrapolation from the observed rate of protein polymorphism. The total amount of the polymorphism and segregational load in populations may not be as great as it is considered to be in many current discussions.

Haldane's Cost of Selection

The substitutional load is closely related to what Haldane (1957, 1960) called the cost of selection, the main subject of this chapter. Haldane's concept of cost of selection refers to the total number of selective deaths involved in one complete gene substitution. Hence cost of selection is synonymous with cumulative substitutional load over a series of generations, and substitutional load per generation is synonymous with incremental cost of selection.

Haldane's two pioneering papers (1957, 1960) contain a mathematical analysis of the total or cumulative substitutional load relative to the population size in any generation. He considered how this relation is affected by selection intensity and other factors. And he considered the bearing of the cost of selection on evolutionary rate. The problem has been restudied and reformulated by various later workers (cf. Crow, 1970; Crow and Kimura, 1970; Flake and Grant, 1974).

The number of selective deaths required for the complete substitution of one allele by another ($\sum D$) will be many times the number of adult individuals in any single generation (N). There is a definite relation between ΣD and N. This relation is expressed by a cost factor (C). In general, if N is constant, $\Sigma D = C \cdot N$.

The formula for C for a single locus in a haploid system (which is much simpler algebraically than a diploid system) is as follows. Cost per generation is the ratio

$$\frac{w_1 sq}{\bar{w}} \quad \text{or} \quad \frac{w_1 - \bar{w}}{\bar{w}}$$

where w_1 is fitness of favored allele, $\bar{w}$ is mean population fitness, and q is frequency of inferior type. The total cost of substitution is then

$$C = \sum \frac{w_1 - \bar{w}}{\bar{w}}$$

The value of C thus depends on the initial frequency of the favored allele and the rate of replacement.

The cost factor is high when the favored allele is at a low initial frequency, and decreases as the initial frequency increases. Crow and Kimura (1970, p. 250) give the following values of C for the following gene frequency changes in a diploid system with no dominance:

q	C
0.001 to 0.999	13.81
0.01 to 0.99	9.19
0.1 to 0.9	4.39
0.2 to 0.8	2.77

If the initial gene frequency is constant, the cost factor varies with the rate of replacement. The rate of replacement depends, in turn, on selection intensity, and therefore s or w enters into the formula for computing cost (C). The total number of selective deaths involved in the gene substitution turns out to be practically independent of selection intensity; but the number of generations over which these selective deaths are spread is measured by selection intensity (s).

Haldane (1957) obtained numerical estimates of the cost of selection in a typical situation. He assumed a large diploid population containing a favorable mutant allele at a low initial frequency. The new allele is supposed to have a moderate selective advantage over the old allele. Under these conditions, the total number of genetic deaths involved in the complete gene substitution is usually 10 to 20 times, or sometimes up to 100 times, the number of breeding individuals in one generation ($C = 10-100$). Taking a single figure as representative, one could expect, on the average, about 30 times as many selective deaths during the course of gene substitution as there are adult individuals per generation ($\sum D = 30\,N; C = 30$).

Selection Cost and Evolutionary Rate

If the rate of replacement (s) is high, the gene substitution will come fast, at least in theory. But in actuality the population may be unable to stand the cost of intense selection. Selective pressures of too great an intensity may exterminate the population.

The number of genetic deaths that can be tolerated by a population in any one generation is strongly limited. The process of gene substitution must be spread out over numerous generations, therefore, if the population is to maintain a sufficient strength in numbers continuously; this restriction places an upper limit on the rate of evolutionary change for a single gene. In the case where $\sum D = 30\,N$, the gene substitution can be accomplished in 300 generations, but obviously not in 30 generations (Haldane, 1957).

The cost of selection for two or more independent genes is still higher. The cost rises exponentially with the number of independent genes that are undergoing selection simultaneously (Haldane, 1957). In a continuously viable population, therefore, the theoretical maximum rate at which evolutionary change can take place in several or many genes concurrently must be much slower than that for any single-gene character alone. Thus, where a single gene is able to undergo substitution in as few as 100 generations, two independent genes being selected for concurrently might require a minimum of 200 generations for substitution.

There is a strong restriction on the rate of gene substitution for any given number of independent genes determining separate single-gene characters, as noted above. This restriction on the evolutionary rate becomes even more severe if the same number of independent genes form components of a new adaptive gene combination, for then there are two interrelated difficulties standing in the way of rapid multiple-gene replacement. To the high cost of selection for multiple-gene substitutions is added the continual breaking-up of the favored gene combination by the sexual process. Both of these factors are especially strong in their restrictive effects when the component alleles of the new gene combination are still rare in the population.

These considerations suggest that evolutionary changes in complex characters and character combinations will normally take place at moderate or slow rates. The empirical evidence concerning evolutionary rates in some low-rate and moderate-rate evolutionary lines is within the theoretical limits imposed by selection cost (Haldane, 1957).

There are, however, some cases of rapid evolution that apparently exceed the theoretical limit. One example will be described.

Certain races of *Mimulus guttatus* (Scrophulariaceae) occupy habitats of post-Pleistocene age in Utah. Their habitats are dated as approximately 4000 years old. Hence racial differentiation in *Mimulus guttatus* has taken place in 4000 years in Utah. This herbaceous plant could pass through 4000 generations in this time (Lindsay and Vickery, 1967). The character differences between the races of another herbaceous plant, *Potentilla glandulosa* (Rosaceae), are due to allelic differences in at least 100 genes (Clausen and Hiesey, 1958). Assume that the genetic differences between races in *Mimulus* are of the same order as those in *Potentilla*. Then we have 100 genes undergoing substitution in 4000 generations, or an average rate of one gene substitution per 40 generations. This estimate errs on the conservative side as regards both the time element and the gene number.

Some organisms have thus evolved at rates that apparently exceed the ceiling imposed by a tolerable cost of selection. The question naturally arises: How have they managed to do so?

Gene Linkage and Gene Interaction in Relation to the Cost Restriction

Haldane's original model postulated a large population size, independence of the two or more genes undergoing substitution simultaneously, and several other conditions. Undoubtedly these conditions are often realized in nature. It is equally true that deviations from the postulated conditions occur frequently in natural populations. Some of these deviations ease the cost restriction on evolutionary rate (cf. Grant and Flake, 1974*b*).

Assume that the separate genes constituting a new adaptive gene combination are closely linked so as to form a supergene. The component genes in the supergene can then undergo substitution at the rate and cost of one gene. The cost of selection is no greater for a supergene than for a single Mendelian gene. If the component genes are closely linked, a population can acquire a new adaptive gene combination within the cost limitations applicable to a single-gene substitution (Grant and Flake, 1974*b*).

Certain modes of gene interaction would also reduce the cost of multiple-gene substitution. This would be the case where the separate genes undergoing selection simultaneously have correlated selective advantages or disadvantages. Different favored alleles may sometimes occur together in the same genotype, and, conversely, different unfavored alleles may also occur together in the same alternative genotype. Then the genetic deaths for numerous separate genes will be pooled in a smaller number of individual deaths. This in turn would make possible a more rapid evolutionary rate for a multifactorial character (Mayr, 1963; Mettler and Gregg, 1969).

Population Structure in Relation to the Cost Restriction

Deviations from the assumption of a single large continuous population can also reduce the cost limitation on the rate of multiple-gene substitution. Let us consider two alternative types of population structure: the series of founder populations and the subdivided or colonial population system (Grant and Flake, 1974a).

It will be recalled that the cost is highest when the favored alleles are at low frequencies. If the favored alleles could be raised above the low level by random non-selective factors, they would escape the most burdensome part of the total substitutional load. A mechanism for such changes exists in founder populations.

In a species with colonizing habits, a large ancestral polymorphic population may give rise to series of daughter colonies founded by one or a few colonizing individuals. Such daughter colonies or founder populations begin with a non-random sample of the ancestral gene pool. Previously rare alleles can be shifted to middle or high frequencies by partly random factors during the establishment of some of the new daughter colonies. These colonies simply evade the heavy part of the substitutional load. And they can support the lighter cost of selection for all subsequent increases in gene frequency, up to complete replacement.

The selective deaths take their inexorable average toll in a population for any given selection intensity. The population has, however, a reproductive potential that allows for the normal proportion of accidental deaths as well as for some quota of selective deaths. In certain population structures the accidental mortality rate can vary upward or downward. Where it varies downward it provides leeway for a higher number of selective deaths.

Take a large continuous population as a standard. Its combined total of selective deaths and accidental deaths sets a limit on the rate of evolution. Now consider a subdivided population of the same total size as the standard population, and subject to the same environmental pressures. In the subdivided population there will be colony-to-colony variation in the accidental mortality rate. Those colonies with low accidental mortality rates can absorb a heavier than average burden of selective deaths and hence can afford a more rapid evolutionary rate than that of the standard large population.

Collateral Readings

Crow, J. F., and M. Kimura. 1970. *An Introduction to Population Genetics Theory.* Harper & Row, New York. Chapter 5, in part.

Flake, R. H., and V. Grant. 1974. An analysis of the cost-of-selection concept. *Proc. Nat. Acad. Sci. USA* **71**: 3716–3720.

Grant, V., and R. H. Flake. 1974*a*. Population structure in relation to cost of selection. *Proc. Nat. Acad. Sci. USA* **71**: 1670–1671.

Grant, V., and R. H. Flake. 1974*b*. Solutions to the cost-of-selection dilemma. *Proc. Nat. Acad. Sci. USA* **71**: 3863–3865.

Haldane, J. B. S. 1957. The cost of natural selection. *J. Genet.* **55**: 511–524.

Haldane, J. B. S. 1960. More precise expressions for the cost of natural selection. *J. Genet.* **57**: 351–360.

Wills, C. 1973. In defense of naive pan-selectionism. *Amer. Nat.* **107**: 23–34.

PART

IV

THE PROBLEM OF ACQUIRED CHARACTERS

15

Phenotypic Modifications

The Theory of the Inheritance of Acquired Characters

The question of the inheritance or non-inheritance of acquired characters has been a center of controversy in evolutionary biology for nearly a century. The traditional theory of the inheritance of acquired characters has been discredited again and again, but refuses to stay down. Some of the reassertions of this theory are scientific anachronisms, but others are not.

The persistence of this controversy is not due primarily to lack of evidence, for ample evidence can be brought to bear on some aspects of the problem, although much more evidence is needed on other aspects. The main source of confusion in the past has been failure, or rather inability, to specify exactly what is meant by the term, acquired character.

Acquired characters may be either hereditary or non-hereditary, as exemplified by hereditary pathogenic diseases and phenotypic modifications, respectively. It is necessary to distinguish between two types of acquired characters: phenotypic characteristics determined by newly acquired genetic material, and characteristics resulting from phenotypic reactions to environmental stimuli, i.e., phenotypic modifications. These two types of acquired characteristics have entirely different roles in heredity and evolution.

The non-inheritance of phenotypic modifications is now so firmly established and widely accepted, and apparently so far removed from the controversial stage, that it is worth reminding ourselves that this principle is of quite recent vintage. The opposite viewpoint—that acquired characters in the sense of phenotypic reactions are inherited—was generally accepted through most of the history of biology, from Hippocrates to Lamarck and Darwin. Indeed, Lysenkoism, which held sway in the Soviet Union from 1939 to 1964, represents an extension of the age-old view of inheritance into modern times.

The modern view that phenotypic modifications are not inherited has a much shorter history, dating from Weismann's *Essays upon Heredity* (1889–1892) and *The Germ-plasm* (1893). However, Weismann's views were not immediately accepted; they were still being debated in American biology in the 1930s, still being ignored in French biology in the 1950s, and still being rejected in official Russian biology in the 1960s.

The controversy between Weismannism and the traditional view of heredity, generally known as Lamarckism, centered at first on Weismann's distinction between soma and germplasm. Weismann recognized a category of somatogenic variations. These are reactions of somatic tissues or of the body as a whole to external influences, including mutilations and the effects of use or disuse. He argued successfully, or almost completely successfully, that, since the soma does not enter the stream of heredity, neither can somatogenic variations.

The distinction between germplasm and soma was soon to be superseded by that between genotype and phenotype (Johannsen, 1911). The former is an anatomical distinction, whereas the latter is genetic and developmental. The genotype-phenotype concept is more refined and has more general applicability. Some somatogenic variations can be hereditary. But phenotypic characters are never inherited. And therefore phenotypic reactions to environmental stimuli are not hereditary.

On the other hand, the principle of the non-inheritance of phenotypic modifications leaves untouched those characters that are determined by acquired genetic material. Such characters can become hereditary and can therefore play a role in evolution.

We will deal separately with the two kinds of acquired characters in Part IV of this book. We will discuss phenotypic modifications in the present chapter, and the effects of acquired genetic material in Chapters 16 and 17.

The Phenotype

The genotype is the sum total of genes of an organism, and the phenotype is its particular set of character expressions. Between genotype and phenotype lie the long and complex processes of gene action and development. These processes take place in an environment and are influenced by that environment. The phenotypic character expression is then a product of two sets of factors: the genotypic determinants and the environmental conditions in which development takes place.

Each genotype has the capacity to give rise to a certain range of phenotypes in different environments. The genotype can be thought of in this regard as a "reaction norm" (Johannsen, 1911). In other words, the action of the genotype is neither rigidly predetermined nor unrestrained; instead, the genotype can determine a series of phenotypic expressions within limits set by the genotype itself (Johannsen, 1911).

A wide range of phenotypic expressions can be clearly demonstrated in plants capable of vegetative reproduction. Take a single individual plant, subdivide it, and propagate the subdivisions under different environmental conditions in a controlled-environment greenhouse or growth chamber.

Good examples are furnished by Clausen, Keck, and Hiesey's (1948) experimental studies of environmental responses in the *Achillea millefolium* group (Compositae). These are perennial herbaceous plants that can be propagated from vegetative subdivisions. The subdivisions were grown in different chambers in a controlled-environment greenhouse in Pasadena, California, and also out of doors in the winter in Pasadena. The controlled environments differed in light duration, day temperature, and night temperature.

In a typical experiment, six subdivisions of one plant of *Achillea borealis* from Seward, Alaska, were grown for 3.5 months in five controlled environments and in a sixth, out-of-doors, environment. The different growth responses at the end of the experimental period are shown in Figure 15.1. The genotype of this plant produced good growth in some environments (long warm days) and slight growth in others (short days). Other genotypes of the same species responded in a different fashion to these environments (Clausen, Keck, and Hiesey, 1948).

Adaptive Properties of Phenotypic Modifications

Phenotypic reactions to normal changes in the environment are usually adaptive. We see this in sun leaves and shade leaves of plants. Shade leaves have a larger surface area, which increases their photosynthetic capacity and compensates for the dimmer light, whereas sun leaves have a smaller surface area, thus cutting down on transpiration and water loss. The same plant genotype has the capacity to produce sun leaves in bright light and shade leaves in dim light. The genotypically determined capacity of a plant to respond phenotypically in this fashion enables that plant to adjust to varying light conditions.

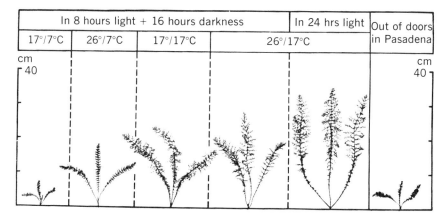

FIGURE 15.1
Growth responses of subdivisions of a single plant of *Achillea borealis* from Seward, Alaska, in six environments. The temperature figures are for days and nights, in that order. The growth period was 3.5 months. (Clausen, Keck, and Hiesey, 1948.)

The range of phenotypic modifiability shows wide differences in different major groups of higher organisms. Higher plants, and particularly herbaceous plants, are characterized by great phenotypic plasticity. Insects exemplify the opposite extreme of phenotypic inflexibility.

These differences in the degree of modifiability of the body are correlated directly with differences in mode of development in the two groups. The plant body develops from growing points that are exposed to and strongly influenced by the environmental factors prevailing when the new parts are forming, so that sun leaves appear in bright light and shade leaves in dim light, and so on. The adult insect body, on the other hand, develops within a hard external skeleton that is formed in a preadult growth stage. The main external features of the insect body are laid down well before they can be functional, and have become set in their mold by the time they can be used.

Beyond these differences in mode of development lie more basic differences in the strategy of individual adjustment to fluctuating environmental conditions. Insects are motile, plants are sedentary. The adult insect can adjust to environmental variables, within limits, by moving to a warm spot in cold weather or to a moist spot in dry weather—in short, by habitat selection. A plant anchored to the ground by its roots does not have this option; for it, phenotypic modifiability is the main means of individual adjustment to variations in the environment.

Both the types of phenotypic modifications and the range of phenotypic modifiability are related to the requirements of the organisms. It is very probable, therefore, that genotypes have been selected for their capacity to make adaptive phenotypic responses to environmental variables.

The Role of Phenotypic Modifications in Evolution

Phenotypic modifications protect the individual organism against environmental stress, and, to the extent that they are successful in this, they constitute a buffer against environmental selection. Theoretically, a population composed of genotypes with a complete range of adaptive phenotypic reactions would not respond to the pressure of natural selection at all.

Phenotypic plasticity has a retarding effect on the action of environmental selection, and this retarding effect is brought to bear on the negative as well as the positive aspects of selection. Phenotypic plasticity retards selective elimination; it permits a population to persist in an environment that has changed in an unfavorable way. This postponement of selective elimination gives the population more chances of acquiring new genetic variations by mutation, gene flow, and recombination, from which it can construct a genotypic response to the new, unfavorable environment.

16

Transduction and Induction

Transformation

The first bona fide case of hereditary change stemming from the acquisition of foreign genetic material was discovered in the pneumonia bacterium, *Pneumococcus*. Transformation was discovered in *Pneumococcus* by Griffith in 1928, and was shown to take place in vitro by Avery, Macleod, and McCarty in 1944. The essential facts are now generally known.

Strains of *Pneumococcus* differ in the nature of the cell surface coat, which may be rough or smooth, and in correlated properties of virulence. These differences are hereditary. When one strain of the bacterium is grown in a sterile extract containing the DNA of another strain, some daughter cells of the former strain acquire the type of cell surface and virulence of the latter strain. The change is hereditary. The hereditary characters of a donor strain are transferred to a receptor strain by the DNA of the donor in a sterile growth medium (Avery, Macleod, and McCarty, 1944).

Hereditary transformations brought about by sterile extracts have been demonstrated in other genera of bacteria, such as *Hemophilus* and *Streptococcus*, and for other characters, such as resistance to penicillin and streptomycin (see Hayes, 1968, Ch. 20).

Bacterial transformation takes place in the laboratory. The question arises: Does any analogous process occur under natural conditions in the bacteria? A natural analogue of transformation is transduction.

Transduction

Under natural conditions, bacterial viruses or phage occasionally transfer genetic material from one bacterial cell to another. This transfer of bacterial genetic material by phage is known as transduction.

The general sequence is as follows. The phage infects one bacterial cell, and then multiplies inside this cell at the expense of the host DNA. A daughter phage particle now infects a second bacterial cell, and, in so doing, introduces genetic material from the first host into the second host. Transduction occurs most successfully with temperature phages that do not completely destroy their bacterial hosts. The transduced bacteria thus live on and perpetuate their altered characteristics.

Transduction has been shown to occur in *Salmonella, Escherichia, Pseudomonas, Staphylococcus,* and other genera of bacteria. Among the characters transferred by transduction are synthetic ability, antibiotic resistance, and motility (see Hayes, 1968, Ch. 21). A case of conversion in *Clostridium* involves the type of toxin produced by the bacterium. A strain of *Clostridium botulinum* that normally causes botulism in man and animals, when it is infected by a certain bacteriophage, is converted to another strain of *Clostridium,* one that causes gas gangrene (Eklund, Poysky, Meyers, and Pelroy, 1974).

Induction of Body Mosaics in *Drosophila melanogaster*

Phenomena analogous to transformation are now known to occur in multicellular organisms. The term induction has come into use to designate such phenomena. Induction in this sense is a hereditary transformation mediated by foreign DNA in multicellular organisms. Our knowledge of induction is still very fragmentary and incomplete.

One of the few known cases involves body mosaics in *Drosophila melanogaster.* Eggs containing young embryos of wild-type *D. melanogaster* were treated with a solution containing DNA from flies with specific mutant characters. Mutant types of body pigmentation, bristle shape, and eye color were among those used as genetic markers (Fox and Yoon, 1966, 1970).

The treated embryos developed into adult flies showing the same phenotypic character as that in the mutant strain that furnished the transforming DNA. Furthermore, the transformed characters proved to be heritable. However, the transformed individuals were mosaics with patches of mutant and normal tissue; the body or body part did not become transformed as a whole (Fox and Yoon, 1966, 1970).

Indirect evidence suggests that the transforming DNA does not become integrated into the chromosomes of the receptor flies. The transforming DNA probably gets into some other parts of the cell in some cell lines and cell sectors. This distribution would account for the mosaicism of the phenotypic effects (Fox, Duggleby, Gelbart, and Yoon, 1970; Fox, Yoon, and Gelbart, 1971).

Induction of Hybrid Sterility in *Drosophila paulistorum*

Drosophila paulistorum is a species group composed of a number of incipient species or semispecies in Central and South America. Hybrid sterility appears between some of the semispecies. Crosses between certain semispecies give sterile male hybrids. Other crosses produce all fertile progeny. The male sterility, where it exists, is determined in part by factors that are transmitted through the cytoplasm of the egg. In the final analysis, the hybrid male sterility is determined by the interaction between these cytoplasmic factors and the chromosomes.

The propensity for sterility is thus passed on by female flies with certain cytoplasmic constitutions to their hybrid male progeny. Consequently, it is possible for a hybrid cross made in one direction (e.g., Santa Marta ♀ × Mesitas ♂) to produce sterile males in F_1, whereas the reciprocal cross (Mesitas ♀ × Santa Marta ♂) yields fertile males in F_1 (Ehrman and Williamson, 1965; Williamson and Ehrman, 1967).

Williamson and Ehrman (1967) made a homogenate of the cytoplasm of eggs of the Santa Marta strain, and injected this homogenate into females of the Mesitas strain. The treated Mesitas females were next crossed with Santa Marta males. The injected Mesitas females (in contrast to untreated Mesitas females) produced sterile male hybrids. Male sterility was induced by injection of the female parent with a particular type of cytoplasm.

What is the nature of the cytoplasmic sterility factors? Electron micrographs of reproductive tissues in sterile males reveal the presence in the cytoplasm of particles resembling *Mycoplasma,* a virus-like organism (Ehrman and Kernaghan, 1971).

A likely hypothesis to account for the various sets of facts is that *Mycoplasma* or similar organisms occur as symbionts in the cells of some semispecies of the *Drosophila paulistorum* group. In these semispecies there is a good adjustment between the chromosomal genes and the infected cytoplasm, so that normal fertility is not upset. However, crosses of these symbiont-carrying semispecies, used as female parents, with symbiont-free strains used as male parents produce combinations of cytoplasm and chromosomal genes that are not mutually adjusted. The phenotypic expression of this cytoplasm-nucleus interaction is hybrid male sterility (Williamson and Ehrman, 1967; Ehrman and Kernaghan, 1971).

Induction in Flax

The response of plants to the quantity of the nutrient elements N, P, and K in their substrate is well known. Good growth and large yield are produced on rich substrates, and poor growth on lean substrates. It is also well known that these responses are phenotypic modifications.

An exceptional case has been found in flax *(Linum usitatissimum),* in which the induced changes have apparently become hereditary. Different flax plants in the original generation were treated with different dosages of N, P, and K, and responded in the usual way. However, the later-generation progeny of induced large plants tended to remain large even though the nutrient level was reduced; conversely, the progeny of induced small plants remained small despite increases in nutrient level. The induced size differences held up through ten generations. Crosses between derived large and small plants yielded intermediate F_1 hybrids, confirming the conclusion that the differences are hereditary (Durrant, 1962a, 1962b; Evans, Durrant, and Rees, 1966).

The hereditary differences between the derived large and small plants apparently do not have a cytoplasmic basis in the flax experiment. There are, however, differences between the lines in the DNA content of the nuclei, the derived large plants having more nuclear DNA than the small ones (Evans, Durrant, and Rees, 1966).

Similar induced size changes correlated with nutrient treatment have been followed through three generations in the wild tobacco species, *Nicotiana rustica* (Hill, 1967).

Many aspects of these two cases are not understood at present. Further research is needed.

Bleaching in *Euglena*

The phenomenon known as bleaching in the green flagellate, *Euglena,* represents still another type of environmentally induced hereditary change. Bleaching is a permanant loss of chloroplasts in a cell line.

The greenness of *Euglena* is due to the photosynthetic pigment chlorophyll in the chloroplasts. The chloroplasts are DNA-containing and ordinarily self-reproducing organelles in the cytoplasm; they constitute a type of cytoplasmic gene. Light and other factors are necessary for the proper development and functioning of the green chloroplasts.

Euglena gracilis, which manufacturers its own sugar in the light, can be maintained artificially in the dark on a growth medium containing sugar. In the dark the chlorophyll does not develop, and later the plastids themselves may fail to reproduce, so that they gradually dwindle in numbers and eventually become lost altogether in a cell line. The loss is permanent. A culture of *Euglena gracilis* that has lost its chloroplasts in a dark environment does not regain them when brought back into a light environment (see Lewin, 1962).

Other environmental factors besides darkness that can cause bleaching in *Euglena gracilis* are heat (32–35°C), UV radiation, and antibiotics (streptomycin, etc.) (Lewin, 1962).

These environmental factors can convert a green photosynthetic strain of *Euglena* into a colorless heterotrophic strain that grows only on a substrate containing sugar. Morphologically, the derived strain of *Euglena* resembles the colorless euglenoid, *Astasia longa.* One hypothesis of flagellate phylogeny holds that the colorless euglenoids are actually derived by bleaching from the green euglenas (Lewin, 1962).

17

The Evolutionary Role
of Induction

Introduction

The evolutionary role of hereditary induction has been a skeleton in the closet of evolutionary biology for many years. The subject is rarely discussed in print; nor is it being researched to any extent. The similarity between hereditary induction and discredited Lamarckian theories of inheritance of acquired characters is probably responsible for this neglect.

Yet hereditary induction does occur, as shown by the fragmentary evidence presented in the preceding chapter, and we must try to follow this evidence wherever it leads. The evidence leads us beyond the limits of the synthetic theory of evolution. This chapter is a preliminary discussion of the problem.

A distinction between types of induced hereditary change is essential to the discussion. The crux of the issue is not environmental induction of hereditary changes per se. High-frequency radiation and mutagenic chemicals are environmental factors, and they induce gene mutations. But the induced mutations are unoriented; the mutagenic agent induces a diversity of mutant types. What we are concerned with here is the environmental induction of *directed* or *oriented* hereditary changes. The orientation of the hereditary change results from the gain (or loss) of preformed genetic material of a particular sort. The term hereditary induction, as used in this discussion, applies to this type of change.

In general, the synthetic theory of evolution begins with hereditary variation originating ultimately as undirected gene mutations, and does not encompass directed hereditary changes. In order to accommodate the latter class of induced changes, therefore, we may have to enlarge our theoretical framework beyond the limits of the synthetic theory.

Possible Examples

The question that concerns us now is not whether hereditary induction occurs occasionally, but whether it has had any important effects in evolution. Some possible examples will be considered.

It will be recalled from the preceding chapter that bleaching in the euglenoid flagellates converts a photosynthetic green *Euglena* into a colorless heterotrophic euglenoid like *Astasia*. On this basis it is hypothesized that the colorless euglenoids have been derived phylogenetically by bleaching from the green euglenoids. This is a plausible but unverified phylogenetic hypothesis.

Although transduction is a firmly established process in bacteria, there does not seem to be any evidence that it has figured in bacterial evolution. On the other hand, the standard processes of mutation and selection are known to govern some microevolutionary changes in bacteria. The development of resistance to phage, for example, depends initially on phage-resistant mutations that appear spontaneously in bacterial colonies. These mutations arise at a given low rate that is not affected by the presence or absence of phage in the environment. The microevolutionary change thus follows the orthodox scheme. The negative evidence regarding the role of transduction in bacterial evolution should not be taken too seriously, however, since transduction has been studied mainly from the molecular rather than the evolutionary standpoint.

The cells of all eukaryotic kingdoms of organisms carry mitochondria in their cytoplasm. The eukaryotic cells of green flagellates and plants also carry chloroplasts as cytoplasmic organelles. Mitochondria and chloroplasts function as centers of cellular respiration and photosynthesis, respectively. Both classes of organelles contain DNA and are self-reproducing and semi-autonomous constituents of the eukaryotic cell.

An old hypothesis, revived and elaborated in recent years, holds that mitochondria and chloroplasts originated from free-living prokaryotes that invaded primitive eukaryotic cells and became established as permanent symbionts in the cytoplasm. Mitochondria are considered to be homologous with and derived from aerobic bacteria, and chloroplasts from blue-green algae. Among the numerous proponents of this hypothesis are Margulis (1970) and Raven (1970); an opposing view is expressed by Uzzell and Spolsky (1974).

The evidence in favor of the theory of a symbiotic origin of mitochondria and chloroplasts consists of structural and biochemical similarities between these cell organelles and the corresponding types of modern prokaryotes. The DNA of mitochondria is double-stranded and circular, like that of bacteria. Chloroplasts and blue-green algal cells both contain a photosynthetic apparatus and bodies that look like ribosomes, in addition to DNA, and both are enveloped by a membrane. Furthermore, blue-green algae have a known tendency to form symbiotic associations with various eukaryotic organisms in the modern world of life (Ris and Plaut, 1962).

The evidence is not decisive. The similarities between the cytoplasmic cell organelles and prokaryotes could be due to common ancestry or to convergence, and convergence has not been ruled out. Nevertheless, the symbiotic hypothesis is an attractive one. If it is true, the acquisition of physiologically useful pro-karyotes by primitive eukaryotes would have been a critical step in the evolution of life, making possible the later development of the plant and animal kingdoms.

A symbiotic and prokaryotic origin has been suggested for various other types of cell organelles. It is suggested that the flagellum of flagellated pro-tistans is derived from a spirochaete (Margulis, 1970). The kappa particles in the cytoplasm of *Paramecium* are bacterial symbionts. The cytoplasmic sterility factors in *Drosophila paulistorum* are probably *Mycoplasma* or similar organisms, as noted in the preceding chapter.

Discussion

The genotype in eukaryotic organisms can be subdivided into two parts on the basis of the degree of internal integration and the susceptibility to invasion by foreign genetic material. The chromosomal genotype is highly structured and highly integrated. It would be very difficult, though not impossible, for a foreign particle of genetic material to become established in a chromosome.

Cytoplasmic genes are, by contrast, a much more loosely integrated lot. Furthermore, the cytoplasm is more exposed to external influences than the chromosome set, and is probably more susceptible to invasion by foreign genetic particles. The cytoplasm is the most likely site for successful hereditary induction in eukaryotes.

The development of complex characters requires the action of well-integrated gene systems. Such gene systems can be assembled and maintained in the chromosome set, which becomes the headquarters and information center of the complex eukaryotic organism. The cytoplasmic genes play a secondary and subservient role. The role of hereditary induction in evolution must therefore also be restricted in scope to the types of characteristics that can be determined by cytoplasmic genes.

The acquisition of foreign genetic material is potentially beneficial to an organism or its descendants, and various mechanisms exist for bringing about genetic exchange. Hereditary induction and transduction are among these mechanisms. These may well be ancient processes. But they are also irregular, hit-and-miss methods.

Eukaryotic organisms possess a regular and precise mechanism for bringing genetic material together from separate lineages. That mechanism is sexual reproduction, which is not hit-and-miss, but channelized within the limits of species populations. The orderliness of the sexual process at the stages of meiosis and fertilization has reference exclusively to the chromosomal genes. It can be argued that sexual reproduction evolved along with eukaryotic organisms and largely supplanted the irregular parasexual methods of genetic exchange in prokaryotes, because of the inadequacy of the more primitive and inefficient methods once a certain stage of organic complexity was reached.

Evolutionary geneticists have followed the classical geneticists in concentrating on the behavior of chromosomal genes and neglecting the effects of the cytoplasm. The Hardy-Weinberg law, the genetical theory of selection, and indeed the whole synthetic theory of evolution are based on the Mendelian behavior of chromosomal genes in sexual organisms. The emphasis on the chromosomal genotype is properly placed. But the picture of evolutionary processes that has emerged is unbalanced and one-sided. We now need to find out more about the evolutionary role of hereditary induction.

PART

V

SPECIATION

(continued)

18

Races and Species

Introduction

In Parts II and III we considered evolutionary forces as they operate within a breeding population. The local population, however, is a component of a more extensive population system; it belongs to a race and to a species. If the single breeding population is a field in which evolutionary changes take place, as discussed in preceding chapters, so too is the interlocking system of populations, the race and the species.

We turn now (in Part V) to a consideration of evolution at the level of races and species. This is the branch of evolution studies known generally as speciation. Our treatment of speciation will build on the foundations laid in Parts II and III.

It is necessary to discuss the nature of races and species before attempting to describe evolutionary processes in these population systems. Three main features of races and species to be considered are the following: (1) A survey of the types of population systems and types of variation patterns in higher plants and animals (this chapter). (2) The barriers to interbreeding and gene exchange within and between population systems (Chapter 19). (3) The ecological relations between species (Chapter 20). With these aspects of population systems in mind, we can then proceed to consider the processes involved in race and species formation in the subsequent chapters of Part V.

Population Systems

The local population is a breeding unit in sexual organisms, but it does not exist in complete isolation. A population system—a race or species—consists of many such local populations, which are linked by breeding relationships. Owing to migration and gene flow, which are normal, a certain amount of interbreeding occurs between populations. Races and species are breeding groups of a more inclusive scope than local populations.

The members of a local population share in a common gene pool. The regional cohort of local populations—the race—also shares in a common gene pool, but it is a more inclusive one. A still more inclusive and diversified gene pool is that of the species, which is the sum total of interbreeding races.

The visible result of interbreeding between populations and between races is intergradation in phenotypic characters that differ regionally within the species. It is by following the course of this intergradation in phenotypic characters over geographical areas that we get our first good indication of the occurrence and channels of interbreeding. When we follow out the intergradations in phenotypes, we come sooner or later to a prominent discontinuity in the variation pattern. This discontinuity marks the boundary line of the species. The interbreeding of populations and of races, in other words, is confined within limits, and these limits circumscribe the species.

Let us note parenthetically that we are using the term species here in the sense in which it is used in evolutionary biology. The same term has come to have other meanings in some branches of taxonomy. We will return to the question of different species concepts later. But here we are discussing biological, not taxonomic, species.

We have stated that the boundary line of the biological species corresponds to the limits of normal interbreeding and intergradation. With these outer limits set, we can go on to describe some types of intraspecific variation patterns. The diverse variation patterns found in nature fall into three main classes, which will be discussed separately. The three modes are: (1) continuous geographical variation, (2) discontinuous geographical variation, and (3) ecological race differentiation. The modes are often mixed in actual cases. A given species may exhibit different variation patterns in different parts of the species area.

Allopatry and Sympatry

The spatial relations between local populations and between population systems are an important factor in determining the amount of interbreeding that takes place between them, and hence the type of variation pattern that develops. The terms allopatric and sympatric are useful to describe these spatial relations.

(1) Allopatry. Populations living in different areas are said to be allopatric. The amount of interbreeding between such populations is a function of the spatial distance between them. Geographical races are allopatric. If the geographical races are contiguous, the intergradation may be more or less continuous; if the races are disjunct, interbreeding is reduced between them and the intergradation is interrupted.

(2) Sympatry. Two or more populations living in the same area are said to be sympatric. Non-interbreeding species coexisting in the same territory are sympatric. The term sympatry covers two different situations, which should be distinguished. In the first case, the populations coexist genetically but not ecologically, while in the second case they coexist both genetically and ecologically. The two situations are designated as neighboring sympatry and biotic sympatry, respectively.

(2a) Neighboring sympatry. One often finds two or more kinds of habitats occupying adjacent positions and having numerous interfaces over a large geographical area. In savanna country, for example, patches of woodland alternate with areas of open grassland in checkerboard fashion; in mountainous country, different altitudinal zones occur in the form of concentric rings. Populations inhabiting the adjacent but different habitats lie within the normal dispersal range of one another, so that interbreeding is possible from a strictly spatial standpoint, but they do not live or grow side by side. Such populations are neighboringly sympatric. Ecological races are usually neighboringly sympatric.

(2b) Biotic sympatry. Now consider two or more populations living in the same habitat, e.g., in the woodland patches in savanna country. Not only do these populations have overlapping zones of dispersal of their gametes, they also come into contact in the non-reproductive or secular phases of their life cycles. Such populations are biotically sympatric.

(3) Parapatry. This is a third and intermediate situation. Two populations are said to be parapatric when they occupy contiguous but non-overlapping areas without interbreeding.

Continuous Geographical Variation

This pattern is typical of outcrossing organisms that form large continuous populations. Examples are furnished by *Drosophila,* man, and many species of forest trees and plains grasses. The local populations are polymorphic. And each local population has its characteristic balance of polymorphic types resulting from gene flow and various modes of selection. Along a geographical transect there is a gradual shift in the frequencies of these polymorphic types.

Consider the ABO blood types in human populations ranging across Europe and Asia. Table 18.1 shows that type B blood, a phenotypic expression of the *I* gene, ranges from frequencies of nearly 40% in Asia in descending order to 5% and 6% in parts of western Europe. The gene frequency of the I^B allele decreases in parallel fashion from 30% in Asia to 4% or 5% in western Europe. The gradients in the allele frequency of I^B are shown again in Figure 18.1. The I^B allele has a high frequency in central Asia and decreases in a regular way across Europe to lows of 0–5% in various parts of western Europe (Figure 18.1A). The I^A and I^O alleles increase in frequency from east to west in Eurasia in a corresponding manner (Mourant, 1954).

The observed gradient has been explained as a result of the repeated migrations of the Mongolians into western Asia and Europe. Interbreeding of the Mongolians with the native races to the west, among whom the I^B allele is believed to have been rare originally, presumably raised the frequency of this

TABLE 18.1
Percentage frequency of type B blood and of the I^B allele in a series of human populations in Europe and Asia. (Data from Mourant, 1954.)

Group	Number of local populations	Blood type B, %	I^B allele, %
Scots	5	8–14	5–9
French	2	6–8	4–6
Swedes	2	10–11	7–8
Swiss	16	5–11	4–8
Poles	1	17	14
Yugoslavs	4	15–18	11–14
Hungarians	1	17	15
Persians	2	22–23	16
Pakistanis	3	32–36	24–26
Hindus of central India	1	39	30

allele in the resident populations. The numbers of Mongolian invaders reaching any region decreased with the distance from Mongolia. The gradient in the allele frequency of I^B across Eurasia thus reflects a gradual decrease in the amount of gene flow from Mongolia to successively more western parts of Asia and Europe (Candela, 1942).

The inversion types in chromosome III of *Drosophila pseudoobscura* have been discussed previously as examples of polymorphism, microevolutionary changes (Chapter 4), and balancing selection (Chapter 11). These inversion types also exhibit continuous geographical variation in western North America.

Figure 18.2 shows the average annual frequencies of the *ST* and *AR* inversions at intervals on a transect through the central Sierra Nevada of California. We see a gradual rise in frequency of *AR* and a decline in frequency of *ST* with rise in elevation in the mountains (Dobzhansky, 1948). This transect is a segment of a much larger pattern of geographical variation, part of which is shown in Figure 18.3. On a transect through the southwestern United States, *ST* has a high frequency in California but declines to the east; *AR* increases in frequency from California to Arizona and New Mexico, but then declines again in Texas. Pikes Peak *(PP)*, on the other hand, has its maximum frequency in Texas, and declines to low levels to the west (Figure 18.3). On a still larger scale, other trends occur to the north in the Rocky Mountain region (where *AR* is high), and to the south in Mexico (where *CH* is high) (Dobzhansky and Epling, 1944).

Geographical racial variation is real in the population systems of *Homo sapiens* in Eurasia and *Drosophila pseudoobscura* in North America. But the drawing of boundary lines between races is an arbitrary procedure, owing to the gradual character of the intergradation; any classification of the population system into races is therefore also arbitrary. How many races should we recognize in *Drosophila pseudoobscura?* One could with equal justification recognize four, or ten times that number.

In practice it is useful to distinguish between local races and geographical races. The former exhibit relatively minor shifts in the frequencies of polymorphic variants (e.g., Jacksonville and Lost Claim in *D. pseudoobscura;* see Figure 18.2), whereas the latter are regional clusters of local races differing significantly from one another in their gene pools (e.g., the California, Arizona-New Mexico, and Texas population systems of *D. pseudoobscura;* see Figure 18.3).

The gradualness of the intergradation is not always uniform. Steep gradients may occur here and there in the variation pattern of the species. Such areas of abrupt change in gene frequencies over relatively small geographical distances provide natural zones in which to draw racial boundary lines and to reduce the arbitrariness of the subdivision of the species into geographical races.

162

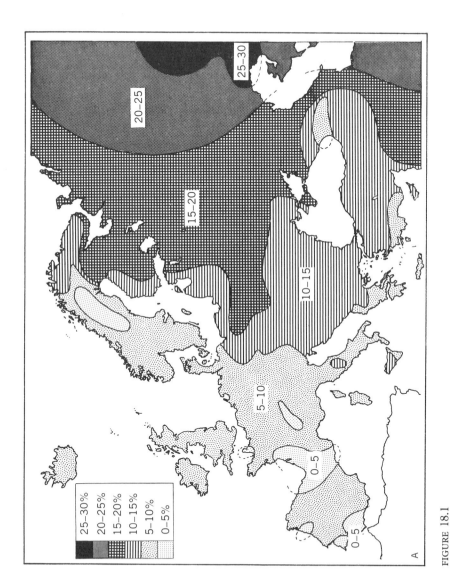

FIGURE 18.1
Geographical distribution of gene frequencies of the blood-group allele I^B in (A) Europe and (B) the world. (Redrawn and reproduced from A. E. Mourant, *The Distribution of the Human Blood Groups*, copyright 1954, Blackwell Scientific Publications, Oxford; by permission.)

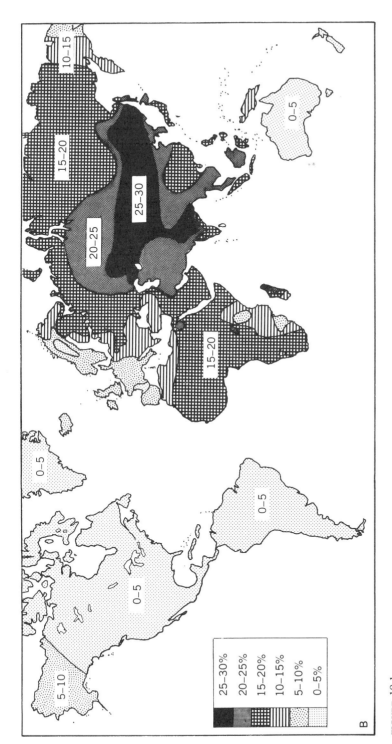

FIGURE 18.1
(continued)

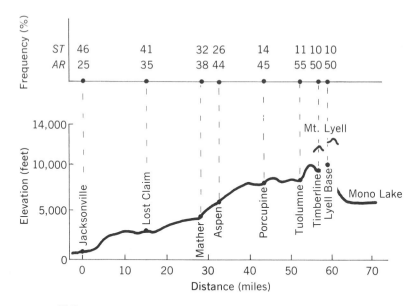

FIGURE 18.2
Average annual frequencies of *ST* and *AR* inversions in populations of
Drosophila pseudoobscura on a transect through the Sierra Nevada, California.
(Redrawn from Dobzhansky, 1948.)

Disjunct Geographical Races

An island-like population system develops where the inhabitable areas of
the species are relatively small and widely spaced. Examples of a discontinuous
distribution of a particular habitat are common: islands in an archipelago for
terrestrial organisms, series of lakes for aquatic organisms, mountaintops
for alpine organisms, rocky outcrops in a grassy plain, etc.

A characteristic variation pattern develops in this situation. The geographical
variation exhibits discontinuities coinciding with the gaps in geographical
variation. The colonies, if sufficiently isolated, tend to become adapted rather
specifically to their local environmental conditions. The species then appears
as a series of disjunct and distinct geographical races. This variation pattern
is the expected result of the reduction of gene flow between colonies and of the
action of selection or selection-drift within each colony.

The annual herbaceous plant *Gilia leptantha* (Polemoniaceae) occurs in
openings in montane pine forests in southern California. The pine forests in
this area occur at middle and high elevations in a series of isolated mountain
ranges separated by many miles of unforested lowlands. The population system

of *Gilia leptantha* accordingly has the disjunct distribution pattern shown in Figure 18.4.

Furthermore, the populations in different mountain ranges group themselves into well-marked geographical races. Four such races, differentiated morphologically on floral characters, are recognized taxonomically as subspecies, as shown also in Figure 18.4. Crossing experiments indicate that these four geographical races are highly interfertile.

The floral characters of the races converge somewhat in their areas of closest proximity, but do not intergrade continuously (Grant and Grant, 1956, 1960).

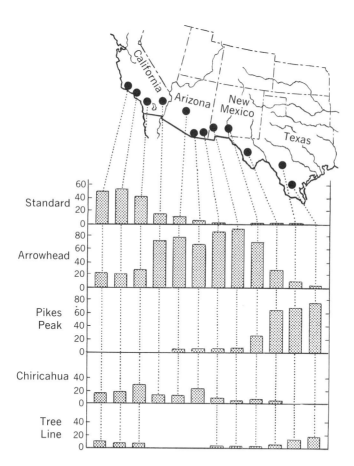

FIGURE 18.3
Frequencies of five inversion types in populations of *Drosophila pseudoobscura* on a transect from southern California to Texas. (Dobzhansky and Epling, 1944.)

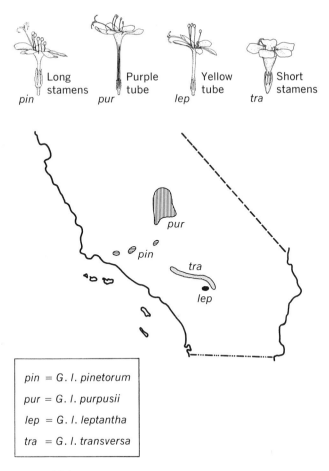

FIGURE 18.4
Disjunct geographical races of *Gilia leptantha* (Polemoniaceae) in separate mountain ranges of southern California. (Drawn from data of Grant and Grant, 1956, 1960.)

A closely related species, *Gilia latiflora*, occurs in arid lowland plains and valleys in the same general area in southern California. It forms continuous or semi-continuous populations over large areas. Its population system consists for the most part of continuously intergrading geographical races (Grant and Grant, 1956, 1960). The contrasting patterns of geographical variation between *Gilia latiflora* and *G. leptantha* are correlated with their respective population structures.

The pattern of geographical variation in the *Nigella arvensis* group (Ranunculaceae) in the Aegean area of southeastern Europe is very interesting. This is a region of complex topography with numerous islands in the Aegean Sea bounded by the mainland areas of Greece and Turkey.

Nigella degenii in this species group is entirely insular in distribution, occurring in the Cyclades Islands. This species consists of a series of more or less discrete local races on the different islands. The local races are grouped into four distinct geographical races with the distributions shown in Figure 18.5. Three other islands on the periphery of the Cyclades archipelago are occupied by two closely related endemic species, *N. icarica* and *N. carpatha* (Figure 18.5). *Nigella arvensis* proper, which is wide-ranging in Europe, shows continuous geographical variation on the mainland, but it too is represented by a distinct race on the islands of Crete and Rhodes (Strid, 1970).

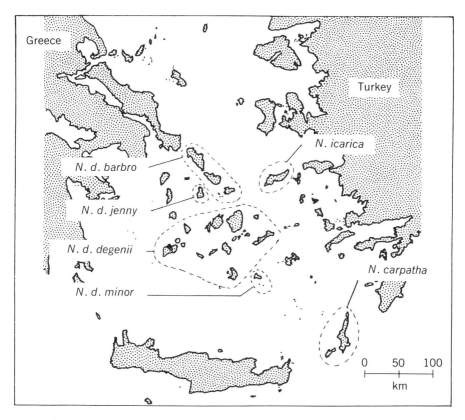

FIGURE 18.5
Geographical races of *Nigella degenii* (Ranunculaceae) in the Cyclades Islands in the Aegean Sea. Also shown are the ranges of two endemic species closely related to *N. degenii*. (Redrawn from Strid, 1970.)

Ecological Races

The third main intraspecific variation pattern is that of ecological racial differentiation. Genetically and phenotypically different but interfertile races are adapted to different habitats in the same territory. These ecological races are consequently sympatric, usually neighboringly sympatric. The races interbreed and intergrade in numerous zones of contact throughout the area of the species. But they preserve their distinctive racial characters in their respective habitats.

Gilia achilleaefolia (Polemoniaceae) occurs in foothills of the Coast Ranges of California. It is a very variable species. The extreme forms are sun races and shade races with the following characteristics. The sun races, occurring in open grassy fields, have large blue-violet flowers clustered in small heads, and are cross-pollinated by bees. The shade races, by contrast, grow in live-oak woods, and have small, pale-colored, solitary, self-pollinating flowers. Shady woods and open fields have a mosaic pattern of distribution in the Coast Ranges, and so do the contrasting ecological races of *Gilia achilleaefolia*. The sun and shade races, which are interfertile, interbreed and intergrade at numerous points of contact between their respective habitats (Grant, 1954).

Ecological races take many forms. Among these are altitudinal races in montane species, host races in insects, and seasonal races in organisms with demarcated breeding systems.

At high elevations in the Sierra Nevada of California several species of woody plants are represented by an erect arboreal race in the subalpine zone and by a low shrubby race in the alpine zone. Adjacent and interbreeding altitudinal races of this sort are found in *Pinus albicaulis, Pinus murrayana, Tsuga mertensiana* (hemlock), and two species of *Salix* (willows). These woody plant species occur at timberline in the Sierra Nevada. Timberline, however, is not the limit of growth of these species, but of their subalpine arboreal races, for their shrubby alpine races extend their range above timberline (Clausen, 1965).

Host races are common in the insects. In many groups of insects the species have narrow host ranges for feeding and breeding. The apple maggot (*Rhagoletis pomonella,* Tephritidae), for example, infests the fruits of two woody members of the Rosaceae in North America, the hawthorn *(Crataegus)* and cultivated apple *(Pyrus)*. The apples and hawthorns harbor different host races of the *Rhagoletis* fruit fly. Since apples and hawthorns grow close together in many places, the host races of *Rhagoletis pomonella* are neighboringly sympatric (Bush, 1969).

The *Rhagoletis* fruit flies mate on their host plants, and this habit reinforces their fidelity to their particular type of host. Host recognition and host preference in *Rhagoletis* and the related genus *Procecidochares* are controlled by a single gene. Hence a mutation in this gene could initiate the formation of a new host race (Bush, 1969; Huettel and Bush, 1972).

Biological Species

The biological species, the sum total of races, is a population system of central importance in nature and in evolutionary biology. It can be observed on every hand—within biotic communities and within natural groups. Thus the deer family (Cervidae) consists of the following well-marked species in North America: whitetail deer, mule deer, American elk, moose, and caribou *(Rangifer tarandus);* in Europe it consists of the fallow deer, roe deer, red deer, moose, reindeer (another race of *Rangifer tarandus*), and, formerly, the Irish elk.

Biological species are kept separate by reproductive isolating mechanisms that prevent or greatly reduce gene exchange between them. The varied forms of reproductive isolation will be discussed later. The point here is that such breeding barriers do operate and are responsible for the prominent discontinuities in the variation pattern that mark the boundary lines of species.

The presence of reproductive isolation between population systems is most clearly evident where those population systems are sympatric and still remain distinct in their phenotypic characteristics. Sympatric population systems of sexually reproducing organisms are, ipso facto, separate biological species. The maintenance of separate character combinations under conditions of sympatry is a natural test, and our best criterion, of the species status of the population systems involved.

Because this situation of distinctness under sympatry is clear-cut, as regards the species status of the entities, it is useful to set it off terminologically and to designate the entities as sympatric species. This case is thereby distinguished from the less clear one of allopatric species.

No one questions the fact that good biological species can live in different territories. Yellow pines and white pines are specifically distinct whether they are growing together in the same mixed-pine forest or in separate forests. Distinct allopatric population systems are not necessarily good biological species, however; they could also be disjunct geographical races that exhibit similar features. The decision as to the rank of distinct allopatric populations depends on the degree of distinctness, which is a relative matter, and on the presence or absence of reproductive isolation, which is usually unknown in this situation without further ad hoc study.

The amount of phenotypic difference between two population systems is less significant for revealing their biological species rank than is the presence of reproductive isolation. In fact, pairs or sets of morphologically very similar but reproductively isolated species are known in many genera of insects, flowering plants, protozoans, and other groups. Such morphologically very similar species are known as sibling species.

For example, the malaria mosquito of Europe, known in the older literature as *Anopheles maculipennis*, turned out on fine analysis to be a complex of six sibling species (segregated and named as *A. maculipennis sens. str., A. melanoon, A. messeae,* and others). These sibling species, though reproductively isolated, are virtually indistinguishable morphologically. Yet they differ in egg color, chromosome morphology, certain physiological traits, behavior, and the medically important character of being or not being a vector of malaria (Mayr, 1963, pp. 35–37, 41).

Semispecies

We occasionally find population systems in nature that are neither good races nor good species. There are intermediate stages of differentiation between disjunct geographical races and allopatric species, or between ecological races and sympatric species. Problematical population systems of this sort are connected by levels of gene flow that are intermediate between the levels typical of races on the one hand and species on the other. Their variation pattern is consequently intermediate between intergradation and discontinuity.

In some cases a combination of race-like and species-like features may be mingled in the same set of population systems. Thus the population systems may intergrade in one part of their total area, but coexist sympatrically without interbreeding in another region (see Figure 18.6). Or the population systems may be sympatric and remain isolated in most places but hybridize in one or a few localities.

The traditional categories of race and species are not sufficient to cover the situation in nature. We need an intermediate category for the intermediate condition. The term semispecies is used to designate population systems that are intermediate between races and species in interbreeding, intergradation, and reproductive isolation.

A set of semispecies constitutes a populational unit of more inclusive scope than the species. A set of allopatric semispecies is known as a superspecies; a set of sympatric or at least marginally sympatric semispecies is a syngameon.

The definitions of the biological species given earlier in this chapter are applicable primarily to sets of races, and require some qualifications when the

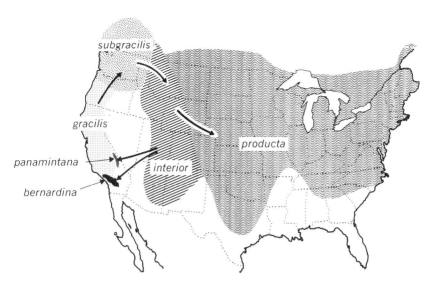

FIGURE 18.6
Overlapping ring of races in the bee, *Hoplitis producta* (Megachilidae). Intergradation occurs from subspecies *H. p. gracilis* to other subspecies, as indicated by arrows. Two derived and terminal subspecies in the series (*H. p. panamintana* and *H. p. bernardina*) coexist sympatrically with *H. p. gracilis* in southern California. (Michener, 1947.)

component units are not races, but semispecies. The biological species is not the most inclusive breeding group in all cases. It is the most inclusive *normally* interbreeding group. But in exceptional cases a more inclusive population system is the syngameon.

The diverse array of population systems in nature requires a more sophisticated classification than the simple dichotomy of race and species. The preceding discussion in this chapter is an attempt to present such a classification (following an earlier treatment by Grant, 1963, Ch. 12). It is desirable to summarize the discussion at this point. A conspectus of population systems is given in Table 18.2.

Species in Uniparental Organisms

Our discussion up to now has been concerned with population systems in sexual and outcrossing organisms. We have been considering aggregations of *breeding* populations. Uniparental organisms, whether they are asexual or parasexual, pose special problems and require separate consideration. The biological species concept as presented earlier in this chapter does not cover, and is not intended to cover, uniparental organisms.

TABLE 18.2
Classification of population systems in sexual organisms.

Level of divergence	Type of population system	Geographical relationships	Phenotypic relationships	Breeding relationships
1	Local races	Microgeographical in extent	Relatively slight differentiation	Interbreeding; intergrading continuously or semicontinuously
2	Contiguous geographical races	Allopatric		As in level 1
	Disjunct geographical races	Allopatric		
	Ecological races	Neighboringly sympatric		
3	Allopatric semispecies	Allopatric		Intermediate between levels 2 and 4
	Sympatric semispecies	Sympatric		
4	Allopatric species	Allopatric		Reproductively isolated; separated by discontinuity
	Sympatric species	Sympatric		
5	Superspecies	Group of allopatric semispecies	Relatively great differentiation	As in level 4
	Syngameon	Group of sympatric semispecies		

Uniparental organisms do form populations. Their colonies consist of genotypically identical or similar individuals, occupying a specific habitat and making particular ecological demands on the environment. They are populations but not breeding populations. And the uniparental populations do not link up into discrete biological species.

The variation pattern in uniparental organisms is quite different from the combination of intergradation and discontinuity that characterizes biparental organisms. What one observes is an array of uniform local (but not breeding) populations built up by the multiplication of one or a few adaptive genotypes. From locality to locality the population genotype is likely to change slightly. If the group occurs in a wide range of environments over a large area, the number of genotypically different kinds of populations may have to be reckoned in the hundreds or thousands, and the extreme forms may differ greatly. The total range of variation within the population system may be comparable to that in a species group in sexual organisms. But the variations are not grouped naturally into race and species units.

The biological species concept obviously breaks down in such cases. And so too, for that matter, does any practical taxonomic species concept, for the variations form a more or less continuous network that is not subdivided into distinct taxonomic units.

Botanists, who encounter this situation in many plant groups, have attempted to deal with it by using a special category, microspecies. A microspecies is a genotypically uniform population or population system with recognizable phenotypic traits of its own in a uniparental group. An array of related microspecies then makes up a particular uniparental complex (see Grant, 1971).

The taxonomic species *Crepis occidentalis* (Compositae) is a very variable group of perennial herbs in western North America. The taxonomic species includes a diploid sexual population system—a biological species—with a localized distribution in the mountains of eastern and northern California. Throughout the remainder of the area of *Crepis occidentalis sens. lat.,* from British Columbia to southern California and from the Pacific Coast to the Great Plains, the populations consist of derived agamospermous types, i.e., plants that reproduce by seeds formed by asexual means (Babcock and Stebbins, 1938).

The agamospermous and hence uniparental populations are uniform. But morphological characters of the leaves, flowering heads, fruits, and other plant parts show recognizable differences from one agamospermous type to the next. The morphological variability in *Crepis occidentalis sens. lat.* is clustered into some 27 recongizable asexual microspecies with their respective geographical ranges (Babcock and Stebbins, 1938). The flowering heads of some of these microspecies are shown in Figure 18.7.

The situation in autogamous plants calls for special comment. Many groups of annual herbaceous flowering plants and some perennial plant groups are autogamous, or automatically self-pollinating and self-fertilizing, and are thus sexual but uniparental. However, the autogamy and inbreeding are usually incomplete and are supplemented by small amounts of cross-pollination. The outcrossing, though constituting only a small fraction of the total reproduction of the autogamous plant, is biologically significant and is often sufficient to link the individuals together into partially breeding populations, and the populations together into species.

The Species in Taxonomy

The species is a basic unit in taxonomy as well as in population biology. The objectives of taxonomy and of population biology are similar but different. Taxonomy is concerned with the formal classification of organisms, and minor systematics, with formal classification at the race and species level. The species

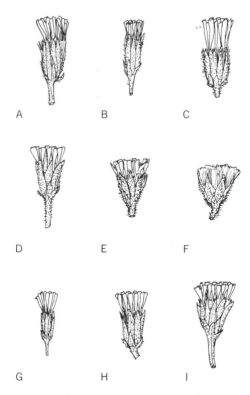

FIGURE 18.7
Flowering heads of nine agamospermous
microspecies of *Crepis occidentalis sens. lat.*
(Compositae). Microspecies names and areas are as
follows: (A) *humilior,* California. (B) *nuttallii,*
Washington. (C) *calyculata,* Colorado. (D) *grayi,*
Utah. (E) *elliptica,* northern California. (F) *deltoidea,*
Alberta. (G) *rydbergii,* Montana. (H) *hamiltonensis,*
northern California. (I) *crassa,* Washington.
(Rearranged and redrawn from Babcock and
Stebbins, 1938.)

in taxonomy is therefore primarily a unit of classification. And the main criterion
for blocking out species units in taxonomy is convenience and workability in
practical classification, identification, and museum filing.

Now the discontinuities in variation patterns that delimit biological species
also serve as convenient boundary markers for the recognition of taxonomic
species. It is convenient taxonomically as well as significant biologically to
subdivide the deer family in North America into the species: whitetail deer,
mule deer, elk, moose, and caribou. In cases such as this, which are widespread

and common in eukaryotic organisms, the species of taxonomy is synonymous with the biological species.

But situations often arise in which the species criterion of taxonomy does not coincide with that of population biology. In some groups it is not convenient, or feasible, or even possible, to recognize biological species in taxonomic practice. There are three fairly common situations in which the taxonomic species necessarily diverges from the biological species.

In the first place, the population systems of sexual organisms do not fall into just two mutually exclusive categories, race and species, but also include intermediate categories, or semispecies. The taxonomic system, however, contains a fixed, artificial hierarchy of categories, i.e., genus, subgenus, section, species, subspecies, and forma. (Section = species group; subspecies = race.) There is no place in the hierarchy for intermediates between species and subspecies; the semispecies of population biology has no counterpart in the taxonomic system. In taxonomy, in order to meet the requirements of a binomial or trinomial nomenclature, a population system at the semispecies level of divergence must be treated as either a species in itself or a subspecies (race) of some other species.

A second set of discrepancies is posed by sibling species. The species units of taxonomy should be identifiable on external morphological characters as preserved in museum specimens. This criterion is not fulfilled by sibling species. Yet sibling species are real biological species, on their breeding relationships, whether they are identifiable in ordinary taxonomic practice or not.

A third area of non-correspondence between the two species concepts occurs in asexual and parasexual organisms. It is a traditional practice in taxonomy, and a requirement of the rules of nomenclature, to classify all organisms into species, whether they are sexual and biparental or not. But biological species simply do not exist in asexual groups.

Conflicts between taxonomic and population-biological systems of classification thus arise in several problem areas. In these situations each field must be true to its own objectives. The resolution is to recognize the legitimacy of both species concepts, that of taxonomy and that of population biology. Semantic confusion can be avoided by designating the two types of species as taxonomic and biological species, respectively, as has been done in the preceding discussion.

Successional Species

The paleontologist dealing with evolutionary changes in geological time is forced to recognize successional species. These are nodal points in a phyletic evolutionary trend that differ enough to warrant the assignment of different species names. For example, a lineage of Late Cenozoic elephants consists of a pair of successional species, *Elephas planifrons* and *E. meridionalis,* which gradually intergrade during the course of time (see Figure 30.2).

Successional species are not coordinate with biological species. Their formation is a product of phyletic evolution rather than of evolutionary divergence. Therefore two successional species do not constitute two biological species.

Collateral Readings

Beaudry, J. R. 1960. The species concept: its evolution and present status. *Rev. Canad. Biol.* **19:** 219–240.

Grant, V. 1963. *The Origin of Adaptations.* Columbia University Press, New York. Chapter 12.

Grant, V. 1971. *Plant Speciation.* Columbia University Press, New York. Chapters 1–4.

Mayr, E. 1963. *Animal Species and Evolution.* Harvard University Press, Cambridge, Mass. Chapters 2, 3, 11–14.

Mayr, E. 1969. *Principles of Systematic Zoology.* McGraw-Hill, New York.

Mayr, E. 1970. *Populations, Species, and Evolution.* Harvard University Press, Cambridge, Mass. Chapters 2, 3, 11–14. These chapters cover the same topics, but in less detail, than those listed above in Mayr, 1963.

Simpson, G. G. 1961. *Principles of Animal Taxonomy.* Columbia University Press, New York. Chapters 1, 2, 5.

19

Isolating Mechanisms

Introduction

Gene exchange between different populations or population systems is reduced or blocked by barriers of various types, known collectively as isolating mechanisms. A classification of the varied array of isolating mechanisms is presented in Table 19.1.

A fundamental distinction is drawn between spatial and reproductive isolation. In spatial isolation, the gametes of different populations do not get together because the populations live in areas separated by distances that are large relative to the dispersal potential of the organisms. In short, spatial isolation is isolation by geographical distance. In reproductive isolation, on the other hand, the barriers to interbreeding stem from inherent characteristics of the organisms themselves. Reproductive isolation permits populations to live in the same territory with little or no gene exchange.

These two modes of isolation enter into the definitions of races and species. Spatial isolation is a characteristic of local populations, local races, and geographical races. Well-developed reproductive isolation is a distinguishing characteristic of biological species. Partial or incomplete reproductive isolation is a characteristic of semispecies.

TABLE 19.1
Classification of isolating
mechanisms.

I. Spatial
    1. Geographical isolation

II. Environmental
    2. Ecological isolation

III. Reproductive
  A. External
    3. Temporal isolation
    4. Ethological isolation
    5. Mechanical isolation
    6. Gametic isolation
  B. Internal
    7. Incompatibility barriers
    8. Hybrid inviability
    9. Hybrid sterility
   10. Hybrid breakdown

Environmental or ecological isolation is a third and somewhat intermediate category. Here the populations differ genetically in their ecological requirements and preferences. The ability of the populations to live in the same territory is determined by the availability of the appropriate habitats or niches and by the force of interspecific competition. If the populations do coexist, hybridization is then controlled by the availability of habitats suitable for their hybrid progeny. The barriers to gene exchange are secular-ecological in nature.

Ecological isolation is a universal feature of species, but it is not a distinguishing feature of species. Ecological isolation is also found between ecological races and between sympatric semispecies. Some ecological differentiation is probably generally present between geographical races.

Reproductive isolation is subdivided in Table 19.1 into external and internal barriers. External barriers are those that prevent the gametes or (in plants) the gametophytes from coming together, whereas internal barriers begin to operate after the gametes or gametophytes do get together.

Ecological and Temporal Isolation

Ecological isolation is a consequence of the secular-ecological differentiation between sympatric species. This latter condition is general and takes numerous forms. *Drosophila* species living in the same area in California or Brazil have different nutritional preferences and feed on different kinds of yeasts. Certain oak species in Texas grow on different soil types, one species *(Quercus mohriana)* occurring on limestone, another *(Q. havardi)* on sand, and still another *(Q. grisea)* on igneous outcrops.

Ecological differentiation in the secular or non-reproductive stages of life reduces the chances of successful hybridization between sympatric species, to one degree or another, and therefore contributes to their isolation. We will discuss ecological differentiation between species more fully in the next chapter (Chapter 20).

Most animals and plants have demarcated breeding or flowering seasons. Mating or cross-pollination takes place during a particular season of the year (e.g., summer or fall), and often at a particular time of day (sometimes by day, sometimes by night). Related species may differ as to the seasonal or diurnal timing of their sexual reproductive periods. Such interspecific differences lead to temporal isolation. Temporal isolation includes seasonal isolation and isolation due to different diurnal periodicities.

Two related species of pines, *Pinus radiata* and *P. attenuata*, come into sympatric contact on the central California coast. Here *Pinus radiata* sheds its pollen early, in February, whereas *P. attenuata* sheds its pollen about six weeks later, in April, and the two species are thus seasonally isolated (Stebbins, 1950, pp. 209–210). *Drosophila pseudoobscura* and *D. persimilis*, which are sympatric over a large area in western North America, breed during the same season of the year but at different times of day, *D. pseudoobscura* being sexually active in the evening and *D. persimilis* in the morning (Dobzhansky, 1951*b*).

Temporal isolation, like any other form of reproductive isolation, may be complete or incomplete. Related species of plants in the same territory often have flowering seasons with different peaks but overlapping ranges, so that the seasonal isolation is incomplete.

Mechanical, Ethological, and Gametic Isolation

Higher animals and plants generally have complex genitalia or flowers consisting of female and male organs that are coadapted structurally so as to facilitate copulation, insemination, or pollination in normal intraspecific matings. Let two such species differ in the configuration of their genitalia or flowers. Then interspecific copulation, insemination, or pollination may be impeded. Such impediments constitute mechanical isolation.

Mechanical isolation occurs between two species of sage, *Salvia mellifera* and *S. apiana* (Labiatae), in southern California. *Salvia mellifera* has typical two-lipped mint flowers with stamens and style standing in the upper lip (Figure 19.1A). It is pollinated by an array of small and medium-sized solitary bees that carry the pollen from flower to flower on their backs (Figure 19.1B). *Salvia apiana* has flowers of a unique conformation with long extended stamens and style and a specialized floral mechanism (Figure 19.1C, D). The floral mechanism of *S. apiana* is worked and pollination is brought about by large carpenter bees and bumblebees that carry the pollen on their wings and other body parts (Figure 19.1E). The normal pollinators of *S. mellifera* are mostly incapable of pollinating *S. apiana*, and vice versa. Mechanical isolation operates. It is not completely effective, however, since some bees of intermediate size can and occasionally do bring about interspecific cross-pollination (Grant and Grant, 1964).

In higher animals copulation is ordinarily preceded by courtship. Courtship comprises a series of stimuli and responses—dances, displays, songs, pheromones, and the like—that prepare the females and males for copulation. The courtship behavior varies from species to species in a group, and the signals are

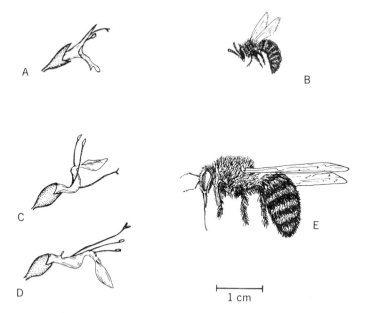

FIGURE 19.1
Two species of sage and their bee pollinators. All drawings to same scale.
(A) Flower of *Salvia mellifera*. (B) *Anthophora*, a common bee pollinator of
S. mellifera. (C, D) Flower of *S. apiana* in untripped (C) and tripped (D)
positions. (E) Carpenter bee, *Xylocopa brasilianorum*, a common pollinator of
S. apiana. (Redrawn from Grant and Grant, 1964.)

species-specific, so the females of one species do not respond to males of a foreign species. The behavioral inhibitions on interspecific mating constitute ethological isolation. Ethological isolation plays a key role in the prevention of interspecific hybridization in many groups of animals, both vertebrate and invertebrate.

Many aquatic organisms release their gametes into the water. External fertilization then depends on the free-living eggs and sperms coming together, and their migration and union may be guided in turn by biochemical substances. These biochemical attractants may be species-specific. Mutual attraction and fertilization then occur between eggs and sperms of the same species, but not between the eggs of one species and the sperms of another in the same water body. This failure of attraction between foreign gametes is gametic isolation.

A classical case of gametic isolation occurs in the sea urchin, *Strongylocentrotus*, in which external fertilization in seawater is the norm. Under controlled experimental conditions, with mixtures of gametes contained in vessels of water, intraspecific fertilizations take place freely in either *S. franciscanus* or *S. purpuratus*, but the interspecific fertilizations (*S. f.* ♀ × *S. p.* and *S. p.* ♀ × *S. f.*) are strongly inhibited.

Incompatibility Barriers and Hybrid Inviability

Let us return to higher animals and plants with internal fertilization and gestation. Assume that the external isolating mechanisms have failed and interspecific copulation or pollination has occurred. A new series of internal barriers to hybrid formation may now come into play in either the parental or F_1 generation.

Many developmental steps occur between pollination and seedling growth in an angiospermous plant. Pre-fertilization steps are: (1) germination of the pollen grains on the stigma, (2) growth of the pollen tube in the style, (3) growth of the pollen tube to the embryo sac in the ovule, (4) release of the sperm nuclei, and (5) their attraction to the female gametes. Fertilization is a step in itself, and in angiosperms it is a two-fold process: (6) fertilization of the egg and (7) fertilization of the endosperm nucleus. The early post-fertilization steps follow: (8) first divisions of the zygote, (9) endosperm development, (10) embryo development, (11) seed formation, and (12) seed germination. And the subsequent stages of the new generation continue: (13) establishment of the young seedling, (14) seedling growth, and (15) development of the adult plant.

Blocks to gene exchange following interspecific crossing can occur at any one of these steps.

A parallel series of stages occurs in mammals: insemination, migration of spermatozoa, conception, implantation, embryo development, birth, and growth and development of the young. And here too, blocks to successful hybridization can occur at any point in the long developmental process.

The blocks are known collectively as incompatibility barriers and hybrid inviability. Where do we draw the line between these two? Let us consider two possibilities, each of which has its advantages and disadvantages. The main point to bear in mind, however we choose to draw the dividing line, is that development is a continuous process that can be blocked at any one of numerous stages.

We can arbitrarily assign the pre-fertilization blocks to incompatibility, and the post-fertilization blocks to hybrid inviability. This is embryologically sound but not always practical. A plant or animal breeder may make a particular cross but not get a viable young F_1 individual, for unknown reasons. The stage of the block could of course be determined by further study, but is somewhat irrelevant as far as the net result of the artificial (or natural) crossing is concerned. It may be more convenient, therefore, to draw the dividing line between incompatibility barriers and hybrid inviability at a later stage.

The second possibility is to define incompatibility so as to include the various blocks after insemination or pollination and up to birth, egg-laying, or seed-ripening. Hybrid inviability would then denote the visible depressions in vigor and malformations in the F_1 individuals after birth, hatching, or germination. Incompatibility blocks in the early stages following insemination or pollination are analogous to gametic isolation in organisms with external fertilization.

Hybrid Sterility

Interspecific crosses in many groups of animals and plants yield F_1 hybrids that are vigorous but sterile, the mule (horse ♀ × donkey ♂) being a well-known example. The sterility phenomena, though confined by definition to the reproductive stage in the F_1's, are still quite heterogeneous. There is variation from case to case in the exact developmental stage affected and in the underlying genetic causes of the infertility.

The development of sex organs and the course of meiosis are complex processes that can easily be upset by disharmonious gene interactions in hybrids. Failure of development of the sex organs can be illustrated by some species hybrids in plants that produce flowers with abortive anthers. In species hybrids in animals the course of cell division in the germ line frequently breaks down owing to genic disturbances. Spermatogenesis may stop before it reaches meiosis, or meiosis may be aberrant, and in either case spermatozoa are not formed. Failure of spermatogenesis in the pre-meiotic stage is the main immediate cause of hybrid sterility in male mules; meiotic disturbances are the cause of sterility in male hybrids of some *Drosophila* crosses (e.g., *D. pseudoobscura* × *persimilis*).

A generalization known as Haldane's rule applies to the sex-limited distribution of hybrid sterility and inviability in dioecious animals. The F_1 hybrids of interspecific crosses in dioecious animals will consist, at least potentially, of a heterogametic sex (carrying the XY chromosome pair) and a homogametic sex (XX). Haldane's rule is that, where sex differences exist in hybrid sterility or inviability, the heterogametic sex is more likely to exhibit these conditions than is the homogametic sex. The heterogametic sex in most animals, including mammals and Diptera, is the male; we have just seen examples of sterile male hybrids in horses and *Drosophila*. There are, however, numerous exceptions to Haldane's rule (see White, 1973, pp. 569ff.).

A third developmental stage in which hybrid sterility can be expressed is the gametophyte generation in plants. In flowering plants the products of meiosis lead directly to gametophytes—pollen grains and embryo sacs—which contain two to several nuclei and house the gametes. Inviability of the gametophytes is a common cause of hybrid sterility in flowering plants. Meiosis is completed but the pollen and embryo sacs fail to develop and function properly.

The causes of hybrid sterility at the genetic level of determination are threefold: genic, chromosomal, and cytoplasmic. Genic sterility is a widespread and basic condition. Unfavorable combinations of the nuclear genes of parental types differentiated at the species level can and do lead to developmental and cytological aberrations in the F_1 hybrids that prevent the formation of gametes. Genetic analysis of genic sterility in *Drosophila* hybrids (*D. pseudoobscura* × *persimilis*, *D. melanogaster* × *simulans*, etc.) shows that sterility genes are located on all or nearly all of the chromosomes of the parental species (see Dobzhansky, 1951*a*, Ch. 8; 1970, Ch. 10). Unfavorable interactions between cytoplasmic genes and nuclear genes also lead to sterility in species hybrids in various plant and animal groups.

Species of plants and animals also differ by translocations, inversions, and other rearrangements that, in the heterozygous condition, cause semisterility or sterility. The degree of sterility is proportional to the number of independent rearrangements; thus heterozygosity for one translocation gives 50% sterility, for two independent translocations, 75% sterility, and so on. The sterility is of the gametophytic type in plants. Meiosis in the chromosome-structural heterozygotes yields daughter nuclei carrying deficiencies and duplications for particular segments; these deficiency-duplication products are unable to form functioning pollen and ovules. Chromosomal sterility of this sort is very common in species hybrids in flowering plants.

The course of meiosis in a hybrid can be disturbed by either genic factors or chromosome-structural differences. Either genic or chromosomal sterility can be expressed in aberrations of meiosis. But the types of meiotic aberration are different. Genic sterility is common in animal hybrids, and chromosomal sterility in plant hybrids. Genetic analysis of some plant species hybrids indicates that chromosomal and genic sterility are often combined in the same hybrid.

Hybrid Breakdown

Let us suppose that a species hybrid is viable enough and fertile enough to reproduce. Its F_2, B_1, F_3, and other later-generation progeny will then generally contain a substantial proportion of inviable, subvital, sterile, and semisterile individuals. These types are the unfavorable recombination products of inter-specific hybridization. The depression of vigor and fertility in hybrid progeny is hybrid breakdown. Hybrid breakdown is the last in the sequence of barriers to interspecific gene exchange.

Hybrid breakdown is invariably found in the progeny of species hybrids in plants, where it can be observed more readily than in most animal crosses. The F_1 hybrid of *Zauschneria cana* $\times$ *septentrionalis* (Onagraceae) is vigorous and semifertile, but most F_2 plants are dwarfish, slow-growing, rust-susceptible, or sterile. In one cross between these species, 2133 F_2's were grown, but not one of these was vigorous (Clausen, Keck, and Hiesey, 1940). Eighty percent of the F_2 generation of *Layia gaillardioides* $\times$ *hieracioides* (Compositae) are inviable or subvital (Clausen, 1951, pp. 108–111). An estimated 75% of the zygotes in the F_2 to F_6 generations of *Gilia malior* $\times$ *modocensis* (Polemoniaceae) were inviable or subvital, and a high percentage (70% or more) of the vigorous fraction were sterile or semisterile (Grant, 1966).

Combinations of Isolating Mechanisms

The isolation of species is rarely if ever dependent on a single isolating mechanism acting alone. It is usual to find several different isolating mechanisms acting in concert.

Consider the species pair *Drosophila pseudoobscura* and *D. persimilis*. The combinations of isolating mechanisms between them include ecological isolation, temporal isolation, ethological isolation, hybrid sterility, and hybrid breakdown. No one of these mechanisms is sufficient in itself to block hybridization. Working in combination, however, they bring about a complete isolation of the two sympatric species in nature (Dobzhansky, 1951*b*, 1955).

Collateral Readings

Dobzhansky, Th. 1970. *Genetics of the Evolutionary Process.* Columbia University Press, New York. Chapter 10.

Grant, V. 1963. *The Origin of Adaptations.* Columbia University Press, New York. Chapter 13.

Gray, A. P. 1954. *Mammalian Hybrids.* Commonwealth Agricultural Bureaux, Farnham Royal, England.

Gray, A. P. 1958. *Bird Hybrids.* Commonwealth Agricultural Bureaux, Farnham Royal, England.

White, M. J. D. 1973. *Animal Cytology and Evolution.* Ed. 3. Cambridge University Press, Cambridge and London. Chapter 15.

20

Ecological Differentiation

Interspecific Competition

Sympatric species interact in their secular or vegetative phase as well as in the reproductive phase. The ecological interactions take different forms. The sympatric species may compete directly for necessary but limited resources — for food, water, space, etc.; they may avoid competition; or there may be an intermediate situation of indirect and subdued competition. These forms of interaction have different evolutionary effects.

Absence of interspecific competition is characteristic of sympatric species that belong to different major groups, have very different ecological requirements, and perform different roles in the biotic community. In fact, such ecologically well-differentiated species, so far from being in competition, may have relations of mutual dependence within their biotic community. A biotic community is an ecological unit that is self-sustaining in energetics, and hence an association of sympatric species playing complementary roles in nutrition and energetics. We will not go into the subject of interdependence in biotic communities here.

Interspecific competition becomes a factor when the sympatric species have similar ecological requirements in general, and, in particular, utilize some resource that is present in a limiting quantity. The species involved in the competition are often more or less closely related, but may also be distantly related but convergent.

Interspecific competition has two broad aspects. The first is passive use by different species of the same resource. Thus different species of shrubs in a desert community are likely to compete passively or non-aggressively for a limited amount of ground water. The second aspect, superimposed on the first, is direct inhibition of one species by a competitor species.

A number of plant species produce chemical substances in the leaves that leach out into the soil and inhibit the germination and growth of other plants in the vicinity (Muller, 1966, 1970). In animals, inhibition may be achieved by aggressive behavior or by the exercise of dominance based on the threat of aggression. Native bighorn sheep *(Ovis canadensis)* and feral burros *(Equus asinus)* compete for water and forage in the Mojave Desert of California and Nevada. Burros are dominant over the bighorn sheep in direct confrontations; when the burros move in on water holes occupied by bighorns, the latter move out and in some cases abandon the area (Laycock, 1974).

Species Replacement

Interspecific selection is the increase in numbers and ecological dominance of one species relative to another, competitor species. Interspecific competition leads to interspecific selection where one species has some inherent advantage over the other in relation to the competition. A great increase in numbers of burros has been observed in Death Valley, California, since 1930, and along with this a marked decline in numbers of bighorn sheep.

The interspecific selection process can go all the way to species replacement, with species A replacing species B completely in a territory, if the environmental conditions in which A has the advantage remain constant. The dingo *(Canis familiaris dingo)* thus replaced the Tasmanian wolf *(Thylacinus)* in most of Australia during historical times.

Species replacement has been studied in laboratory experiments with *Paramecium, Tribolium,* and other organisms. In his classical experiments with *Paramecium,* Gause (1934) put *P. aurelia* and *P. caudatum* together in a medium of water, salts, and food bacteria *(Bacillus pyocyaneus)* in glass containers, and kept the temperature and composition of the medium constant. *Paramecium aurelia* replaced *P. caudatum* completely in a few weeks. Different replications of this experiment carried out under identical environmental conditions always gave the same result. But when the conditions were changed, by using a different strain of *Bacillus* as a food organism, *P. caudatum* replaced *P. aurelia*. In general, whenever two species of *Paramecium* were forced to compete in a homogeneous culture medium, one species always replaced the other eventually.

If, however, the laboratory environment is heterogeneous for a species mixture of *Paramecium*, complete species replacement does not necessarily occur. This is illustrated by combinations of *Paramecium aurelia* and *P. bursaria* placed in tubes containing vertically stratified suspensions of food yeasts. Here *P. aurelia* feeds mainly in the upper layers and *P. bursaria* mainly on the bottom. Given different food niches of this sort, and different feeding preferences of the two species, the two species can continue to exist in equilibrium indefinitely (Gause, 1935).

Species replacement undoubtedly occurs commonly in nature. The process is usually difficult to observe and interpret correctly in the modern world of nature, apart from a few clear-cut cases like that of the Tasmanian wolf and dingo in Australia. But the indirect evidence for frequent species replacements is inescapable when we consider evolutionary changes on a geologic time scale. An estimated 98% of the living families of vertebrate animals are descended from about eight species in the early Mesozoic, and eight is only a very small fraction of the thousands of vertebrate species living at that time (Wright, 1956).

Competitive Exclusion

The experiments of Gause, the observations of field naturalists (especially Grinnell and Lack), and mathematical equations (of Volterra and Lotka) have led to a generalization known as the competitive exclusion principle (also known variously as Gause's law, Gause's principle, and the Lotka-Volterra principle; see Hardin, 1960).

The competitive exclusion principle (as we shall refer to it) has served as a focal point in species ecology since the 1930s and has been extensively discussed by many authors. Different formulations of the principle are given by different authors. It would take us too far afield to trace the various formulations here; instead we will quote one passage from Lack (1947), which expresses the essential idea, and then give a general statement of the principle. Lack states (1947, p. 62):

> My views have now completely changed, through appreciating the force of Gause's contention that two species with similar ecology cannot live in the same region (Gause, 1934). This is a simple consequence of natural selection. If two species of birds occur together in the same habitat in the same region,

eat the same types of food and have the same other ecological requirements, then they should compete with each other, and since the chance of their being equally well adapted is negligible, one of them should eliminate the other completely. Nevertheless, three species of ground-finch live together in the same habitat on the same Galápagos islands, and this also applies to two species of insectivorous tree-finch. There must be some factor which prevents these species from effectively competing.

The above considerations led me to make a general survey of the ecology of passerine birds (Lack, 1944a). This has shown that, while most closely related species occupy separate habitats or regions, those that occur together in the same habitat tend to differ from each other in feeding habits and frequently also in size, including size of beak. In a number of the latter cases it is known that the beak difference is associated with a difference in diet, and this correlation seems likely to be general, since it is difficult to see how otherwise such species could avoid competing.

The competitive exclusion principle makes two general statements about sympatric species. (1) If two species occupy the same ecological niche, one species is practically certain to be superior to the other in this niche, and will eventually replace the inferior species. Or, more briefly, "complete competitors cannot coexist" (Hardin, 1960). The second statement is a corollary of the first: (2) If two species coexist in a stable equilibrium, they must be ecologically differentiated so that they can occupy different niches.

There are two ways of looking at the competitive exclusion principle: as an axiom and as an empirical generalization. Considered as an axiom or law of nature, it is logical, coherent, and has been very heuristic. Considered as an empirical generalization, it is widely but not universally valid, as will be pointed out next.

Species Coexistence

Complete replacement of one species by another is not the only outcome of interspecific competition, as is evident from the fact that related species with similar ecological requirements do frequently coexist in nature. Complete species replacement can be circumvented in any one of several ways. Four such ways are mentioned below (see also Grant, 1963, pp. 403ff.).

First, species replacement is a time-consuming process. It is to be expected that observations made at any given moment of time will find some pairs of competing species in an uncompleted stage of replacement.

Second, ecologically similar species may coexist without ever reaching a stage of direct interspecific competition. This would be the case if their numbers were kept in check by some factor other than direct competition. For instance, populations of herbivores in natural communities are often held down by predators, and predation then prevents interspecific competition between herbivores for food from becoming an important factor affecting herbivore coexistence.

Third, the environmental conditions may change reversibly during the course of interspecific selection so as to give a selective advantage to species A in one phase and to species B in another phase. The two species would then coexist in a cyclical equilibrium.

A fourth situation, which is very important in nature, is that where the environment is heterogeneous and contains a range of conditions for some critical factor. Species A may be superior to species B in one facet of the environment while species B is superior in another facet. Species A and B can coexist in this situation by living partly or mainly in their preferred respective facets of the environment. Laboratory experiments with *Paramecium, Drosophila,* and other organisms are relevant in this connection (Gause, 1935; Ayala, 1969). It will be recalled that certain pairs of *Paramecium* species form stable mixtures when their food is dispersed in different feeding zones corresponding to the differential feeding preferences of the species. Natural habitats are, of course, always heterogeneous, and are often if not usually heterogeneous in ways that permit some ecological segregation of sympatric species, and hence a continuous coexistence.

Selection for Ecological Differentiation

We find that interspecific selection can lead to complete species replacement in some situations and to ecological divergence and coexistence in others. A model of the latter process, where ecological differentiation is the outcome of interspecific selection, will be presented here.

The model contains the following assumptions: (1) There are two ecologically similar and competitive species, A and B. (2) Their environment is heterogeneous, with facies E_1, E_2, and E_3. (3) Each competing species contains polymorphic types ($A_1, A_2, A_3; B_1, B_2, B_3$) fitted to the three environmental facies (to E_1, E_2, and E_3, respectively). (4) Polymorphic type A_1 is superior in E_1, while B_3 is superior in E_3.

The setup at the beginning of the interspecific selection process is as follows:

Environment	E_1	E_2	E_3
Species A	A_1 (superior)	A_2	A_3
Species B	B_1	B_2	B_3 (superior)

Letting interspecific selection run its course, and ignoring various complicating factors such as interbreeding between polymorphic types in each species, the final result will be

Environment	E_1		E_2	E_3
Species A	A_1		A_2	
Species B			B_2	B_3

We see that ecological divergence develops between competing species A and B. Each species becomes more narrowly specialized for the aspects of the common environment in which it has the selective advantage. Furthermore, the process of specialization is accompanied by a loss of genetic variability in each species.

Considerable evidence from natural populations is in agreement with this model.

Lack (1947) points out that the White-eye bird *(Zosterops palpebrosa)* in southeast Asia has a different altitudinal range in areas where it occurs alone from that in areas where it is sympatric with other species of *Zosterops*. In Burma, where *Z. palpebrosa* is the sole member of its species group, it breeds over a wide altitudinal range from sea level to the high mountains. But in Malaya and Borneo, where a related species of *Zosterops* occupies the middle and higher zones, *Z. palpebrosa* is restricted to the lowlands. And in Java, Bali, and Flores, where two other related species occur in the higher and coastal zones, respectively, *Z. palpebrosa* is restricted to the middle altitudinal zone.

A second line of evidence comes from the phenomenon of character displacement, to be described next.

Character Displacement

Character displacement can be observed in some pairs of species that have overlapping areas, being sympatric in one part of their range and allopatric elsewhere. Let allopatric and sympatric races of the two species be compared with respect to their morphological, ecological, or behavioral characteristics. It frequently turns out that the sympatric races of two such species are more distinct than the allopatric races. The enhanced degree of differentiation in the sympatric races of overlapping sympatric species is known as character displacement (Brown and Wilson, 1956).

Character displacement is observed in a number of animal groups and in a few plant groups. Examples are found in nuthatches *(Sitta)* in Eurasia for bill size and body size, in ground finches *(Geospiza)* in the Galápagos Islands for bill size and body size, in *Lasius* ants in North America for various characters, and in other groups of vertebrates and insects (Brown and Wilson, 1956).

Characters exhibiting character displacement may owe their differentiation to different modes of selection. Character displacement for reproductive characteristics may be a product of selection for enhanced reproductive isolation. This process does not concern us here (see Chapter 24). But sympatric divergence of characters affecting success in the secular-ecological phase of life, such as bill size and body size in birds, is likely to be a result of interspecific selection for ecological differentiation (Brown and Wilson, 1956).

Ecological Niche

"Ecological niche" is a key term in the competitive exclusion principle. How do we define this term? And can we define it in a way that is independent of the competitive exclusion principle?

Much effort has been devoted to the search for a satisfactory formal definition of niche in either verbal or mathematical terms (see Connell and Orias, 1964; Hutchinson, 1965; Levins, 1968; Krebs, 1973; Whittaker, Levin, and Root, 1973; Pianka, 1974; Rejmanek and Jenik, 1975). The subject is an elusive one, and the search has not been entirely successful.

Hutchinson (1965, pp. 32ff.) begins by distinguishing between the so-called fundamental niche and the realized niche. The fundamental niche, according to Hutchinson, is the volume in multiple dimensions occupied by a species, where each dimension corresponds to one variable factor necessary for the life of the species. The multidimensional volume of the fundamental niche is considered to be undistorted by the presence of competitor species. The realized niche is then the fundamental niche as restricted by the presence of competitor species.

However, Hutchinson's (1965) definition of the fundamental niche seems to be a definition of what plant ecologists and plant geographers have long called the "tolerance range" of a species. It specifies the potential species area, not the niche. Hutchinson's "realized niche" *is* the ecological niche. The inclusion of the effects of interspecific competition in the definition of the realized niche is important.

The field naturalist knows what an ecological niche is in practice, whether he can define it or not. Deep groundwater and topsoil water represent different ecological niches for the root systems of plants. Large seeds and small seeds are different food niches for seed-eating birds. A good example of niche diversity concerns the feeding zones of warblers in spruce forests in the northeastern United States (MacArthur, 1958). Five sympatric species of *Dendroica* feed on insects in three different parts of the crowns. Their characteristic feeding zones are as follows:

> Upper crown: Cape May warbler *(Dendroica tigrina)* and Blackburnian warbler *(D. fusca)*
>
> Mid part of crown: Black-throated green warbler *(D. virens)* and Bay-breasted warbler *(D. castanea)*
>
> Lower part of tree: Myrtle warbler *(D. coronata)*

There are finer degrees of ecological differentiation between warbler species that feed in the same zones. Thus in the middle zone *D. castanea* gets much of its insect food by hawking, whereas *D. virens* engages only rarely in hawking (MacArthur, 1958).

Examples such as the foregoing, which could be multiplied endlessly, suggest that the niche can be characterized in terms of two sets of factors: (1) the habitat or broad environmental field for which a species is adapted; and (2) the restrictions on exploitation of the habitat imposed by interspecific competition and interspecific selection. Hence the niche can be regarded as a facet of the habitat for which a species is particularly specialized.

Following logically from this is a recognition of differences in niche breadth. Ecological niches may be relatively narrow or relatively broad. And the niche breadth presumably correlates with the degree of specialization of the species occupying the niche in question. Various ecologists have pointed out that niche breadth diminishes with increase in interspecific competition. The presence of a number of ecologically similar species in a community tends to promote narrow specialization and reduced niche breadth (Lack, 1945; Connell and Orias, 1964; Levins, 1968).

Influence of Ecological Demands

Interspecific competition for essential environmental resources sets in motion the process of interspecific selection, which can result eventually in either ecological divergence or complete replacement. It is logical to assume that interspecific selection will run its course more rapidly where the interspecific competition is intense. A general rule, subject to some exceptions, is that large-bodied animals and plants make heavier demands on the environment, and more quickly exhaust the resources present in limiting amounts, than do small-bodied organisms. These differences between large and small organisms can be expected to lead to correlated differences in the intensity of interspecific competition and the rate of interspecific selection.

Ross (1957) studied the geographical and ecological distribution of six related species of leafhoppers in Illinois (*Erythroneura lawsoni* group, Jassidae, Hemiptera). These small insects feed and breed on sycamore leaves *(Platanus occidentalis)* and thus have a narrow ecological niche. Nevertheless, biotic sympatry is common in the *E. lawsoni* group, and several species are frequently found coexisting in the same niche. Signs of interspecific competition were looked for in such sympatric associations but were not found.

Ross (1957) attributes the apparent absence or weakness of interspecific competition to the small size of the leafhoppers, which makes it possible for them to maintain their populations in a restricted niche without depleting the food supply. He remarks dryly that "It is difficult to visualize half a dozen species of elephants all maintaining reproductive units if restricted to sycamores for food."

He goes on to state the general conclusion that "The number of species which [can] occupy the same niche [is] inversely proportional to the food requirements, hence usually to the absolute size, of the organism."

Distribution patterns in a number of animal groups are in agreement with Ross's generalization. It is significant that related species of large mammals are nearly always allopatric, e.g., the African and Asiatic elephants, the species of big cats, and bears in most species combinations. The Black rhinoceros and White rhinoceros occur together in the same general area in South Africa, but they are neighboringly sympatric there and do not occupy the same niche; furthermore, one species is rare. On the other hand, biotically sympatric assemblages are quite common in small rodents.

The opposite extreme among higher animals, however, is found in small insects such as *Erythroneura, Drosophila,* and *Anopheles,* in which biotically sympatric assemblages composed of several or many species are normal.

Parallel trends are seen in higher plants in a temperate-zone flora. Related species of dominant trees in genera such as *Quercus* and *Pinus,* and of perennial herbs in genera like *Iris* and *Polemonium,* mostly have allopatric or neighboringly sympatric distributions. But small annual herbs like *Gilia* and *Mentzelia* commonly form sympatric assemblages of several species.

The larvae of butterflies are heavy feeders on plant herbage, whereas the adult butterflies are light-to-moderate flower feeders. It is significant that a given species of butterfly often has a broad food niche in the adult stage but a narrow and specialized food niche in the larval stage. The adult butterflies feed on a fairly wide range of flower species, but lay their eggs on specific host plants, where the larvae later complete their development.

Different groups of butterflies have different larval host plants. Thus for the pipevine swallowtail *(Battus philenor)*, the larval food plant is *Aristolochia*; for the monarch butterfly *(Danaus plexippus)*, milkweed *(Asclepias)*; and for the alfalfa butterfly *(Colias eurytheme)*, alfalfa, clover, and vetch (Fabaceae). But the adults of these same species of butterflies feed on broadly overlapping arrays of flower species. Parallel differences between the food-niche breadths of larval and adult stages occur in various groups of moths. Ecological differentiation has thus progressed further in the heavy-feeding larvae of many butterflies and moths than in the moderate- or light-feeding adults.

Collateral Readings

Andrewartha, H. G., and L. C. Birch. 1954. *The Distribution and Abundance of Animals.* University of Chicago Press, Chicago. Chapter 10.

Brown, W. L., and E. O. Wilson. 1956. Character displacement. *Syst. Zool.* **5:** 49–64.

Grant, V. 1963. *The Origin of Adaptations.* Columbia University Press, New York. Chapter 14.

Hardin, G. 1960. The competitive exclusion principle. *Science* **131:** 1292–1297.

Hutchinson, G. E. 1959. Homage to Santa Rosalia or why are there so many kinds of animals? *Amer. Nat.* **93:** 145–159.

Hutchinson, G. E. 1965. *The Ecological Theater and the Evolutionary Play.* Yale University Press, New Haven, Conn.

Lack, D. 1947. *Darwin's Finches.* Cambridge University Press, Cambridge.

Lack, D. 1971. *Ecological Isolation in Birds.* Harvard University Press, Cambridge, Mass.

Pianka, E. R. 1974. *Evolutionary Ecology.* Harper & Row, New York. Chapters 5, 6.

21

Geographical Speciation

Evolutionary Divergence

Divergence between related evolutionary lines, if gradual and long-continued, goes through a series of stages with time. A common ancestral population gives rise, in succession, to two or more local races, geographical races, semispecies, biological species, and species groups (Figure 21.1). The divergence can continue to higher levels from genus to class and phylum.

The populational units and taxonomic categories are stages in evolutionary divergence. Considered as evolutionary products, the higher taxonomic categories are problems in macroevolution. Our concern here is with the process of divergence in the lower middle range involving races and species. Divergence at these levels constitutes problems of speciation.

Divergence at the race and species levels is manifested in three main sets of characteristics. Progression from lower to higher levels on the branching phylogeny is accompanied by (1) increasing differentiation in genotype; (2) increasing differentiation in morphology, physiology, ecology, and behavior; and (3) stronger isolation (Figure 21.1).

It must be recognized that we are speaking in very general terms here. The three sets of characteristics listed above may evolve at different rates in different groups of organisms. In some groups a relatively small amount of genotypic differentiation is amplified into striking phenotypic differentiation, while other groups have extensive genotypic but small phenotypic differentiation. Reproductive isolation is only loosely correlated with degree of phenotypic differentiation when a broad sample of plant and animal groups is considered.

Variation also exists in the sequence of stages. Some of the middle stages of speciation may be bypassed in some cases. Reversals in the process of divergence are also possible. Such variant modes can be considered later. We ruled out these variations at the beginning of this section by postulating a gradual and long-continued process of divergence.

A common mode of divergence, known as geographical speciation, follows the sequence: (1) common ancestral population, (2) geographical races, (3) allopatric semispecies, (4) biological, and eventually (5) sympatric, species. This mode of speciation is not universal, as we shall see later. However, it is well known in some groups, and furthermore, being a gradual process, it passes

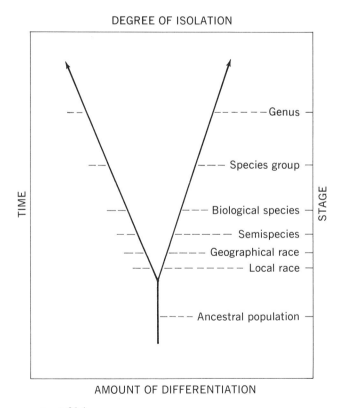

FIGURE 21.1
Stages of divergence.

through all the intermediate stages on the way to species status. For these reasons it is useful to begin our survey of speciation processes with geographical speciation.

Factors in Race Formation

The formation of races—be they local or geographical, contiguous or dis-junct—is brought about by the same evolutionary forces as those that control microevolutionary changes within populations. It will be recalled that the basic variation-producing processes are mutation, gene flow, and recombination, and that the variation-sorting processes are selection and drift (Chapters 4 and 7). Race formation takes place when these processes follow different courses, that is, produce different gene pools, in different geographical subdivisions of a species area.

The Role of Selection

The chief effective force involved in race formation is undoubtedly natural selection. Two groups of phenomena furnish evidence for the role of selection in race formation. First, racial traits are commonly seen to be adaptive for the environment inhabited by the race in question. Second, parallel racial variation is often found in different species that occur throughout a similar range of environments.

A generalization in animal systematics known as Bergmann's rule applies to species of warm-blooded vertebrates that occur in both warm and cold areas. Bergmann's rule states that such species tend to be represented by large-bodied races in areas with cold climates and by small-bodied races in warm areas. This rule holds true for about 75–90% of the bird species and 60–80% of the mammal species in various faunas. It is thus a valid generalization, subject to some exceptions (see Mayr, 1963, pp. 319–321).

The adaptive value of the body-size differences described by Bergmann's rule is a function of the thermodynamic properties of warm bodies. In a cold climate a large warm body has an advantage in respect to heat conservation because of the low ratio of surface to volume. This advantage tends to disappear in a moderately warm climate. And in a hot climate a small body may have a positive advantage for temperature regulation because of a high surface-to-volume ratio, which promotes heat dissipation.

If racial variation in body size along the above lines were known in only a few isolated cases, the interpretation that such variation is adaptive, and hence controlled by selection, would rest on relatively weak foundations. But the fact that the same body-size trend occurs in many distinct species in different faunas, so as to warrant recognition as Bergmann's rule, makes the case for a controlling role of natural selection compelling.

A related ecogeographic rule, Allen's rule, pertains to the length of protruding body parts—ears, nose, bill, tail—in species of warm-blooded animals ranging through different climatic zones. According to Allen's rule, the races of a species of bird or mammal inhabiting a cold area generally have short protruding body parts, while the races in warm areas have long extremities.

This general trend in racial variation, like that in body size, is related to the problems of temperature regulation in warm-blooded vertebrates. Long protruding parts give off body heat. Therefore, for a warm-blooded animal, long extremities are usually selectively disadvantageous in a cold climate but advantageous in a warm climate.

Many other regular patterns of racial variation, which are probably adaptive, have been recognized in various groups of animals. For reviews see Rensch (1960a; 1960b, pp. 43–46) and Mayr (1963, pp. 318ff.).

Parallel racial variation correlated with environmental features is also found in many wide-ranging plant species. Here the subject is subsumed under the heading of "ecotypes" (or adaptive races) rather than codified as ecogeographic rules.

Plant species ranging along a coastline may be represented by prostrate ecotypes on sand dunes and by bushy ecotypes on cliffs. Species of woody plants in high mountains may be represented by arboreal ecotypes in the subalpine zone and by shrubby elfinwood ecotypes in the alpine zone (see discussion of altitudinal races of *Pinus*, etc., in Chapter 18). Such ecotypes or races are undoubtedly produced by environmental selection. Good experimental studies of plant ecotypes are summarized by Turesson (1922, 1925) and Clausen, Keck, and Hiesey (1940, 1948).

The Roles of Drift and Introgression

Haphazard racial variation is found in some species that have colonial population systems. By "haphazard" we mean that the racial differences are not obviously correlated with environmental differences in transects through the species area. An illustration is found in the New Guinea kingfisher *(Tanysiptera galatea)* in the New Guinea territory (Mayr, 1942, pp. 152–153).

New Guinea proper is a very large island, 1500 miles long and 100 to 500

miles wide. It spans a wide range of climates and vegetation types, from ever-
green rainforest at one end to seasonally dry monsoon forest at the other. New
Guinea is surrounded by numerous medium-sized and small islands with climates and vegetation similar to those of the neighboring parts of the main
island.

The mainland of New Guinea proper is occupied by three races of *Tanysiptera
galatea,* namely, *T. g. galatea, T. g. meyeri,* and *T. g. minor.* These races extend
with gaps from one end of the large main island to the other (Figure 21.2). They
are very similar. Five additional races occur on some of the small neighboring

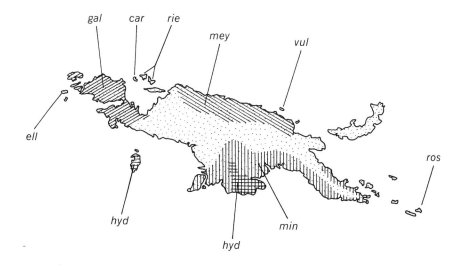

gal	=	T. g. galatea
mey	=	T. g. meyeri
min	=	T. g. minor
ell	=	T. g. ellioti
car	=	T. g. carolinae
rie	=	T. g. riedelii
vul	=	T. g. vulcani
ros	=	T. g. rosseliana
hyd	=	T. hydrocharis

FIGURE 21.2
Distribution of races of the kingfisher (*Tanysiptera galatea* group) in the New Guinea Territory.
(Redrawn from Mayr, 1942.)

islands (Figure 21.2). These insular races are quite distinct from one another and from the adjacent mainland races from which they probably descended. A closely related species, *T. hydrocharis,* occurs on two islands as well as on part of the mainland (Figure 21.2).

Mayr's (1942, 1954) interpretation of this pattern of racial variation is as follows. The orthodox racial variation on the New Guinea mainland is a product of selection in a large and widely interbreeding population system. The island populations were founded by small numbers of colonizing individuals migrating from the mainland. Drift (or founder effect in Mayr's terminology) occurred during the early generations of colonization, and its effects were perpetuated by spatial isolation from the large mainland population. These factors enabled the small-island races to diverge markedly from the neighboring and probably ancestral mainland population. One insular population system, *T. hydrocharis,* has diverged to the species level.

It is not necessary to assume that selection is not involved in race formation in the *Tanysiptera galatea* group. Drift and selection usually operate hand in hand in natural populations, as pointed out in Chapter 13. The biologically realistic distinction regarding factors responsible for race formation is that between selection in large populations, on the one hand, and the selection-drift interaction, on the other.

Many cases of racial variation are known that conform to the *Tanysiptera* pattern. An example in the *Nigella arvensis* group (Ranunculaceae) in the Aegean Islands and surrounding mainland was described earlier (see discussion in Chapter 18 and Figure 18.5).

Introgression is a special mode of natural hybridization known particularly in plants. Introgression consists of the formation of natural F_1 hybrids and their backcrossing to one or both parental species or semispecies. The backcrossing often occurs repeatedly in successive generations. The result is a flow of certain genes from the donor species or semispecies into the recipient population system. Introgression is an extension of gene flow and recombination to the species level.

Where two species or semispecies, A and B, are marginally sympatric and engage in introgression, races develop in the zone of overlap that are convergent in various morphological characters and ecological preferences. Species A gives rise, by introgression, to a race in the area of species B that approaches B in various characteristics, and vice versa. Examples have been studied in *Iris, Tradescantia, Helianthus, Gilia,* and many other plant groups (for review see Grant, 1971; Heiser, 1973*b*).

Here again, as in the case of drift, we should not assume that selection is not involved. Introgression generates the variations, but natural selection sorts them out, eliminating some and preserving others. Introgressive races are products of both introgression and selection.

Incipient Reproductive Isolation

Different geographical races of the same species are often partially isolated reproductively. Races living in different geographical areas are likely to have different blooming seasons or breeding seasons. They are likewise apt to have different ecological requirements. Partial internal barriers may also exist between geographical races. Artificial interracial F_1 hybrids in herbaceous flowering plants are usually semisterile, and their F_2 progeny usually show some hybrid breakdown. All of the known reproductive isolating mechanisms surveyed in Chapter 19 can be found, in partial or incipient form at least, between certain pairs of geographical races.

Reproductive and environmental isolation can and does develop as a byproduct of evolutionary divergence. Consider two geographical races of a plant species living in different areas, one area being cool and moist and the other warm and dry.

The two races will be exposed to selection for different climatic and perhaps different edaphic conditions; this is the first stage of ecological isolation. They are likely to come into flower at different seasons; partial seasonal isolation develops. Different groups of bees are apt to be on the wing in the cool moist and warm dry areas, and as the indigenous plant races become adapted to their normal bee visitors and pollinators, their floral characteristics undergo changes. We now have the beginnings of mechanical or ethological isolation or both. Some changes in flower structure, such as style length, affect the ease of crossing, and can initiate an incompatibility barrier. Finally, the large number of gene differences accumulated by the two races in the course of their differentiation may well result in the formation of some disharmonious recombination products in their hybrids. Partial hybrid inviability, semisterility, and partial hybrid breakdown come into the picture.

Divergence at the racial level thus inevitably brings about some reproductive isolation as a side effect. This sets the stage for possible future speciation.

The Geographical Theory of Speciation

The geographical theory of speciation was developed by a series of zoologists including M. Wagner in the late nineteenth century and K. Jordan, D. S. Jordan,

Rensch, Dobzhansky, Mayr, and others in the first half of this century. It was later extended to plants by Clausen, Stebbins, and others. The arguments and evidence are clearly set forth by Mayr in *Systematics and the Origin of Species* (1942).

According to this theory, biological species are normally derived from geographical races. Spatial isolation is a normal prelude to the development of reproductive isolation. The sequence of stages in species formation is as shown in Figure 21.3. A species composed of contiguous geographical races (Figure 21.3B) breaks up into a series of disjunct races (Figure 21.3C). Reproductive isolation develops as a by-product of divergence in the disjunct races. When

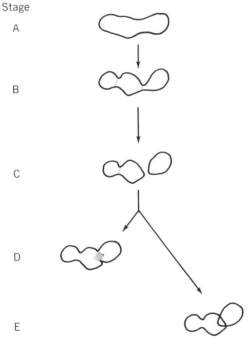

FIGURE 21.3
Stages of geographical speciation. (A) Wide-ranging uniform species. (B) Polytypic species with contiguous races. (C) Polytypic species with one disjunct and distinct race. (D, E) Migration leading to reestablishment of contact between the disjunct race and the main body of the species. (D) Interbreeding and secondary intergradation occur. (E) Interbreeding does not occur; the formerly disjunct populations behave as young species. (Rearranged from Mayr, 1942.)

this reproductive isolation reaches a certain critical point, the divergent populations can live in the same territory without interbreeding, and have thus attained the stage of species (Figure 21.3E).

All of the stages of speciation required by the geographical theory – the intermediate as well as terminal stages – are actually found in various groups of animals and plants. This is the strongest line of evidence in favor of the geographical theory.

A set of examples from a single natural group, the Cobwebby gilias (Polemoniaceae), is shown in Figure 21.4. These are diploid, annual, predominantly outcrossing plants of desert and mountain regions in the American Southwest. In this group we find the following series of stages (Grant and Grant, 1956, 1960). (1) A species composed of contiguous geographical races (Figure 21.4A). (2) Disjunct and distinct geographical races (Figure 21.4B, also Figure 18.4). (3) A syngameon composed of three marginally sympatric and sporadically hybridizing semispecies (Figure 21.4C). (4) A pair of intersterile allopatric species (Figure 21.4D). (5) A pair of intersterile sympatric species (Figure 21.4E).

Borderline cases between geographical races and full-fledged species are particularly significant in this connection. Such borderline cases, or semispecies, are not uncommon, as every animal and plant systematist knows. Mayr (1942, p. 165) estimates that 12.5% of the taxonomic species of birds of North America are in a borderline condition between species and subspecies, so that ornithologists sometimes treat them in one taxonomic rank and sometimes in the other. Examples are the Red-shafted flicker and Yellow-shafted flicker *(Colaptes)* and the Audubon warbler and Myrtle warbler *(Dendroica)*.

Comparable examples are found in many other animal and plant groups. As a plant example we can mention the eastern North American white pine *(Pinus strobus)* and the western white pine *(Pinus monticola)*.

An especially interesting type of borderline case is the overlapping ring of races. Here we have a chain of intergrading races – A, B, C, D, E – stretched out in a great circle with overlap between A and E. Races A and E represent the morphological and ecological extremes in the population system. They intergrade in the series A–B–C–D–E. Considering this pattern alone, the population system is a normal species composed of geographical races. But the terminal and extreme members of the series, A and E, coexist sympatrically, and the relationship between A and E, therefore, is that of sympatric species.

FIGURE 21.4
Stages of geographical speciation in the Cobwebby gilias (Polemoniaceae). Further explanation in text. (Drawn from data of Grant and Grant, 1956, 1960.)

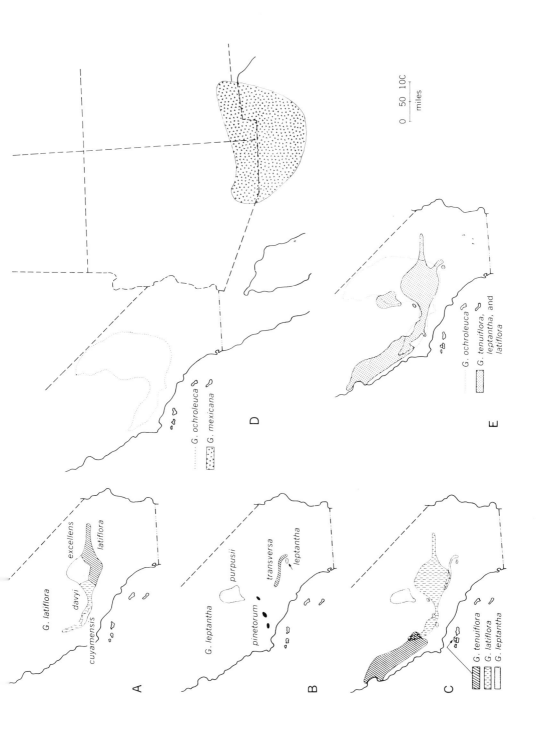

A

G. latiflora

davyi
excellens
latiflora
cuyamensis

B

G. leptantha

purpusii

pinetorum

transversa
leptantha

C

G. tenuiflora
G. latiflora
G. leptantha

D

G. ochroleuca
G. mexicana

E

G. ochroleuca
G. tenuiflora, leptantha, and latiflora

0 50 10C
miles
0 50

Among the classical cases of overlapping rings of races are the titmouse *(Parus major)* in Eurasia, warbler *(Phylloscopus trochiloides)* in central Asia, gulls *(Larus argentatus* group) in northern Eurasia and North America, and deermouse *(Peromyscus maniculatus)* in western North America. In some of these older cases, e.g., *Parus major,* reinvestigation reveals less complete isolation between the terminal races than was formerly believed. Other cases have been described more recently in various genera of birds, amphibians, insects, and angiosperms (for review see Mayr, 1963, pp. 507–512). Circular overlap of races in the North American solitary bee, *Hoplitis producta* (Megachilidae), is illustrated in Figure 18.6.

Reversals

Geographical speciation is not an inexorable process. Evolutionary divergence from geographical races to biological species continues only so long as the evolutionary forces producing that divergence continue to act. The process of divergence may be arrested at any stage. Geographical races in some species are very ancient. Divergence may be not only arrested, but reversed at any stage up to the formation of complete sterility barriers between the related species. Natural hybridization, which is common in higher plants and well known in many groups of animals, represents a reversal in the process of divergence.

Collateral Readings

Mayr, E. 1942. *Systematics and the Origin of Species.* Columbia University Press, New York.

Mayr, E. 1963. *Animal Species and Evolution.* Harvard University Press, Cambridge, Mass. Chapters 11–13, 15, 16. See also the abridged chapters with the same numbers in Mayr, 1970:

Mayr, E. 1970. *Populations, Species, and Evolution.* Harvard University Press, Cambridge, Mass. Chapters 11–13, 15, 16.

22

Modes of Speciation

Introduction

Geographical speciation as described in the preceding chapter is one of several known modes of speciation in higher organisms. There are in addition certain modes of speciation, not involving geographical isolation in the intermediate stages, that are theoretically possible but unverified, and that are the subject of considerable controversy at present. The purpose of this chapter is to review briefly and in synoptical form the various modes of speciation, both known and controversial.

There is some redundancy in the terminology of speciation now. This is due to the fact that different students often apply different terms to what is essentially the same process or mode. The duplication of terminology obscures the unity of the processes in such cases. In this chapter we will use a set of identifying terms for the modes of speciation that we consider to be preferable, and will indicate the synonyms in passing.

Quantum Speciation

In geographical speciation, the pathway is: local race–geographical race–allopatric semispecies–species (as in Figure 21.1). In quantum speciation, by contrast, the pathway runs directly from local race to new species. Quantum speciation thus represents a shortcut method of species formation. The differences in pathway are associated with differences in controlling forces. The divergence is controlled by selection in the case of geographical speciation, and by the selection-drift combination in the case of quantum speciation.

The starting point for quantum speciation is a large, polymorphic, outcrossing, ancestral population. Now let a small daughter colony be founded by one or a few migrants somewhere beyond the ancestral species border. The daughter colony will be spatially isolated in its peripheral location.

The founding individuals bring only a small and non-random sample of the gene pool of the ancestral population to this daughter colony. Drift therefore takes place during the founding of the new colony. Furthermore, since the founder individuals will have to interbreed with one another at first, inbreeding and more drift occur during the early generations in the new colony.

Inbreeding in the small isolated daughter population leads to the formation and establishment of new homozygous gene combinations that would ordinarily be swamped out by outcrossing in the ancestral population. This is a source of phenotypic novelty in itself. But there is an additional source of novelty.

Habitually outcrossing populations tend to develop homeostatic buffering systems that ensure that "normal" phenotypes develop from highly heterozygous genotypes. To put it in other words, selection in outcrossing populations favors alleles that interact with one another in heterozygous combinations so as to produce normal phenotypes.

The corollary is that enforced inbreeding and homozygosity remove these homeostatic buffers and permit the formation of novel phenotypes, often radically different from the norm. Such novel phenotypes resulting from enforced inbreeding in normally outcrossing populations are referred to as phenodeviants. Phenodeviants are well known from experimental studies in *Drosophila*, chickens, primroses, *Linanthus*, and other animals and plants. They are predictably formed in small isolated daughter colonies derived from large outcrossing ancestral populations.

We have, then, two levels of novel phenotypes to distinguish: first, simple homozygous segregates, which are swamped out in the ancestral population but can be fixed in the small isolated daughter colony; and second, phenodeviants.

Now there is no assurance that the novel phenotype in the daughter colony, whether it is a simple homozygous segregate or a phenodeviant, will be adaptively valuable in its environment. If it is not, the colony will soon become extinct. In fact, most new phenotypes will probably be misfits and the colonies harboring them will probably not persist. But in a large series of trials, as where numerous daughter colonies are founded by a large ancestral population, over periods of time and in different parts of the species border, some adaptively valuable genotypes and phenotypes are likely to arise once or a few times.

Such adaptively valuable genotypes, when they do arise, are quickly fixed in the derivative colony by the combination of inbreeding, drift, and selection. The colony, with its novel phenotypic characters preserved by selection, can then go on to increase and spread as a divergent race or species.

The foregoing process of speciation was described by Mayr in a classic paper (1954). The ideas involved here are related to the earlier concepts of genetic drift (Wright, 1931) and quantum evolution (Simpson, 1944), discussed elsewhere in this book.

Several examples are now known of probable quantum speciation. Let us recall the variation pattern in the New Guinea kingfishers (*Tanysiptera galatea* group). As noted in Chapter 21, the insular races of *T. galatea,* derived probably from small numbers of migrants from the mainland population, are markedly different. Their degree of differentiation stands in contrast to the conservative racial variation of the intergrading mainland populations of the same species. One insular population has diverged to the species level *(T. hydrocharis)* (see Figure 21.2).

Some species of *Drosophila* in the Hawaiian Islands appear to have arisen by quantum speciation. Probable phylogenies in the Hawaiian drosophilas can be traced by similarities and differences in the inversion types as seen in salivary-gland chromosomes. Also, the known ages of the islands inhabited by the flies help to establish the relative ages of the species.

The *Drosophila planitibia* group contains three closely related species (*D. planitibia, D. heteroneura,* and *D. silvestris*) with the same chromosomal formula; that is, they carry the same set of inversion types in homozygous condition. The three species occur on two islands with the following distribution: *D. planitibia* on Maui, and *D. heteroneura* and *D. silvestris* on Hawaii. Now the island of Hawaii is only 700,000 years old, whereas Maui is older. This fact, combined with the chromosomal homologies, suggests that *D. planitibia* is ancestral and *D. heteroneura* and *D. silvestris* are derived (Carson, 1970).

The sequence of events as regards one speciational change, e.g., *D. planitibia–D. heteroneura,* is probably as follows. (1) Colonization of Hawaii by one or a few migrant individuals of *D. planitibia* from Maui, perhaps by a single fertilized female. (2) Fixation of formerly polymorphic genes in homozygous condition in the new daughter colony on Hawaii. (3) This fixation is accompanied by rapid divergence of the Hawaiian colony to the level of a new derivative species, *D. heteroneura* (Carson, 1970).

The second daughter species on Hawaii, *D. silvestris,* could have arisen by an independent event of colonization from the same ancestral population on Maui, or, alternatively, by divergence from the same original foundation colony on Hawaii that produced *D. heteroneura* (Carson, 1970).

It is significant that parallel cases of apparent quantum speciation occur in two other species groups of *Drosophila* on the same two islands (Carson, 1970).

A case in annual flowering plants is the species pair *Clarkia biloba* and *C. lingulata* (Onagraceae) in the Sierra Nevada of California. The wide-ranging *Clarkia biloba* is clearly ancestral on chromosomal and ecological evidence; the narrowly endemic *C. lingulata,* which occurs on the southern periphery of the area of *C. biloba,* is just as clearly derived. The species differ with respect to two independent translocations, which bring about a chromosomal sterility barrier, and by floral characters associated with one of the translocations. A chromosomally polymorphic population of *C. biloba* on the southern edge of its species area probably became fixed by drift and selection for a set of translocations and genes, to give rise to the new deviant species *C. lingulata* (Lewis and Roberts, 1956; Lewis and Raven, 1958; Lewis, 1962; Gottlieb, 1974). This mode is probably common in annual plants.

The term quantum speciation was proposed for the mode of speciation described here by Grant (1963, p. 459; also 1971). Synonymous terms used by other students are: speciation by genetic revolutions (Mayr, 1954); speciation by catastrophic selection (Lewis, 1962); speciation by population flush-crash cycles (Carson, 1971); and, as a near synonym, homosequential speciation (Carson, 1970, 1971). The term quantum speciation is preferable, in my opinion, since it ties in with the longer-known and more general phenomenon of quantum evolution. (See also a short discussion in Chapter 30).

Geographical and Quantum Speciation Compared

Geographical speciation and quantum speciation both start with large sexually reproducing populations. Both modes involve passage of the divergent lines through a period of spatial isolation on their way to attainment of species status. Here the similarities cease.

In geographical speciation, the allopatric precursor of the new species is a geographical race; in quantum speciation it is a local race. In the former, the forces involved are regional environmental selection in large outcrossing populations; in the latter, they are selection combined with inbreeding and drift in small daughter colonies. In geographical speciation the changes are gradual, slow, and conservative, whereas quantum speciation occurs at a rapid rate and may bring about drastic changes.

The difference in controlling forces between the two modes of speciation has some ramifications. Directional selection for new gene combinations, as involved in geographical speciation, entails a high substitutional genetic load. Such selection is costly in numbers of genetic deaths, and places a burden on the reproductive potential of the population. This burden, in turn, restricts the evolutionary rate to levels of selective mortality that can be tolerated by the population (see Chapter 14).

If, however, the parental population breaks away from reliance on directional selection in large populations, and instead exposes some small populations to the joint action of selection and drift, it can avoid some of the restrictive cost of selection (see again Chapter 14). And therefore it can change more rapidly. The cost-of-selection factor thus operates in a restrictive manner on evolutionary rates during geographical speciation; but quantum speciation is a way out of this same cost-of-selection restriction, and, in theory at least, can proceed more rapidly than geographical speciation.

Allopolyploid Speciation

The formation of new species by allopolyploidy has long been known in angiosperms and other land plants. This is indeed the mode of speciation that is best documented experimentally in any major group of organisms. Let us define allopolyploidy in stages. Polyploidy is the occurrence of three or more chromosome sets (or genomes) in a cell, individual, or population; we are of course interested here in polyploidy in populations. Allopolyploidy (also known as amphiploidy) is hybrid polyploidy, or, more specifically, the polyploid derivative of an interspecific hybrid.

Assume that the chromosome sets or genomes of two diploid species differ by several or many chromosomal rearrangements, such as translocations and inversions. Their genome constitutions can be written as AA and BB, respectively, where A and B represent the structurally differentiated haploid sets. The interspecific F_1 hybrid AB will have aberrations of meiosis and consequently will be chromosomally sterile.

Now suppose that the sterile hybrid AB undergoes chromosome-number doubling. This happens spontaneously in plants. Doubling can occur in somatic cells to give rise to a tetraploid branch and, eventually, tetraploid flowers with the genomic constitution $AABB$. Or, the diploid hybrid AB produces unreduced gametes *(AB)* as an end result of the aberrant meiosis, and union of two such unreduced diploid gametes yields tetraploid zygotes in F_2 with the constitution $AABB$.

The tetraploid plant *(AABB)*, having two sets of A chromosomes and two sets of B chromosomes, has normal chromosome pairing in bivalents at meiosis in the combinations A/A and B/B, and normal separation of chromosomes to the poles, so that it forms chromosomally and genically balanced gametes *(AB)*.

The allotetraploid is fertile. It has its own hybrid combination of morphological, physiological, and ecological characteristics, a combination that is different from that of either parental diploid species. And it breeds true for its intermediate hybrid constitution because of the intragenomic chromosome pairing and separation (A/A and B/B) at meiosis.

There is a chromosomal sterility barrier between the allotetraploid and its diploid parental species. Their F_1 hybrid, if it should be formed at all, would be triploid, and triploids are sterile in plants, owing to irregularities of meiosis. Our new allotetraploid *AABB* is fertile, possesses its own particular character combination, for which it breeds true, and is intersterile with its most closely related diploid species. It is therefore a new biological species of hybrid origin.

Allopolyploidy is a very common mode of speciation in angiosperms, ferns, and some other groups of plants. Recent estimates indicate that 47% of the species of angiosperms and 95% of the species of pteridophytes (ferns and fern allies) are polyploids. Most of these polyploids are allopolyploids (see Grant, 1971, pp. 233–234, 241ff.).

Some common economic plants are allopolyploids. Examples are the New World cottons (*Gossypium hirsutum* and *G. barbadense*; $2n = 4x = 52$), bread wheat (*Triticum aestivum*; $2n = 6x = 42$), and tobacco (*Nicotiana tabacum*; $2n = 4x = 48$). ($4x$ = tetraploid, $6x$ = hexaploid, etc.) *Nicotiana tabacum*, with the genome constitution *SSTT*, is an allotetraploid derivative of two South American diploids ($2n = 24$) close to the present species *N. tomentosiformis* (*TT*) and *N. silvestris* (*SS*) (Clausen, 1941).

Hybrid Speciation in Plants

Allopolyploidy is one of several modes of hybrid speciation in plants. By hybrid speciation we mean the formation, in the progeny of a natural hybrid, of a new, true-breeding line that is isolated from the parental species and from its siblings in the hybrid population. The new line must circumvent the obstacles of hybrid sterility and hybrid breakdown. Allopolyploidy accomplishes this feat. Some other mechanisms achieve a similar end result without a change in the number of chromosome sets. Two of these will be described briefly here (for a more comprehensive treatment see Grant, 1971).

Assume that the parental species differ by two or more independent translocations (*P, Q, . . .*). The chromosomal arrangement will be *PP QQ* in one parental species and *pp qq* in the other. Their F_1 hybrid is a double (or multiple) translocation heterozygote (*P/p Q/q*), and is consequently chromosomally sterile to some degree. It can, however, segregate several classes of structurally homozygous, fertile types in F_2 or later generations.

Two such classes of fertile progeny are the parental types, but they do not represent new lines. In addition, the hybrid can produce the homozygous recombination types (*PP qq* and *pp QQ*). These are fertile and new. Furthermore, they are separated from the parental species and from one another by a chromosomal sterility barrier, albeit a weak one in this case.

With larger numbers of independent translocations the chromosomal sterility barriers surrounding the new homozygous recombination types become stronger and the new line accordingly becomes better isolated. The model has been presented here in terms of translocations. Sets of independent inversions or transpositions have similar effects on sterility and, with appropriate modifications, can be used in the model.

The process described above has been termed recombinational speciation. It can be defined formally as the formation, in the progeny of a chromosomally sterile species hybrid, of a new, structurally homozygous recombination type that is fertile within the line but isolated from other lines and from the parental species by a chromosomal sterility barrier.

The process of recombinational speciation has been followed in experimental hybrid progenies in *Nicotiana* (Solanaceae), *Elymus* (Gramineae), and *Gilia* (Polemoniaceae). Its role in nature is still poorly understood. It probably occurs occasionally but less frequently than allopolyploidy.

Internal isolating mechanisms are determined by genic factors—sterility genes, gamete lethals, incompatibility genes, etc.—as well as by segmental rearrangements involved in chromosomal sterility. There is no apparent theoretical reason why a process parallel to recombinational speciation could not be carried out with sets of genic sterility factors rather than with chromosomal rearrangements. However, this possible mode of hybrid speciation has not yet been investigated thoroughly.

In some plant groups the species are interfertile and are isolated mainly by external barriers. Ecological, seasonal, and floral isolation are the main species-separating barriers. The morphological, physiological, and behavioral differences between the species that lead to these barriers are, of course, gene-controlled. The progeny of natural hybrids between the species, if such arise, will segregate for the gene differences and corresponding character differences determining the external isolation. This opens the possibility of the formation of interspecific recombination products with new character combinations so as to set up new, externally isolated subpopulations. If the external isolation persists, these could develop into new species of hybrid origin.

Probable examples of hybrid speciation by segregation and recombination of external barriers have been described in several plant groups. The California foothill species of larkspur, *Delphinium gypsophilum,* is probably derived from *D. recurvatum* × *D. hesperium* (Ranunculaceae) in this manner (Lewis and Epling, 1959). Other parallel cases are found in *Amaranthus* (Amaranthaceae), *Carex* (Cyperaceae), and other genera.

We don't know whether hybrid speciation plays any significant role in animal evolution. Most animal evolutionists consider that it does not (e.g., Mayr, 1963; White, 1973). However, reports are occasionally published of the probable hybrid origin of new species in various animal groups [e.g., Miller (1954) on woodpeckers, and Ross (1958) on leafhoppers], and the possibility should be kept in mind. The problem requires much further study.

The Problem of Sympatric Speciation

One of the controversial topics in evolutionary biology is sympatric speciation. This controversy has generated a rather large number of journal articles and book discussions over the years. One question is whether sympatric speciation is theoretically possible; this would have to be answered in the affirmative now. The next question is whether sympatric speciation is a real process in nature; on this, opinions are rather sharply divided at present. There are elements of confusion in the discussions pro and con of this latter question.

It is necessary to bring the issue of sympatric speciation into clearer focus. We should specify the circumstances in which sympatric speciation is supposedly going on; then we can debate the feasibility of the process under these restrictive conditions. Three sets of conditions are very relevant to the discussion.

First, are we talking about sympatric speciation as a product of primary divergence, or of hybridization, or of either one interchangeably? Allopolyploid speciation is sympatric and is common in the plant kingdom. Other modes of hybrid speciation in plants are also sympatric and occur occasionally, at least, indicating that allopolyploidy is not a special, exceptional case. The real problem, then, centers on the possibility of sympatric speciation during primary divergence.

Second, what is the breeding system of the organism involved in sympatric speciation? If it is self-fertilizing there is no particular problem. The problematical situation is sympatric speciation in outcrossing organisms, because of the swamping effect of cross-fertilization.

Third, what microgeographical field do we have in mind when we use the term "sympatric"? Sympatry may be either neighboring or biotic (see Chapter

18), and these two conditions should be distinguished in discussions of the problem. Sympatric speciation is much more feasible in a neighboringly sympatric field than in a biotically sympatric field.

Neighboringly Sympatric Speciation

We are concerned here with primary divergence in an outcrossing population where the divergent lines occupy neighboringly sympatric zones. The divergence is brought about by disruptive selection in the neighboring zones, which differ environmentally, and is protected from swamping to some extent by the partial spatial separation of the zones.

A number of actual cases are known, in both plants and animals, of divergence under these conditions to the *race* level. In plants, one good example, among others, is the outcrossing grass, *Agrostis tenuis,* in Great Britain, which has developed different edaphic races on different adjacent soil types (Antonovics, 1971). An interesting case in insects involves the fruit fly *Rhagoletis* (Tephritidae).

The larvae of the apple maggot, *Rhagoletis pomonella,* live in and feed on the fruits of cultivated apples *(Pyrus)* and native hawthorns *(Crataegus)* in North America. The populations on apples and on hawthorns are racially different. The flies mate on the host plant, where the females then deposit the eggs in the fruits. The larvae develop in the fruits and the next generation of adults emerge and mate again on the host plant, completing the cycle (Bush, 1969*a*).

Hawthorns and apples grow close together in some apple-growing districts of the northern United States. And here the host races of *Rhagoletis pomonella* also maintain themselves in close proximity. Their habit of mating on their respective host plants helps them to preserve their racial differentiation in their neighboring habitats (Bush, 1969*a*).

Now hawthorns are native in North America, whereas apples are introduced. The original host plant of *Rhagoletis pomonella* in North America was therefore probably *Crataegus.* The fruit fly is known to have infested apples in North America since 1866. It probably switched over from hawthorns to apples sometime in the nineteenth century (Bush, 1969*a*).

The formation of the new host race was probably an event taking place in a neighboringly sympatric field. Host-plant recognition and preference in a related genus, *Procecidochares,* and probably also in *Rhagoletis,* are controlled by a single gene. It will be recalled that mating takes place on the host plant. Therefore a mutation in the gene governing host recognition could initiate a process of neighboringly sympatric race formation (Bush, 1969*a*; Huettel and Bush, 1972).

This brings the divergence to the stage of host races on the basis of plausible circumstantial evidence. But the problem before us is sympatric species formation.

Rhagoletis pomonella is one of four sibling species. These species are sympatric but infest fruits of different plant families. The plant hosts of the *R. pomonella* group are as follows:

	Certain genera of
R. pomonella	Rosaceae
R. mendax	Ericaceae
R. cornivora	Cornaceae
R. zephyria	Caprifoliaceae

This host distribution could have arisen by shifts from one plant family to another in a sympatric field (Bush, 1969*b*).

The case for neighboringly sympatric speciation in fruit flies rests on an extrapolation from race formation to species formation. Such extrapolations are hazardous. Nevertheless, the pattern of distribution in the fruit flies is suggestive of neighboringly sympatric divergence at different taxonomic levels. The problem is still under investigation.

Narrow, species-specific food niches are common in many groups of insects. It is tempting to think that abrupt shifts from one food niche to another, like those postulated for fruit flies, have constituted a common pattern of speciation in the insects.

Biotically Sympatric Speciation

The formation of a genetically different population within a biotically sympatric field in an outcrossing organism is beset with serious difficulties. The new population is vulnerable to swamping and reabsorption in the parental population. Ordinarily the process cannot occur in nature.

Sympatric divergence in a cross-fertilizing population requires intense disruptive selection. The intense selection, however, is accompanied by a heavy genetic load that the population may not be able to bear.

Strong disruptive selection has been carried out in experimental populations of *Drosophila melanogaster* and *Musca domestica*. The experiments were successful in that divergences developed between the high selection lines and the low lines for particular characters. Partial reproductive isolation also developed between the diverging lines (Thoday and Gibson, 1962, 1970; Streams and

Pimentel, 1961; Pimentel, Smith, and Soans, 1967; Soans, Pimentel, and Soans, 1974).

These results are promising as far as they go. However, it must be admitted that the extracted lines in the disruptive selection experiments fall far short of species status.

The role of disruptive selection would be easier if combined with assortative mating. This combination of factors would operate if the character undergoing disruptive selection were to induce positive assortative mating. The disruptive selection could then bring about a given amount of divergence with a reduced selection cost.

The problem of biotically sympatric speciation is at present in the theoretical and experimental stages of inquiry. There is no evidence as yet to indicate that biotically sympatric primary speciation plays any role in nature.

Collateral Readings

Dobzhansky, Th. 1970. *Genetics of the Evolutionary Process.* Columbia University Press, New York. Chapter 11.

Grant, V. 1963. *The Origin of Adaptations.* Columbia University Press, New York. Chapter 16.

Grant, V. 1971. *Plant Speciation.* Columbia University Press, New York. Chapters 8, 13–16.

Mayr, E. 1963. *Animal Species and Evolution.* Harvard University Press, Cambridge, Mass. Chapters 15, 17, 18.

Mayr, E. 1970. *Populations, Species, and Evolution.* Harvard University Press, Cambridge, Mass. Chapters 15, 17, 18.

23

General Theory
of Speciation

Introduction

We have considered several modes of speciation in the preceding two chapters. Let us now look for the common denominators in the various modes. We will follow an earlier treatment (Grant, 1963, pp. 448–454).

The speciation process can be broken down into four basic components: (1) the goal, (2) the steps for reaching the goal, (3) the pathways composed of different series of steps, and (4) the fields. We can then consider the various modes of speciation as different combinations of these basic components.

The Goal

Sympatric species are adapted for different niches or habitats in a common territory. The adaptations of each species for its niche or habitat are based on character combinations and on the underlying gene combinations. The species is a field for the process of gene recombination, or, more particularly, a field for the generation of adaptively useful gene recombinations.

Not all the recombination products that can be generated by the sexual process are adaptively useful. In fact, many recombination products are inviable or subvital and represent a wastage of the reproductive potential of the population. Reproductive isolating mechanisms function to keep the interbreeding and the generation of recombinations within adaptively useful limits. Those limits are normally the boundaries of biological species.

Speciation is therefore essentially the formation of different reproductively isolated sets of adaptive gene combinations.

The Steps

The process of speciation requires a series of consecutive steps. The steps are: (1) production of multiple-gene variation, (2) formation of the new allele combination, (3) fixation of the new allele combination in a derivative population, and (4) protection of the new allele combination by reproductive isolating mechanisms.

Let us consider each step in turn. It will be helpful to think in terms of specific though oversimplified genotype changes. Thus an ancestral species with a genotype $A/A\ B/B$ gives rise to a daughter species $a/a\ b/b$.

(1) Production of multiple-gene variation. The ancestral population A/A B/B must acquire some a alleles and b alleles. Three main sources of such variation in natural populations are relevant to this discussion: first, normal existing variability in polymorphic populations, variability accumulated over long periods of time; second, special new variability resulting from events of hybridization; and third, special new variability resulting from sporadic bursts of mutation, either genic or chromosomal.

(2) Formation of the new allele combination. The sexual process now produces some $a/a\ b/b$ genotypes in the ancestral $A/A\ B/B$ population. But these are rare at first, and furthermore, they are broken up again by the same sexual process.

(3) Fixation of the new allele combination in a derivative population. The fixation of the new adaptive gene combination $aa\ bb$ in a daughter population is a critical step in the speciation process. There are two main ways.

The first is selection under wide outcrossing in a geographical race of the ancestral species. Directional selection for the genotype $a/a\ b/b$ in a large outcrossing population containing predominantly A and B alleles in its gene pool gradually increases the frequency of the a and b alleles, and eventually brings about the substitution of the *aa bb* genotype. The process has a high selection cost and is slow. But a population that is able to afford numerous selective deaths and has a long time to effect the substitution will eventually reach the new adaptive goal.

The other way is inbreeding. If the rare *aa bb* individuals can cross *inter se* or self-fertilize they can multiply quickly in numbers. The fixation of gene combinations can be more rapid and more economical of selective deaths in the case of inbreeding than it is with selection and outcrossing.

Two modes of inbreeding are known to be significant in natural populations, namely, crossing in small populations, which amounts to crossing between sibs and cousins in the main, and self-fertilization. A third mode, assortative mating, or preferential mating of genetically similar individuals, could be significant also, but its role in natural populations is not well known as yet.

(4) Protection of the new allele combination by reproductive isolation. The reproductive isolating mechanisms arise in two ways: as by-products of the divergence in genotype (see Chapter 21), and as products of selection for reproductive isolation per se (see Chapter 24).

The Pathways

Steps 1 and 3 have different variant forms. These forms can be combined in different ways to give a variety of alternative pathways.

Let us assign letter symbols to the forms of variation production and variation fixation for convenience of reference.

> Step 1
>> Normal existing variability (v)
>> Hybridization (h)
>> Novel mutations (u)
>
> Step 3
>> Selection (with wide outcrossing) (s)
>> Inbreeding (and selection) (i)
>>> In small populations (i_p)
>>> By self-fertilization (i_f)
>>> By assortative mating (i_a)

We can now link the variant forms of steps 1 and 3 together, as in Figure 23.1, to reveal the alternative pathways. Thus there is pathway $v-s$, pathway $v-i_p$, and so on. The number of possible pathways shown in Figure 23.1 is nine or twelve, depending on whether assortative mating (i_a) is counted or not.

VARIATION-PRODUCING FACTORS   VARIATION-FIXING FACTORS

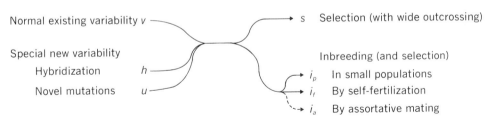

Normal existing variability v s Selection (with wide outcrossing)

Special new variability   Inbreeding (and selection)

     Hybridization h i_p In small populations

     Novel mutations u i_f By self-fertilization

                                   i_a By assortative mating

FIGURE 23.1
Pathways of speciation. Further explanation in text. (Redrawn from V. Grant, *The Origin of Adaptations,* copyright 1963, Columbia University Press, New York; by permission.)

The Models

The speciation process can take place in three types of spatial fields, namely, allopatric, neighboringly sympatric, and biotically sympatric fields. The three types of fields can be combined with the nine or twelve pathways to give a larger number of possible models. To be sure, most of these possible models are unlikely or unknown to occur in nature; but a fair number of models remain that are either real or worthy of serious consideration.

We will list here eight models corresponding to modes of speciation discussed earlier in Chapters 21 and 22. The modes of speciation at the top of the list are known; those at the bottom of the list are possibilities currently under investigation.

$v-s$, allopatric.	Geographical speciation.
$v-i_p$, allopatric.	Quantum speciation.
$h-i$, biotically sympatric.	Recombinational speciation, and hybrid speciation with segregation of external barriers.
$h-u-i$, biotically sympatric.	Allopolyploid speciation.
$v-i_p$, neighboringly sympatric.	Neighboringly sympatric speciation in outcrossing insects and plants (a definite possibility).
$v-i_f$, biotically sympatric.	Biotically sympatric speciation in self-fertilizing plants and invertebrates (a definite possibility).
$v-s$, biotically sympatric.	Biotically sympatric speciation in outcrossing organisms under disruptive selection. (Regarded as a real possibility by some workers, as a remote possibility by others, including myself.)
$v-i_a$, biotically sympatric.	Biotically sympatric speciation in outcrossing organisms with assortative mating (a possibility).

24

Selection for Isolation

Introduction

Reproductive isolation is believed, on good evidence, to develop in two ways. First, reproductive isolating mechanisms develop as by-products of evolutionary divergence from the race to the species level, as noted in preceding chapters. This enables the divergent species to coexist sympatrically.

But the primary, by-product isolating mechanisms may not prevent hybridization completely. Some hybrids may be produced. These are likely to be inviable, sterile, weak, or poorly adapted to one extent or another. Their production accordingly represents a wastage of the reproductive potential of the parental species.

Under these conditions, it may be selectively advantageous for the parental species to reinforce the existing primary isolating mechanisms with new barriers that do effectively block hybridization. Selection can act to build up special reproductive isolating mechanisms for the sake of reproductive isolation itself, if and when such isolation is selectively advantageous.

This second mode of formation of reproductive isolation will be described briefly in the present chapter. Selection for reproductive isolation per se has also been referred to in the literature as the Wallace effect or Wallace process, after Darwin's great contemporary, A. R. Wallace, and as character displacement in the reproductive phase.

The Process

Selection for reproductive isolation per se comes into operation when all of the following conditions are fulfilled. (1) The species populations are in sympatric contact. (2) They engage in hybridization. (3) The hybrids and hybrid progeny are inviable or sterile or otherwise adaptively inferior. (4) The loss of reproductive potential in the parental species due to the hybridization is selectively disadvantageous; in other words, prevention of hybridization has a definite selective advantage over the status quo. (5) Individual variation exists within one or both parental species with respect to characteristics affecting reproductive isolation.

Condition 5 goes to the heart of the selective process itself. We have to assume that a species population is polymorphic for ease of hybridization with a foreign species and that the hybridization has some deleterious effects. Under such conditions, those individuals in the species population that possess barriers to hybridization contribute more viable and fertile progeny to future generations of their population than do sister individuals that hybridize freely. And the genes of the former will therefore increase in frequency and spread throughout the population.

The conditions listed above vary greatly from one species pair to another and from one group of organisms to another. Biotic sympatry (condition 1), for example, is generally more common among insects and annual plants than among large land vertebrates and large woody plants. And the selective disadvantage of a loss of reproductive potential (condition 4) is much greater in ephemeral organisms, such as annual plants and small insects, than in long-lived perennial plants with a high excess fecundity.

The conditions promoting selection for isolation are sometimes realized and sometimes wanting in an array of species groups. Selection for isolation is not expected to be a universal process.

When the process of selection for isolation does come into play, it has reference primarily to isolating mechanisms that operate in the parental generation. The adaptive goal is the prevention of hybridization. And this goal is achieved most efficiently by the building up of mechanisms, such as ethological isolation or incompatibility, that act prior to fertilization.

This rule can be stretched in the case of organisms where the maternal parent carries and nurtures the embryo, as in mammals and most seed plants. Here it is theoretically possible for selection to build up barriers that operate in the early stages of the F_1 generation. The rationale is that an abortive young hybrid embryo that blocks hybridization, although not as efficient a barrier as ethological isolation, nevertheless represents an improvement in reproductive efficiency over the production of sterile or inviable adult hybrids.

The Evidence

One approach to the study of the efficacy of selection in natural populations makes use of a species pair with overlapping sympatry. The setup is as shown in Figure 24.1, top diagram. Two species A and B have races that are allopatric with respect to one another (A_a and B_a) and other races that are sympatric (A_s and B_s). Now compare the strength of ethological or other pre-fertilization barriers in the allopatric combination of races with that in the sympatric combination.

This comparison has been made for partially sympatric species pairs of *Drosophila (D. pseudoobscura* and *D. miranda)*, tree frogs *(Microhyla carolinensis* and *M. olivacea, Hyla ewingi* and *H. verreauxi)*, phlox *(Phlox pilosa* and *P. glaberrima)*, and some other organisms. In each case, ethological isolation itself or some measurable component of ethological isolation is more strongly developed in the sympatric areas of the two species than in the allopatric areas (Dobzhansky, 1951a; Blair, 1955; Littlejohn, 1965; Levin and Kerster, 1967; Levin and Schaal, 1970).

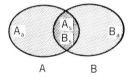

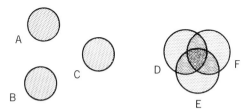

FIGURE 24.1
Two types of distribution patterns in species groups in which it is possible to compare allopatric and sympatric populations with respect to parental-generation isolation. Top: partially sympatric species pair. Bottom: species group containing allopatric species and sympatric species.

Thus in the southern United States, the mating calls of *Microhyla caroli-nensis* and *M. olivacea*, which function in ethological isolation, are very different in their zone of sympatric overlap but are not well differentiated when allopatric races of the two species are compared (Blair, 1955). The same pattern of racial variation for call notes is found in the partially sympatric species pair *Hyla ewingi* and *H. verreauxi* in southeastern Australia (Littlejohn, 1965).

More extensive comparisons of the same sort were made in the superspecies *Drosophila paulistorum*. This tropical American superspecies consists of six semispecies (Amazonian, Andean, etc.), which are partially sympatric in various combinations. The strength of ethological isolation between the semispecies can be measured in mixed laboratory populations and expressed quantitatively as an isolation coefficient (I). This coefficient varies from 0 to 1, where 0 is random mating and 1 is complete isolation (Ehrman, 1965).

The isolation coefficient was computed for allopatric race combinations and for sympatric race combinations representing eight semispecies pairs (e.g., Amazonian $\times$ Andean, Amazonian $\times$ Guianan, etc.) in the *D. paulistorum* superspecies. The values of I for the race combinations in the different semispecies pairs were as follows:

	I, range	I, average
Allopatric race combinations	0.46–0.76	0.67
Sympatric race combinations	0.68–0.96	0.85

The isolation coefficient is consistently higher between sympatric races of a semispecies pair than between allopatric races of the same semispecies pair throughout the large and complex *D. paulistorum* superspecies (with one exception) (Ehrman, 1965).

Another setup in which it is possible to detect the occurrence of selection for isolation is that shown in Figure 24.1, bottom diagram. A species group contains some allopatric biological species (A, B, C) and some sympatric species (D, E, F). The strength of parental-generation isolating mechanisms is then compared in two classes of interspecific combinations, allopatric $\times$ allopatric and sympatric $\times$ sympatric.

A favorable group of organisms in which to make this comparison is the Leafy-stemmed gilias (Polemoniaceae). This natural group of annual herbs consists of nine biological species in Pacific North America and South America. The species fall into two classes on geographical distribution. Five species in the foothills and valleys of California are extensively sympatric. Four other species on the coastline and offshore islands are entirely allopatric with respect to one another and only rarely sympatric with outlying populations of the foothill-and-valley species. The artificial F_1 hybrids between the nine species are highly sterile, with chromosomal sterility in all combinations (Grant, 1966*b*).

Since the plants are annual herbs, and flower during only one season, seed set is critical for them. By the same token, the production of seeds that develop into sterile hybrids represents a serious loss of reproductive potential.

An effective block to hybrid formation in *Gilia* is incompatibility. The strength of incompatibility barriers between species can be determined quantitatively in artificial interspecific crosses. A good measure is the average number of hybrid seeds produced per flower cross-pollinated for a large number of flowers in a given species cross (S/Fl) (Grant, 1966*b*).

The value of S/Fl was determined for 20 interspecific combinations of the nine species of Leafy-stemmed gilias. Incompatibility was found to differ greatly between the two geographical classes of hybrid combinations, as the following results show:

	S/Fl, range	S/Fl, average
Allopatric species combinations	7.7–24.8	18.1
Sympatric species combinations	0.0–1.2	0.2

Very strong incompatibility barriers occur between the sympatric species but not between the allopatric species (Grant, 1966*b*).

Ethological Isolation

Species-specific recognition marks and courtship behavior are well developed in most groups of animals. One can think of courtship plumage, displays, and songs in birds, of courtship dances and olfactory signals in insects, and so on. The varied types of recognition signals usually differ from species to species. They function to promote intraspecific matings and to discourage interspecific matings.

The origin of species-specific courtship characteristics should probably be attributed mainly to selection for isolation, although other selective factors may well enter in as contributing causes. Thus the courtship characteristics act as stimuli and releasers of sexual behavior, and could be products of selection for reproductive output per se, to some extent at least; but this mode of selection does not provide an explanation of the species specificity of the courtship characteristics. Again, *some* differences in courtship characters and behavior may develop as by-products of divergence; but it is very difficult to explain the more elaborate species-specific paraphernalia of courtship as side effects of differentiation in respect to food-getting and other secular aspects of life. The best hypothesis is that these courtship characteristics, which are so widespread in the animal kingdom, are mainly products of selection for ethological isolation.

Collateral Readings

Dobzhansky, Th., L. Ehrman, O. Pavlovsky, and B. Spassky. 1964. The superspecies *Drosophila paulistorum*. *Proc. Nat. Acad. Sci. USA* **51:** 3–9.

Ehrman, L. 1965. Direct observation of sexual isolation between allopatric and between sympatric strains of the different *Drosophila paulistorum* races. *Evolution* **19:** 459–464.

Grant, V. 1966. The selective origin of incompatibility barriers in the plant genus *Gilia. Amer. Nat.* **100:** 99–118.

Koopman, K. F. 1950. Natural selection for reproductive isolation between *Drosophila pseudoobscura* and *Drosophila persimilis. Evolution* **4:** 135–148.

Levin, D. A. 1970. Reinforcement of reproductive isolation: plants versus animals. *Amer. Nat.* **104:** 571–581.

Paterniani, E. 1969. Selection for reproductive isolation between two populations of maize, *Zea mays* L. *Evolution* **23:** 534–547.

25

Geological Time

The Nature of Macroevolution

Macroevolution involves changes of far greater magnitude than those seen in microevolution and speciation. Changes of macroevolutionary extent occur in the development of the characteristics that distinguish major groups such as genera, families, orders, classes, and phyla. Such developments take place on a scale of geological time.

The methods of research in macroevolutionary studies are necessarily different from those employed in studies of microevolution and speciation. Genetics, ecology, and minor systematics are the key approaches to the latter fields, whereas our knowledge of macroevolution has been gained by paleontology and comparative morphology. Work in population genetics and related fields has yielded the direct evidence concerning the evolutionary forces. But we owe our knowledge of the broad sweep of organic evolution to work in paleontology and comparative morphology.

Different groups of research workers are necessarily engaged in the macro and non-macro areas of evolutionary biology, and they often differ in their perspectives on experimentalism, the time scale, and other matters. Many older paleontologists and morphologists failed to appreciate the significance of genetic findings for evolutionary theory; conversely, neontological population biologists are apt to have only a vague feeling for the time element involved in macroevolution. Many controversies in evolutionary biology can be traced to characteristic differences between macroevolutionists and microevolutionists in research methods, habits of thinking, and background knowledge.

One of the questions in evolutionary biology has been whether macroevolution is simply an extension of microevolution, or whether there is some essential difference between the two. Let me state two extreme positions. (1) The only difference between macroevolution and microevolution is the length of time involved, i.e., the former is a simple extension of the latter. (2) Macroevolution involves special factors or processes that we do not find operating at the microevolutionary level, e.g., saltations, systemic mutations, orthogenetic tendencies, racial senescence, etc. Although each position has had its adherents, many adherents for position 1 and a small handful for position 2, we are not interested in the personalities here, but rather in the central issue and how to resolve it.

We note that position 1 is on the right track as far as economy of hypotheses is concerned. If we cannot discover an evolutionary force or factor on the microevolutionary level, where experimentation is possible, we are not entitled to invent it for the more recondite phenomena of macroevolution. Position 2 has never been able to gain wide acceptance among evolutionists for this and other reasons.

But let us also note a possible weakness in position 1. Position 1 assumes that we can safely extrapolate from 10 or 50 or 100 generations in a laboratory experiment to millions or tens of millions of years in nature. It does not allow sufficiently for the possibilities latent in geological time.

A third position that reconciles knowledge about microevolution and speciation with the geological time scale is possible. Such a position has been formulated (Simpson, 1944), and is now rather widely accepted. Position 3, as we will call it here, accepts the evolutionary forces and factors worked out by population biology, but rejects *simple* extrapolations from microevolution to macroevolution.

The known evolutionary forces and factors may well have operated in different ways at different times in the history of a group. Perhaps the mutation rate was different in the past as compared with the present. The environment might have been changing rapidly in the past, but is stable today, or vice versa. The group could have been geographically widespread at one stage and narrowly endemic at another. Or perhaps it consisted of numerous sympatric species at one stage but was monotypic at another.

When we consider the long time available for macroevolutionary changes, we also have to consider the possible occurrence of unique events that have far-reaching consequences. A terrestrial organism may colonize a distant vacant island only once and then embark upon a new course of evolution in its adopted territory. A primitive microorganism may acquire a symbiont by infection that enables it to carry out photosynthesis.

There is considerable evidence to support this third view of macroevolution. Macroevolution *is* different from microevolution, for macroevolution is a historical process.

Main Stages of Earth History

It is convenient to divide earth history into four main stages according to the dominant mode of cosmic developmental process at each stage. (1) Atomic evolution, in which nuclear reactions occurred to form hydrogen atoms, and, from hydrogen, to build up the other elements. (2) Chemical evolution, in which the atoms combined into chemical compounds of various degrees of complexity, including non-living organic molecules. (3) Organic evolution, encompassing the developments from the origin of life through the formation of higher animals. (4) Cultural evolution, the accumulation and transmission of a cultural heritage that commenced with the rise of mankind above the animal level.

It is possible to assign approximate dates to these stages. The earth as a solid sphere is between 4.5 and 5.0 billion years old. The dating is based on the ratios between uranium and lead and between different lead isotopes in the earth's crust; these ratios are products of radioactive decay that occurs at known rates. The oldest known igneous rocks are 3.0–3.5 billion years old (see Dott and Batten, 1971, p. 108), and the oldest sedimentary rocks are 3.4–3.8 billion years old (Cloud, 1974). Atomic evolution prevailed in the formative period of Earth and graded into chemical evolution later.

Chemical evolution eventually reached a stage in which complex organic molecules were formed. Such non-living organic compounds paved the way for the origin of life. The oldest known fossil organisms are bacteria and unicellular blue-green algae about 3.2 billion years old, in the pre-Paleozoic Figtree formation of South Africa (Barghoorn, 1971; Schopf, 1974). Therefore the origin of life occurred sometime before 3.2 billion years ago but after chemical evolution had reached an advanced phase. Organic evolution thus spans over 3 billion years of earth history (see Figure 25.1).

The stage of cultural evolution belongs to the latter phase of human evolution and is thus a very recent phenomenon, on the planet Earth at least. The genus *Homo* is about 1.3 million years old. However, the evolution of *Homo* has involved a mixture of organic and cultural evolution, with the proportions

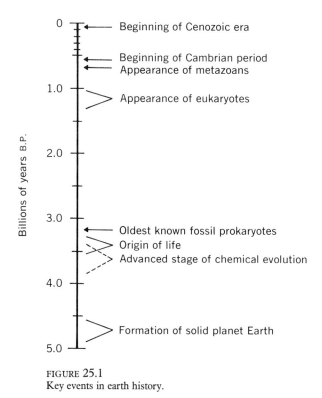

FIGURE 25.1
Key events in earth history.

of the two components varying through time. The transition from predominantly organic to predominantly cultural evolution in man was gradual and cannot be dated exactly. Cultural evolution will be discussed in Part VII.

The Origin of Life

The essential step for the origin of a simple form of life was the formation of DNA or DNA-like macromolecules possessing the properties now found in genes. These properties are the ability to direct the synthesis of proteins and other substances that help them maintain themselves as entities, for awhile at least, and the ability to replicate themselves.

The carrying out of these processes requires food as an energy source. The accumulation of energy-rich carbon compounds during the previous period of chemical evolution provided a store of potential food.

The accumulation of organic compounds in the course of chemical evolution depended on two favorable conditions that existed then but ceased to exist later. The two conditions were the absence of living organisms and the absence of free oxygen in the atmosphere. These are factors that bring about the breakdown of organic molecules in the modern world. In the pre-biotic world the absence of these factors meant that such molecules, once formed by spontaneous chemical reactions, would persist and accumulate.

The chemical reactions that produce organic molecules do take place spontaneously, as has been confirmed in experiments. Set up an atmosphere composed of hydrogen, water, ammonia, and methane. Expose it to lightning discharges, natural radioactive emissions, cosmic rays, or ultraviolet rays. Various organic compounds, including amino acids, are then formed spontaneously.

Therefore it is logical to infer that an "organic soup" containing stored energy would form and accumulate during the stage of chemical evolution. And at some point the processes of chemical evolution must have produced nucleic acid molecules with genetical activity. These first living particles would be "naked genes" living on the stored energy in the organic soup.

Genes have the capacity to replicate, multiply, and mutate. Replication spins off a fraction of mutant types. Multiplication goes on to the physical limits of the environment. Consequently a competition sets in sooner or later among the genes or their carriers for limiting environmental resources, and in this competition some types will be more successful reproductively than others. Thus the tendency to evolve is one of the basic attributes of life in a world of limited resources. With the origin of life, organic evolution was on its way.

Main Stages of Organic Evolution

Let us now divide the 3.2–3.5-billion-year span of organic evolution into stages. We will recognize five stages, on the basis of the nutrition and structural organization of the most advanced forms in each stage. We can then discuss the stages separately in later sections.

The stages are: (1) Simple heterotrophic particles; virus-like or bacterium-like. (2) Prokaryotic autotrophic cells; photosynthetic or chemosynthetic; exemplified by blue-green algae and autotrophic bacteria. (3) Eukaryotic cells; aerobic; exemplified by unicellular green algae and flagellates. (4) Simple multicellular organisms; aerobic; exemplified by multicellular green algae, liverworts, and sponges. (5) Complex animals, e.g., Metazoa, and land plants.

Stages 1 through 4 occupy most of the time span of organic evolution, from 3.2–3.5 billion years B.P. to the beginning of the Paleozoic era about 0.6 billion years B.P. (see Figure 25.1). Ancient fossil prokaryotes, both bacteria and blue-green algae, are found in the following formations (Barghoorn, 1971):

	Age in billions of years
Figtree formation, South Africa	ca. 3.2
Gunflint chert, Ontario, Canada	1.6–1.9
Bitter Springs formation, Australia	ca. 1

Green algae and possible fungi also appear in the Bitter Springs formation (Barghoorn, 1971). Primitive eukaryotes are also reported from the 1.3-billion-year-old Beck Spring flora in California (Cloud, 1974). Simple eukaryotes thus appear on the scene at least by 1.0–1.3 billion years B.P.

Stage 5 begins in the Late Precambrian and picks up momentum in the Cambrian period. Ancient metazoan fossils—coelenterates and annelids—occur in the Ediacara fauna of Australia, dated as about 0.7 billion years old in Late Precambrian time (Simpson, 1969). The major groups of aquatic metazoan animals became abundant later in the Cambrian (see Table 25.1 and Figures 25.1 and 25.2).

Primitive Evolution

Several major advances in nutrition and structural organization took place during the first 2 billion years, more or less, of organic evolution, advances that were essential for the later evolution of multicellular organisms. The important developments were: autotrophic nutrition, particularly photosynthesis; aerobic respiration; eukaryotic cellular organization; and sexual reproduction. These developments cannot be pinpointed on the time scale in Figure 25.1; they must have occurred during the long gap between 3.5 and 1.3 billion B.P., and they probably occurred in the sequence listed above.

The continued growth of the primitive heterotrophic particles necessarily led to the gradual depletion of the original "organic soup." The depletion of the food reserves would have put a premium on the ability of organisms to synthesize their own foods out of inorganic raw materials such as water and carbon dioxide. The development of photosynthesis and chemosynthesis in primitive prokaryotes was a major advance in early evolution. It paved the way for a second important advance, aerobic respiration.

The primitive atmosphere contained hydrogen but not free oxygen; it was a reducing rather than an oxidizing atmosphere. Under these conditions the

TABLE 25.1
Ages of the geological periods. (Kulp, 1961;
Dott and Batten, 1971.)

Era	Period	Millions of years B.P.
		0
	Quaternary	
		2.5
Cenozoic	Late Tertiary (Neogene)	
		27
	Early Tertiary (Paleogene)	
		65
	Cretaceous	
		130
Mesozoic	Jurassic	
		185
	Triassic	
		230
	Permian	
		265
	Carboniferous	
		355
	Devonian	
Paleozoic		413
	Silurian	
		425
	Ordovician	
		475
	Cambrian	
		600
	Late Precambrian	
		700
Precambrian		

primitive heterotrophic particles and cells would have had to get the energy out of the organic soup by fermentation. Now fermentation is metabolically inefficient in that it leaves most of the energy of the carbon compounds untapped. Aerobic respiration, involving a more complete breakdown of the carbon compounds, releases far more energy. Organisms obtaining their energy by cellular respiration can operate at a much higher metabolic rate than organisms relying on fermentation.

Aerobic respiration requires an oxygen-containing atmosphere. Oxygen is a by-product of photosynthesis. Students of early earth history believe that the change from the original oxygen-free atmosphere to an oxygen-containing atmosphere was a product of the activity of primitive photosynthetic organisms. The transition to an oxidizing atmosphere permitting respiration must have been gradual and very slow. Its slowness may account for the long lag of 2

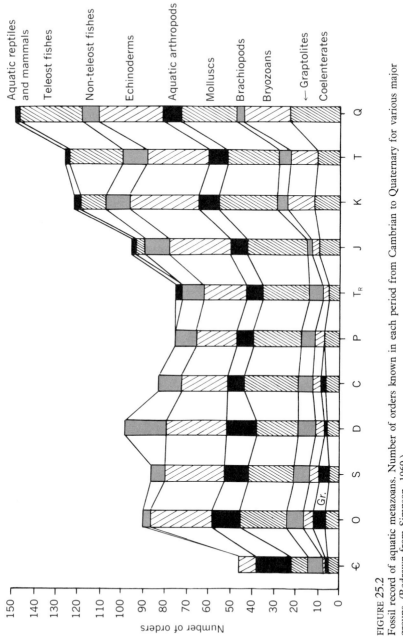

FIGURE 25.2
Fossil record of aquatic metazoans. Number of orders known in each period from Cambrian to Quaternary for various major groups. (Redrawn from Simpson, 1969.)

billion years or more between the origin of life and the appearance of eukaryotes and metazoans (see Figure 25.1).

The eukaryotic cellular organization, with its grouping of numerous gene centers onto true chromosomes, its division of labor between nucleus and cytoplasm, and its capacity to harbor organelles such as chloroplasts and mitochondria, represented another big step forward in complexity and in the ability to carry out diverse self-sustaining life processes.

The two basic components of sexual reproduction—fertilization and meiosis—were made possible by the grouping of genes onto true chromosomes and of chromosomes into a nucleus in eukaryotes. Sexual reproduction, as an orderly and symmetrical method of generating recombinational variability (see Chapter 7), enters the picture at the level of unicellular eukaryotes.

The Stage of Complex Multicellular Organisms

The stage of complex multicellular organisms occupies the last 700 million years of earth history. Such organisms make rare appearances in the fossil record in the Late Precambrian, and become abundant and varied in the Cambrian. Animals and plants were aquatic through the Cambrian and Ordovician, emerged onto land in the Silurian and Devonian, and gradually became better adapted for terrestrial and aerial life during and since the Carboniferous. Animals and plants of a modern cast belong to the Cenozoic era, which began 65 million years ago (see Table 25.2).

TABLE 25.2
Chronology of the Cenozoic era. (Raven and Axelrod, 1974.)

Period	Epoch	Millions of years B.P.
Quaternary	Recent	
		0.01
	Pleistocene	
		2.5
Late Tertiary (Neogene)	Pliocene	
		10
	Miocene	
		27
Early Tertiary (Paleogene)	Oligocene	
		38
	Eocene	
		54
	Paleocene	
		65

The increasing diversification of metazoans through the Phanerozoic is summarized in graphs in Figures 25.2 and 25.3. The successive rise of various groups to dominance in the biotic world is shown graphically in Figure 25.4.

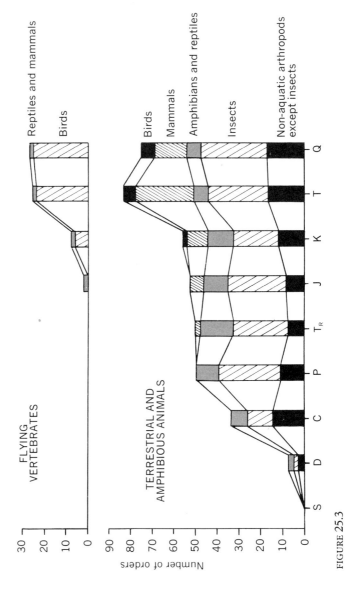

FIGURE 25.3
Fossil record of non-aquatic animals. Number of orders known in each period from Silurian to Quaternary for various major groups. (Redrawn from Simpson, 1969.)

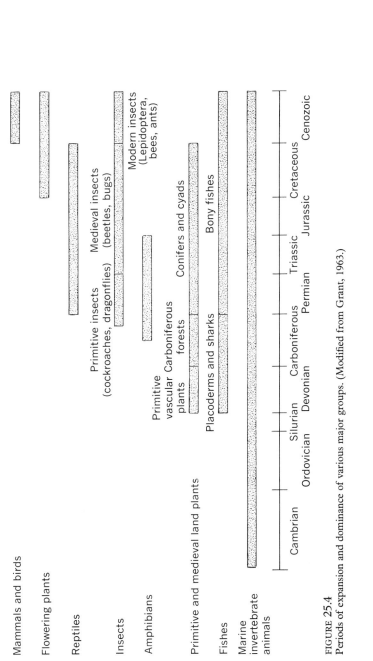

FIGURE 25.4
Periods of expansion and dominance of various major groups. (Modified from Grant, 1963.)

Collateral Readings

Barghoorn, E. S. 1971. The oldest fossils. *Sci. Amer.* **224**(5): 30–42.

Colbert, E. H. 1961. *Dinosaurs: Their Discovery and Their World.* E. P. Dutton & Co., New York.

Margulis, L. 1970. *Origin of Eukaryotic Cells.* Yale University Press, New Haven, Conn.

Oparin, A. I. 1962, 1964. *Life: Its Nature, Origin and Development.* Oliver & Boyd, Edinburgh (1962); Academic Press, New York (1964).

Romer, A. S. 1966. *Vertebrate Paleontology.* Ed. 3. University of Chicago Press, Chicago.

Simpson, G. G. 1967. *The Meaning of Evolution.* Ed. 2. Yale University Press, New Haven, Conn. Chapters 1–9.

Simpson, G. G., and W. S. Beck. 1965. *Life: An Introduction to Biology.* Ed. 2. Harcourt, Brace & World, New York. Chapters 29–31.

26

Evolutionary Trends

Introduction

The fossil record of many animal groups and some plant groups provides clear evidence for the occurrence of long-term oriented changes or evolutionary trends. By evolutionary trend we mean a series of progressive changes in a phyletic line, changes that take place in a given direction and continue to take place over a long period of time. Such evolutionary trends are a fact of paleontology. They can also be inferred with reasonable confidence from the evidence of comparative morphology in many contemporaneous plant groups that lack a fossil record but that have preserved an array of primitive and advanced forms.

The facts regarding evolutionary trends are now well established, but their interpretation has been a matter of controversy, which is still simmering in some quarters. The two contending theories are orthogenesis and orthoselection. The evidence is overwhelmingly in favor of the theory of orthoselection. We will return to the question of interpretation later in this chapter.

Examples

A good example of a trend is the increase in size of the "horn" in titanotheres (Brontotheriidae, Perissodactyla) from Eocene to Oligocene (Osborn, 1929; Simpson, 1967, pp. 155–158; Stanley, 1974). This trend is illustrated in Figure 26.1. The Early Eocene titanothere, *Eotitanops,* had no horn (Figure 26.1A). Small bony protuberances were present on the front of the face of later Eocene titanotheres (Figure 26.1B, C). Moderately large, blunt, horn-like protuberances occurred in one Early Oligocene titanothere, *Megacerops,* and very large "horns" in another Oligocene genus, *Brontotherium* (Figure 26.1D).

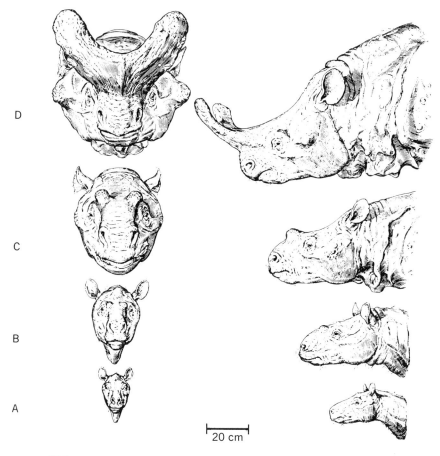

FIGURE 26.1
Evolutionary trend in the horn of titanotheres. (A) *Eotitanops borealis,* Early Eocene.
(B) *Manteoceras manteoceras,* Middle Eocene. (C) *Protitanotherium emarginatum,* Late Eocene.
(D) *Brontotherium platyceras,* Early Oligocene. (Stanley, 1974, after Osborn.)

Another classical example of an evolutionary trend occurs in the freshwater snail *Viviparus* (formerly known as *Paludina,* Viviparidae) in the Pliocene of Europe. Figure 26.2 shows a series of shells of successively younger age in the Late Pliocene. The shells gradually became more tightly coiled and more definitely sculptured with the advance of time (see Abel, 1929).

The most famous and best documented example of all known evolutionary series involves characters of the teeth, feet, and other body parts in horses during the Tertiary period. The trends in the horse family (Equidae) will be described in the following sections.

Some trends occur independently and in parallel in a number of separate groups. Thus a trend toward increased body size occurs in many groups of

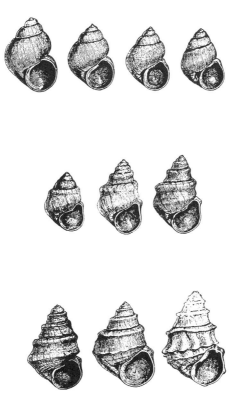

FIGURE 26.2
Evolutionary trend in coiling and sculpturing of the shell
in the snail *Viviparus* (= *Paludina*) during the Late
Pliocene. The series runs from an ancestral form close to
Paludina unicolor (upper left) through intermediate
stages to *P. hoernesi* (lower right). (Abel, 1929,
after Neumayr.)

animals. This trend is found in many groups of land mammals during the Tertiary period, e.g., in titanotheres (Figure 26.1), horses (Figures 26.4, 26.6), and proboscidians; in dinosaurs during the Mesozoic; and in some marine animals such as sea snails in the Paleozoic. So widespread among animals is the trend to increased body size that it has become codified as Cope's rule. For discussions of Cope's rule see Rensch (1960*b*, pp. 206ff.) and Stanley (1973).

Many independent groups of flowering plants exhibit parallel trends to reduced size and simpler morphology. Such trends are known collectively as reduction series. A common type of reduction series runs from flowers with numerous petals, stamens, and/or carpels to flowers with only a few of these organs. Another common reduction series is that from large perennial herbs to small annual herbs. This trend often continues within annual herbs from large-flowered cross-pollinating forms to small-flowered self-pollinating forms. Trends such as these in plants are recognized mainly on comparative morphological rather than paleontological evidence.

Trends in the Equidae

The very good fossil record of the Equidae spans a time period of about 60 million years since the Paleocene. The extreme members of the family are the small four-toed *Hyracotherium* (or *Eohippus*) of Late Paleocene and Eocene age and the modern genus *Equus,* consisting of horses, onagers, and zebras. A nearly continuous series of gradations is found between these extreme forms in the fossil record.

The Equidae developed out of a still earlier Paleocene stock, the condylarths, which are small dog-like animals with five-toed padded feet. The fossil series extends downward from *Hyracotherium* to the condylarths.

The phylogeny of the horse family has been worked out from the fossil evidence. A greatly simplified version of the phylogeny is shown in Figure 26.3. The geological ages and geographical distribution of the principal equid genera are also listed for reference purposes in Table 26.1. The main theater of equid evolution was North America, but branch lines also evolved in Eurasia, Africa, and South America at various times during the Tertiary and Quaternary.

We will not attempt to review horse evolution in general here, for which see Simpson (1951), but instead will concentrate on some of the trends included in this evolution. The main trends involve: (1) body size; (2) tail length; (3) foot mechanism; (4) leg length; (5) head shape; (6) brain size and complexity; and (7) cheek teeth. The following account is based mainly on Simpson (1951, 1953).

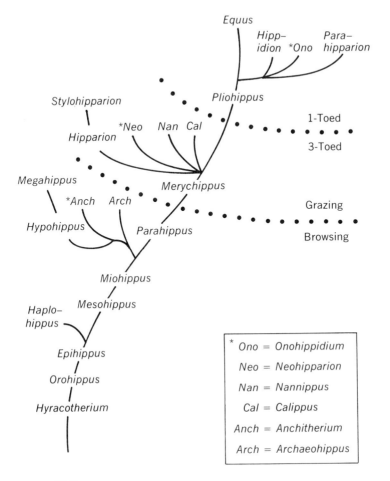

FIGURE 26.3
Phylogeny of the horse family, greatly simplified. (Redrawn from Simpson, 1953.)

As regards body size, *Hyracotherium* was small, like the earlier condylarths, standing 25–50 cm tall in different species, in contrast to the well-known large size of *Equus*. A general increase in body size occurred in many lines, including the equine horses from Oligocene to Pliocene time. Milestones in this trend are indicated by the scaled drawings of heads and feet in Figures 26.4 and 26.5. It should also be noted that the size-increase trend, though real enough, did not occur at an even rate through time, was not universal throughout the horse family, and was subject to reversals in various horse genera, including *Equus* itself.

TABLE 26.1
Ages of genera of the horse family. (Based on Simpson, 1951, and personal communication.)

Epoch	North America	Old World	South America
Recent	*Equus*	*Equus*	*Equus*
Pleistocene	*Equus*	*Equus* *Stylohipparion*	*Equus* *Hippidion* *Onohippidium* *Parahipparion*
Pliocene	*Equus?* *Pliohippus* *Calippus* *Nannippus* *Neohipparion* *Hipparion* *Megahippus* *Hypohippus*	*Stylohipparion* *Hipparion* *Hypohippus*	
Miocene	*Merychippus* *Archaeohippus* *Megahippus* *Hypohippus* *Parahippus* *Anchitherium*	*Anchitherium*	
Oligocene	*Miohippus* *Mesohippus* *Haplohippus*		
Eocene	*Epihippus* *Orohippus* *Hyracotherium*	*Hyracotherium*	
Late Paleocene	*Hyracotherium* (= *Eohippus*)		

The tail changed from long relative to body size in *Hyracotherium* to relatively short with long hairs in *Equus*.

The most conspicuous trend in horse feet is reduction in the number of toes. The starting point of this trend is in the ancestral condylarths, which were five-toed. *Hyracotherium* was four-toed and three-toed in the forefeet and hindfeet, respectively (Figure 26.6A). Oligocene, Miocene, and most Pliocene horses were three-toed (Figure 26.6B, C). Among three-toed horses a difference is apparent between earlier genera, in which the two lateral toes are well developed (Figure 26.6B), and later forms, in which they are shortened (Figure 26.6C). The trend culminates in one-toed feet in the Pliocene genus *Pliohippus* and its later descendants, including *Hippidion* and *Equus* (Figure 26.6D).

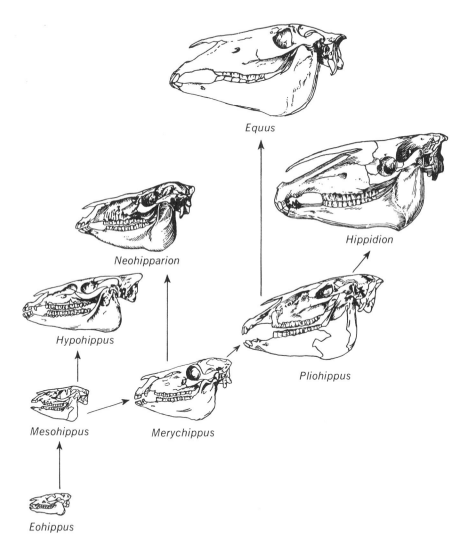

FIGURE 26.4
Evolution of horse skull. Shown to same scale. (From G. G. Simpson, *Horses,* copyright 1951, Oxford University Press, New York; reproduced by permission.)

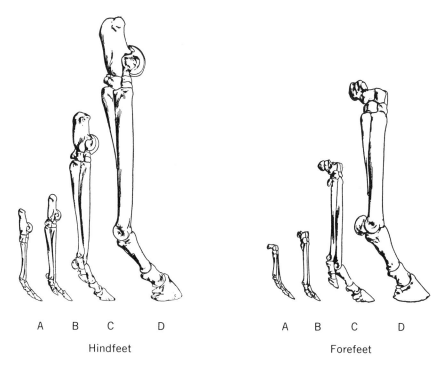

A B C D

Hindfeet

A B C D

Forefeet

FIGURE 26.5
Evolution of horse feet. Shown to same scale. (A) *Hyracotherium.* (B) *Mesohippus.* (C) *Merychippus.* (D) *Equus.* (Rearranged from G. G. Simpson, *Horses,* copyright 1951, Oxford University Press, New York; by permission.)

The reduction in toe number accompanies a change in the foot mechanism. The feet of condylarths and *Hyracotherium* were padded, and the animals walked on these pads, as do dogs. The earliest three-toed horses of the Oligocene and some of their direct descendants in the Miocene were also pad-footed (Figure 26.6B). In another branch of three-toed Miocene horses, however, and in the later one-toed horses, the foot has a spring action. The body weight is borne not on pads, which are now absent, but on the hoof of the central toe (Figures 26.5C, D; 26.6C, D).

In the three-toed spring-footed horses the two short side toes may have functioned as shock absorbers, according to Simpson, to relieve some of the strain on the main central toe (Figures 26.5C, 26.6C). These side toes eventually disappeared in the one-toed horses. Here the body weight rests entirely on the hooves of the central toes. Powerful ligaments attached to the bones of these toes produce a spring action in running.

The changes in foot structure were episodic rather than slow and gradual; these changes came in certain lineages only and at relatively rapid rates.

Increase in leg length in equid evolution is partly a function of increase in body size. But the change in foot mechanism from padded feet to hooved spring-feet added to the effective length of the legs, for spring-footed horses walk and run on their tiptoes.

Inspection of Figure 26.4 reveals numerous and complex changes in proportions in the skull as well as the change in size. The muzzle becomes longer and the jaw bones wider. These alterations in the head are related to changes in dentition, to be considered next.

Other changes in the equid head are correlated with a trend in brain capacity. *Hyracotherium* had a small brain, like those of reptiles and primitive mammals (Figure 26.7). In the subsequent history of the family the brain became larger and more complex, with a development of regions associated with intelligence in modern mammals. Progressive development of the equid brain and intelligence took place gradually throughout the Tertiary and Quaternary (Figure 26.7).

Trends in Horse Teeth

The evolutionary trends in horse teeth are very complex and cannot be fully described without recourse to a highly technical terminology. It is sufficient

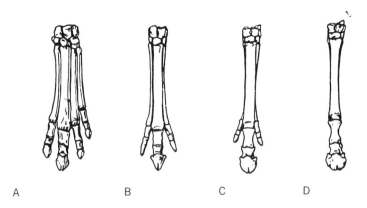

A   B   C   D

FIGURE 26.6
Evolution of horse feet and toes. Forefeet. Not shown to same scale.
(A) *Hyracotherium.* (B) *Mesohippus.* (C) *Merychippus.* (D) *Pliohippus.* (Redrawn from Simpson, 1951, and Romer, 1966.)

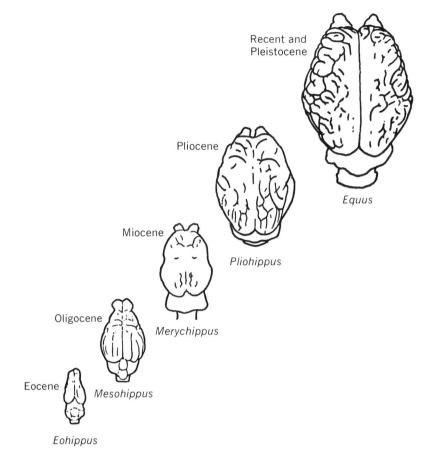

FIGURE 26.7
Evolution of horse brain. Casts of cranial cavities. Shown to same scale. (From G. G. Simpson, *Horses*, copyright 1951, Oxford University Press, New York; reproduced by permission.)
NOTE ADDED IN PROOF: A recent report (Radinsky, 1976) indicates that the brain of *Hyracotherium* was more advanced than that shown here.

for our present purposes to present the main outlines of horse-tooth evolution with a minimum of terminology.

The teeth of *Equus* exhibit the following advanced features. (1) Differentiation of chisel-like front teeth for biting and rear cheek teeth for grinding; (2) large-sized cheek teeth; (3) long (or high-crowned) cheek teeth; (4) an increased number of functionally grinding cheek teeth (six on each side of each jaw); (5) highly developed ridges (crests) composed of enamel on the grinding surfaces; and (6) development of hard cement in pockets between the crests.

The corresponding primitive features in the teeth of *Hyracotherium* are: (1) less differentiation between front and rear teeth; (2) smaller teeth; (3) short (low-crowned) cheek teeth; (4) no grinding teeth (three molars on each side of each jaw, but these were not true grinders); (5) a simple pattern of low enamel crests on the grinding surfaces; and (6) absence of cement. See Figures 26.4 and 26.8.

These dental differences reflect differences in diet. *Hyracotherium* was a forest-inhabiting browsing animal, whereas *Equus* is a grazing animal of grassy plains.

The various dental character differences between *Equus* and *Hyracotherium* were brought about by a series of trends, e.g., the grinding teeth became larger (trend 2), higher-crowned (3), with more elaborate crest patterns (5), and so on. The increase in number of grinding teeth was achieved by the gradual transformation of three of the premolars of *Hyracotherium* into molar-like teeth in *Equus,* a process known as molarization.

The changes in the various tooth characters were not simultaneous but occurred at different times. Molarization (trend 4) took place mainly during the Eocene and Oligocene and leveled off thereafter. Increase in height of crown (trend 3) took place mainly in the Miocene, and, moreover, exclusively in the then new grazing groups such as *Merychippus,* while the Miocene browsing horses remained low-crowned.

The changes in tooth characters were gradual, with one type grading into the next. The intergradation can be seen in cases where a phyletic line is represented by fossil teeth in a series of geological horizons of closely spaced ages. Consider the lineage *Parahippus–Merychippus–Nannippus* of Miocene-Pliocene age in Figure 26.8. Figure 26.9 portrays crown height and grinding surface in a more complete series of stages in this same lineage. A gradual and nearly continuous increase in crown height is evident.

We can next amplify the Late Miocene segment of the series shown in Figure 26.9. Downs (1961) has made a detailed quantitative study of the differences in teeth between two populations of *Merychippus* in the Mascall and Coalinga formations, respectively. The two populations are separated by 1–2 million years of time. The similarities and slight differences in their enamel crest patterns can be seen in Figure 26.10.

Downs (1961) measured a number of tooth characters in population samples of adult individuals of *Merychippus* from the Mascall and Coalinga formations, and treated the data statistically. Some of his results are given in Table 26.2. We note first that there is much individual variation within each population; and second, there are slight differences between populations in means and ranges. Most of the differences in means (including some not shown in Table 26.2) are

254

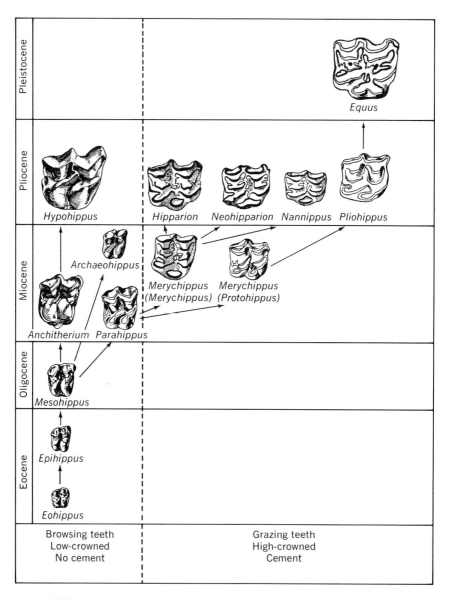

FIGURE 26.8
Evolution of horse teeth. Crest patterns on grinding surfaces of upper molars are shown in the drawings; other features are indicated in the legends. All drawings to same scale. (From G. G. Simpson, *Horses,* copyright 1951, Oxford University Press, New York; reproduced by permission.)

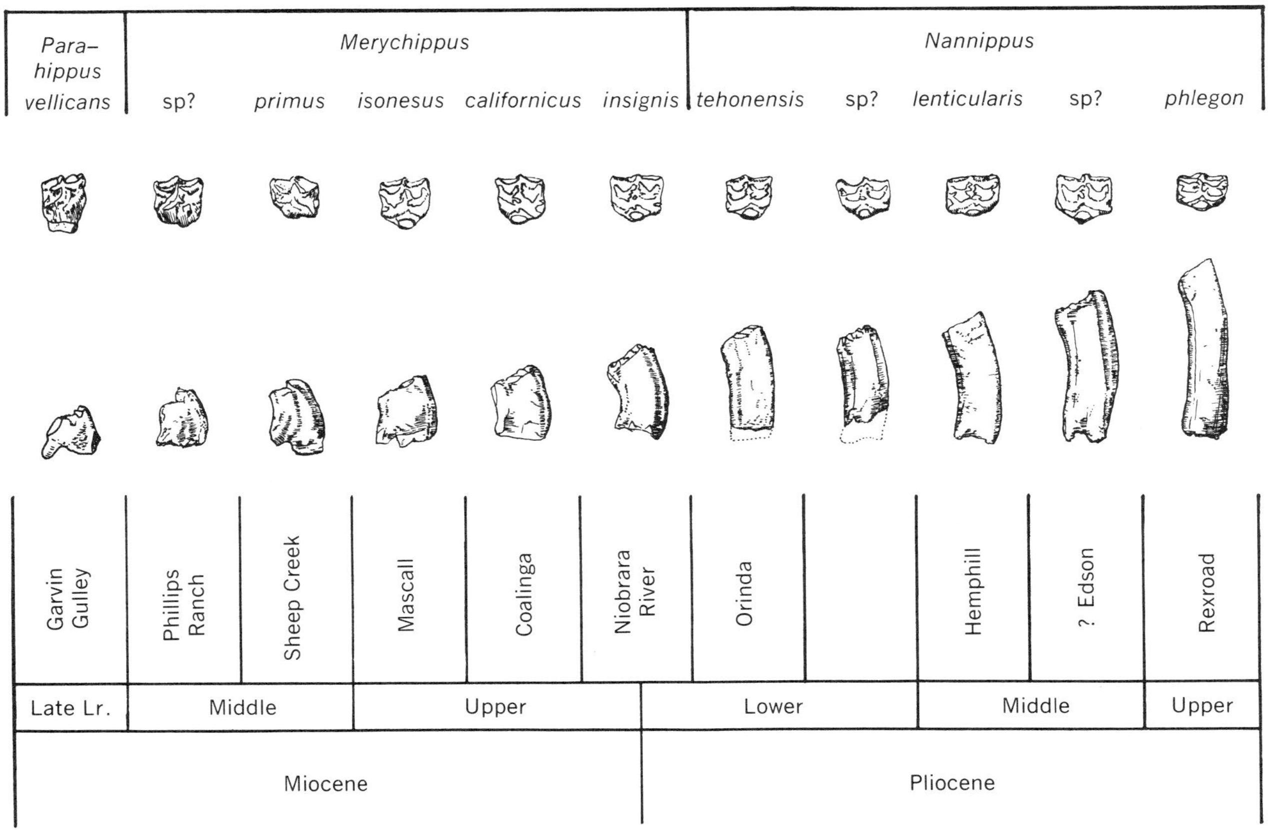

FIGURE 26.9
Evolutionary trend in height of crown of cheek teeth in a lineage of horses in the Miocene and Pliocene. (Redrawn from Stirton, 1947.)

significant statistically. Nevertheless, there is a broad overlap between the two populations, as shown by the ranges and standard deviations in Table 26.2. The statistical nature of the differences between the Mascall and Coalinga populations bridges the gap between macroevolutionary and microevolutionary changes.

Some interesting cases are known in which a tooth character appears as a polymorphic variant in an older fossil population, and becomes fixed as a constant character in a later population. Here again a macroevolutionary trend is resolved into a microevolutionary change.

One such case concerns a particular crest (known as the crochet) on the upper cheek teeth in the *Mesohippus–Equus* line. This crest was usually lacking in *Mesohippus*. But some populations of *Mesohippus* contained a few mutant individuals with tiny crochet crests. Some derivative species of *Miohippus* in the Oligocene were relatively constant for the presence of a tiny crochet; other later derivative species of *Parahippus* in the early Miocene possessed a small crochet as a constant feature. The crochet later increased to a prominent size

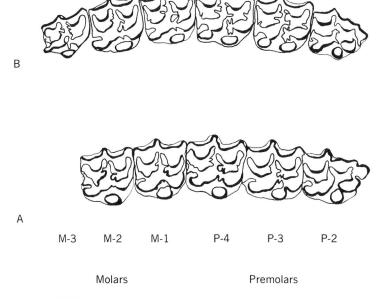

FIGURE 26.10
Enamel crest pattern in upper cheek teeth of two Late Miocene species of
Merychippus. (A) *Merychippus seversus*, Mascall formation. (B) *M. californicus*,
North Coalinga formation. (Redrawn from Downs, 1961.)

TABLE 26.2
Variation in individual upper cheek teeth of adult horses *(Merychippus)* from two Late
Miocene formations. (Downs, 1961.)

Character and formation	Range	Mean	Standard deviation	Number of teeth	Statistical significance (*P*)
Height of crown of unworn premolars and molars, mm					
N. Coalinga (younger)	30.5–39.6	34.9	2.23	39	<0.01
Mascall (older)	24.6–32.3	29.4	2.45	18	
Height of crown of worn molars, mm					
N. Coalinga	23.4–33.9	27.9	2.53	57	<0.01
Mascall	18.1–27.5	22.9	2.37	16	
Total plications on molars, mm					
N. Coalinga	2–17	7.6	2.02	66	>0.05
Mascall	5–11	7.3	1.52	19	

in the line leading to *Equus* (Simpson, 1944, pp. 59–60; 1953, pp. 105–106).
Tooth cement makes its first appearance in a similar way in the *Parahippus–
Merychippus* line. Cement is typically lacking in *Parahippus,* but occurs as a
polymorphic variant in some populations of *Parahippus,* and is later fixed in
monomorphic condition in *Merychippus* (Simpson, 1953, pp. 106–107).

Adaptive Nature of Evolutionary Trends in the Horse Family

The adaptive nature of the trends described above can be inferred with reasonable confidence from the circumstances of the animals. Habitat, diet, and
defense were undoubtedly important factors in equid evolution. Indirect as
well as direct effects of these and other factors must be taken into account.

Forests were very widespread in the early Tertiary, but climatic changes
during the Miocene caused these extensive forests to break up and be partially
replaced by open savannas and prairies. In the Pliocene the area of open grassland continued to spread at the expense of forest.

The Early Tertiary Equidae were forest-dwelling herbivores with a browsing
mode of feeding, as noted earlier. In the Miocene, however, a branch line of
the horse family tree moved into the open savanna and grassland habitats and
adopted a grazing mode of feeding. This change in habitat and food habits took

place in North America. It must have exposed the branch line to strong new selective pressures, which can account for many of the observed evolutionary trends.

Grass is a harsh, abrasive type of food material that wears out teeth. Most of the changes in tooth characters in Miocene and Pliocene horses are initial adaptations and improved adaptations for grazing. The front biting teeth, the long rows of cheek teeth, and the enamel crests and cement on the grinding surfaces of the cheek teeth are all functional for nipping off and chewing grass. The long crowns of the cheek teeth give these teeth, and their owners, a longer life span.

Some of the changes in the head also appear to be related to the adoption of a grazing habit. Elongation of the muzzle helps to accommodate differentiated sets of front nippers and rear grinders. Wide jaw bones are needed to house high-crowned cheek teeth. Strong wide jaws are required to grind large quantities of grass.

Grazing animals in open savannas and plains are more exposed visually to predators than are forest animals. Increase in body size and strength is one effective means of defense against predators in land mammals. Intelligence is another. And speed in running is still another. Trends in these characteristics in the horse family can probably be related in large measure to the requirements for defense in a plains animal.

Large body size solves some problems but creates other new ones. A large grazing horse needs bigger, harder, more durable teeth for its support than does a small grazing horse. Therefore, as Simpson (1951) points out, the trends in tooth characters are probably correlated adaptively with increase in body size as well as with change in diet.

Increase in body size also has side effects on running speed. The heavy body of a large horse is a handicap for fast running, as compared with a smaller horse, and calls for improved leg and foot mechanisms to overcome this handicap. *Hyracotherium* was built for fast running. Changes in the proportions of the legs, feet, and toes were necessary in the later, larger-bodied horses just to maintain the speed of the ancestral *Hyracotherium*, let alone to surpass it (Simpson, 1951).

Adaptive Aspects of Cope's Rule

Many groups of animals exhibit a phylogenetic increase in body size. This common, but not universal, trend is known as Cope's rule, as noted earlier (see Cope, 1896; Newell, 1949; Rensch, 1960; Stanley, 1973). Among mammals the size-increase trend is found in the marsupials, carnivores, perissodactyls, artiodactyls, primates, and other groups; parallel trends occur in reptiles, arthropods, molluscs, and other major groups.

Large body size is selectively advantageous in several ways, in land animals at least. Four advantages will be listed; others could be added.

Body size is an important factor in predator-prey interactions. A predator must possess the size, weight, and strength to bring down its prey animal; the prey, conversely, must have the size and strength to ward off its predator. Some predator-prey combinations probably become locked into armament races that lead to ever-increasing body size in both parties. Horses and big carnivores may have interacted in this manner and with this effect.

Intelligence depends upon a large and complex brain, and the latter requires a large head, which in turn calls for a large body. Selection for intelligence may entail selection for increased body size as a correlated character.

Strength in combat is another correlated aspect of body size. In struggles between males for possession of females, such as occur in many species of ungulates, there is a selective premium on superior size and strength. Some size trends may be products, in part, of sexual selection.

Large animals have still another advantage over small ones in the retention of body heat. Large bodies are more efficient than small ones for heat retention because of a smaller and hence more favorable surface-to-volume ratio in the former.

The size-increase trend has never been universal. Some phyletic lines have increased in size, while other related lines decreased in size, during the same time span. And the same phyletic line may exhibit a size-increase trend during one time period and a reversed, size-decrease trend in another period. Both deviations from Cope's rule are found in the horse family, as well as in elephants, deer, and other groups.

This is a reflection of the relativity of the selective advantage of body size. Increase in body size is advantageous – under certain conditions. It is not necessarily advantageous, and may indeed be disadvantageous, under other conditions. Island animals generally become smaller than their mainland ancestors as a result, in part, of selection for ability to survive on a strictly limited food supply.

Orthogenesis vs. Orthoselection

The old theory of orthogenesis holds that evolutionary trends are oriented by some force within the organisms themselves. According to one version of orthogenesis, the internal directing force is the "need" of the organism, an inner urge, an *élan vital,* or the like (Lamarck, 1815–1822; Bergson, 1911; Osborn, 1934; Teilhard, 1959; etc.). Among the names assigned to this mysterious inner guiding force are: *besoin* (Lamarck), *pouvoir de la vie* (Lamarck), *élan vital* (Bergson), aristogenesis (Osborn), entelechy (Driesch), telefinalism (du Noüy), and *orthogénèse* (Teilhard). This is, of course, no scientific explanation at all; it is mystical name-calling. According to another version of orthogenesis, the evolutionary trend is oriented by an alleged tendency of mutations to be oriented in the direction of the trend (Cronquist, 1951; Werth, 1956; etc.). This is at least a testable scientific hypothesis, and it has been tested and found to be incorrect (see Chapter 5).

There is, in short, no scientific evidence for orthogenesis in any of its versions. Orthogenesis did have a pioneering status in the early nineteenth century. In the mid-twentieth century, with the development of the well-founded synthetic theory of evolution, it is an anachronism, one of the unsuccessful ventures of past generations of thinkers.

We bring up the dead issue of orthogenesis here because it is not, in fact, dead in the minds of many persons. It should be remembered that orthogenesis was a widely held viewpoint among paleontologists and morphologists until the 1930s and 1940s. Some older biologists today still cling to an orthogenetic interpretation of macroevolution, while others maintain an agnostic position, refusing to accept the synthetic theory or to drop orthogenesis entirely. Furthermore, orthogenesis under one name or another (*élan vital*, entelechy, aristogenesis, telefinalism, etc.) has always had an unfailing appeal for literary intellectuals and religious philosophers interested in evolution (see, for example, Teilhard, 1959; Koestler and Smythies, 1969). It is comforting to believe that evolution has a purpose. Orthogenesis, then, is very much a live issue when we consider the views on evolution held in various popular and semi-professional circles.

Orthogenesis can be placed in opposition to a corollary of the synthetic theory known as orthoselection. According to the synthetic theory, the orientation of evolution comes not from within the organism, but from the organism-environment interaction. If environmental selection continues to operate in a given direction for a long time, we can call it orthoselection, and it will produce an evolutionary trend.

Orthogenesis and orthoselection can be treated as alternative hypotheses, giving the former the benefit of the doubt for purposes of debate. The paleontological evidence can then be reexamined with both possible explanations, and not just orthogenesis, in mind. When the paleontological evidence is carefully examined in a group that is well known both paleontologically and neontologically, so that the unknown factors are reduced to a minimum, the evidence is found to fit the hypothesis of orthoselection better than that of orthogenesis (Simpson, 1944, 1951, 1953, 1967).

In the first place, many evolutionary trends are clearly adaptive. This is the case for size-increase trends subsumed under Cope's rule, for increase in grinding capacity and durability of horse teeth, for many reduction series in plants, and so on.

Second, if an evolutionary trend is due to orthogenesis, it should be general and undeviating within a phyletic group; but if it is due to orthoselection, we would expect the trend to exhibit signs of opportunism, continuing while conditions are favorable, but shifting direction when conditions change.

The evidence in the Equidae, presented earlier, as well as in other groups, definitely fits the expectations of orthoselection. There were no trends in the Equidae that continued throughout the whole history of any phyletic line, or that took place in all lines at any one time. Body-size increase was a general trend in the Equidae, but it was subject to reversals in some lines and at some times. Horse phylogeny was actually highly ramified with many side branches, and some of the trends, including those in the line culminating in *Equus,* occurred in side branches (Figure 26.3) (Simpson, 1951).

Third, certain factors of individual variation in fossil populations, as observed under favorable conditions, are difficult to account for on the orthogenesis hypothesis, but are just what one would expect on the synthetic theory. Some trends in tooth characters in horses can be traced back in time to polymorphisms in ancestral populations, as we described in a preceding section.

Collateral Readings

Colbert, E. H. 1969. *Evolution of the Vertebrates.* Ed. 2. John Wiley & Sons, New York.

Cronquist, A. 1968. *The Evolution and Classification of Flowering Plants.* Houghton Mifflin, Boston. Chapters 2, 3.

Simpson, G. G. 1951. *Horses.* Oxford University Press, New York. Reprinted by Doubleday & Co., Garden City, N.Y.

Simpson, G. G. 1953. *The Major Features of Evolution.* Columbia University Press, New York. Chapter 8.

Simpson, G. G. 1967. *The Meaning of Evolution.* Ed. 2. Yale University Press, New Haven, Conn. Chapter 11.

Stebbins, G. L. 1974. *Flowering Plants: Evolution Above the Species Level.* Harvard University Press, Cambridge, Mass. Chapters 7, 11, 12.

27

Specialization

Introduction

Although all organisms are specialized to some extent, the degree of their specialization varies over a range from broad to narrow. Specialization in biology is an aspect of adaptation, and, like the latter, may be relatively narrow or relatively broad. Indeed, specialization may be viewed as the set of adaptations and tolerances of an organism for its adaptive zone in nature, or, more concretely, for its normal habitat or niche. If the organism is adapted for a narrow zone, habitat, or niche, it is relatively highly specialized; if it is adapted for a wider range of environmental conditions, its specializations are relatively broad. Thus an omnivorous raccoon has a broad food niche in comparison with a koala restricted to a diet of *Eucalyptus* leaves; the dietary specializations of the two mammals are broad and narrow, respectively.

I avoid using the terms "unspecialized" and "overspecialized" here, although these are often seen in writings about evolution, for both terms have misleading connotations. Simpson and others have pointed out that an unspecialized organism (if such can be imagined) could scarcely make a living. What constitutes overspecialization, on the other hand, depends on the pragmatic test of survival during unpredictable future changes. No group of organisms is overspecialized as long as it is able to survive; any group can be called overspecialized after it has become extinct. It is preferable to think in terms of relative degrees of specialization.

Narrow specialization is promoted by interspecific competition. A biotic community containing several or many species with similar ecological requirements will tend to develop into an array of narrow specialists, as explained in Chapter 20.

A narrowly specialized organism is more vulnerable to extinction during a change in the environment than an organism with broad specializations, other factors being equal. If one type of food should disappear, koalas would become extinct, whereas raccoons would not. Vulnerability to extinction is a weakness inherent in narrow specialization.

Self-Perpetuation in Evolutionary Trends

Evolutionary trends often involve an increase in specialization. In the preceding chapter we concluded that trends are not directed by internal evolutionary forces. There is a sense, however, in which a specialization trend, once started, may become self-perpetuating as a result of internal factors. The mechanism responsible for the self-perpetuation is fundamentally different from anything envisioned by the theory of orthogenesis.

An evolutionary trend involving a complex character or character combination, once well under way, often tends to become self-perpetuating. The older evolutionists used the term irreversible for such trends, and went on to recognize a general principle of irreversibility in evolution (also known as Dollo's law); but this is an overstatement of the case, since reversals do sometimes occur. Nevertheless, a tendency for self-perpetuation in trends can be discerned.

This self-perpetuating tendency can be explained as a joint effect of orthoselection and specialization. Orthoselection is the primary force guiding the phyletic line in a certain direction. But as the group advances in the pathway of its trend it accumulates specialized characters that restrict the range of functionally useful new mutations. New mutations are selectively valuable only if they produce phenotypic changes that fit in harmoniously with the existing character combination. And that range of selectively valuable mutant types becomes restricted with an increase in specialization. Specialization reinforces orthoselection to keep the trend "on the track."

Escape from Specialization

If specialization trends have a tendency to become self-perpetuating, and also increase the chances of eventual extinction, the long-term survival of a group already well advanced in such a trend might depend upon its ability to escape from specialization. We are interested in how a group can get out of the potential blind alley of a specialization trend.

Specialized characters develop during ontogeny as well as phylogeny. Juvenile stages usually lack the specializations found in adult organisms. Changes in rates of development, which are gene-controlled and susceptible to selective control, can lead to reproductive maturity in a previously juvenile stage (a condition known as paedomorphosis or neoteny). Selection for paedomorphosis, therefore, could give a group that is narrowly specialized (in the adult stage) a new lease on life. This possible way of escape from specialization has been suggested by de Beer (1951), Hardy (1954), Takhtajan (1959), and others.

Hardy (1954, p. 167) has expressed the thought clearly as follows:

> However highly specialized a race of animals may have become in its typical adult condition, provided it has a less or differently specialized young or larval form (which naturally will already be well adapted to its particular mode of living), and has a gene-complex which may sooner or later produce neoteny, then given sufficient time it stands a chance of escape from its path to extinction. In the great majority of stocks the end must come before this rare opportunity of paedomorphosis can intervene; but in a very small minority the chance comes earlier, before it is too late, and such lines are switched by selection to new pathways with fresh possibilities of adaptive radiation. So vast is the span of time available, that, rare as they may be, these escapes from specialization seem likely to have provided some of the more fundamental innovations in the course of evolution.

I might add that we know very little about the balance of factors that determines whether a group can or cannot succeed in escaping from a specialization trend.

28

Evolutionary Rates

Measurement of Rates

In this chapter we will cite evidence for differences in evolutionary rates, present Simpson's classification of rates, and discuss briefly the factors determining different rates. Before we can discuss these questions, however, we must define our units of measurement of rates, and this is by no means a simple task.

The measurement of evolutionary rates is beset with many technical and practical difficulties. Evolutionary rate is the amount of evolutionary change per unit of time. The first question to arise is: What time scale should be used? Should we use geological time (i.e., million-year intervals) or biological time (number of generations)? If we want to compare rates in different groups we have to use geological time.

The next question then concerns the type of evolutionary change to be considered and measured: whether it is phyletic, speciational, or a mixture of both. Practical necessity provides an answer to this question. The obtainable evidence pertains to phyletic evolution in some groups and to speciation and branching in others. We have to work with the evidence available, and since it includes a mixture of indicators, we should try to keep the rate comparisons between groups within one category of change or another, as much as possible.

A third problem is to find a measure of the amount of morphological change that permits rate comparisons between, say, horses and opossums. Every systematic biologist recognizes the difficulty of comparing similarities and differences in groups belonging to different families and possessing different sets of variable characteristics. Simpson (1949, 1953, 1967) takes, as an admittedly imperfect but practically useful measure, the number of new genera originating per million years.

Evolutionary rates measured by this criterion are known as genus origination rates. For example, the average genus origination rates in three groups are as follows (Simpson, 1953, p. 33):

	Genera/10^6 years
Ammonites	0.05
Chalicotheriidae	0.13
Equidae *(Hyracotherium–Equus)*	0.13

Biochemical methods can give a more precise estimate of the amount of change in a fraction of the group's internal phenotype and genotype. Thus a diverse natural group can be assayed for some homologous protein such as hemoglobin so as to determine the similarities and differences in amino acid sequence. The number of amino acid substitutions made during the course of diversification of the group is revealed by such assays. And this combined with other evidence concerning the time element leads to estimates of the rate of molecular evolution in the group (see Chapter 29).

It is possible to estimate the age of a living group without a fossil record if that group is both endemic and autochthonous in an area whose geological history is known. Thus a particular race or species may be endemic in a local area that became habitable at some recent date; for example, an isolated mountain range covered with ice until several thousand years ago. A terrestrial group of larger taxonomic rank may be confined to an isolated large island or island continent that the group could have reached only by migrating over land connections that existed at some known previous time, say, several million years ago. In such cases the age of the autochthonous and endemic race, species, genus, or family can be inferred.

The age of the endemic group sets limits for estimates of evolutionary rate, but does not necessarily indicate the maximum rate actually attained, for the group could have diverged rapidly from the ancestral stock during the early period of colonization and undergone little or no change thereafter.

Differences in Rates

Paleontological evidence shows conclusively that some groups have evolved rapidly and others have remained static during the same time span. Thus during the Tertiary there was a succession of eight genera in the equid line, from *Hyracotherium* to *Pliohippus* (for an average origination rate of one new genus per 6.25 million years), and other hoofed mammals show similar rates in the Tertiary. But during a longer time span from the Late Cretaceous to the Recent, the opossum lineage from *Alphadon* to *Didelphis* underwent relatively little change, the Cretaceous opossums being very similar to the modern ones (Simpson, 1967, p. 100).

Some classic examples of evolutionary conservatives, or "living fossils," are the opossum *(Didelphis)*, the rhynchocephalian reptile *Sphenodon*, the crocodiles *(Crocodylus)*, the oyster *(Ostrea)*, the horseshoe crab *(Limulus)*, and the marine brachiopod *Lingula*. These groups have survived into the Recent from earlier geological periods with relatively little change.

The various conservative groups have shown relatively little change during the geological time spans indicated below. We will list separately the lineages consisting of two similar genera and the ancient single genera. The information is from Simpson (1967, pp. 191–193, and personal communication).

Alphadon–Didelphis	Late Cretaceous to Recent
Homoeosaurus–Sphenodon	Jurassic to Recent
Limulitella–Limulus	Triassic to Recent
Crocodylus	Late Cretaceous to Recent
Ostrea	Cretaceous to Recent
Lingula	Silurian (or Ordovician?) to Recent

One does not have to be reminded of all the evolutionary waters that have flowed over the dam while these conservative groups have plodded on in their age-old ways.

The paleontological evidence also points to the conclusion that evolutionary rates have not remained constant in any one phyletic line throughout its whole known history. This is clearly indicated by Simpson's graphs for the brachiopods (Figure 28.1) and for selected orders of mammals (Figure 28.2). Figure 28.2 also shows group-to-group differences in rates.

Simpson (1967, p. 102) points out that living fossils such as the opossum

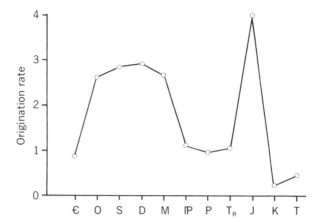

FIGURE 28.1
Variation of evolutionary rates in brachiopods through geological
time. Time span is Cambrian to Tertiary. Ordinate indicates
number of new genera per million years (origination rate).
(Redrawn from G. G. Simpson, *The Meaning of Evolution*, ed. 2,
copyright 1967, Yale University Press, New Haven, Conn.;
by permission.)

and horseshoe crab, though conservative today, were progressive when they
first appeared in the geological record. Prior to this first appearance, when the
groups were evolving from their ancestral stocks in a limited time period, their
evolutionary rate must have been relatively high. And the rate then leveled off
to low or nil after the initial period of formation.

Classification of Rates

A few groups of organisms, such as carnivores, pelecypods, and diatoms,
are diverse enough and well enough known paleontologically to permit a
meaningful plotting of the frequency distribution of their evolutionary rates.
When the frequency distribution of rates is plotted in such cases, the histogram
or curve turns out to deviate from a normal distribution by being leptokurtotic
and strongly skewed. There is a high peak, and this peak lies near the fast end
of the range of rates. The leptokurtotic and asymmetrical pattern of distribution
of rates may well be general in many or most large groups (Simpson, 1944;
1953, pp. 314–319).

The pattern of rate distribution provides a natural basis for classifying rates.
Three categories of rates are recognized: normal rates (horotely, horotelic),
low rates (bradytely), and high rates (tachytely) (Simpson, 1944, 1953).

The high peak in the distribution histogram or curve sets the norm for evolutionary rates in any given group. The leptokurtosis means that most members of a group have similar rates, rates that are characteristic of the group in question. These are the horotelic rates. The skewness of the distribution indicates further that the horotelic rates are relatively rapid.

A smaller number of members of the group have bradytelic rates, and a still smaller number, tachytelic rates. Living fossils would of course fall at the low end of the bradytelic range. Rapid changes at particular stages of horse evolution would exemplify tachytelic rates.

It should be noted that paleontological methods and evidence, by their nature, are best suited for dealing with bradytelic and horotelic rates. Direct paleontological evidence of tachytelic rates requires a very good fossil record, such as

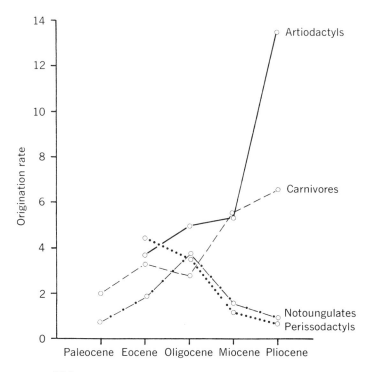

FIGURE 28.2
Variation in evolutionary rates in four orders of mammals during the Tertiary. Ordinate indicates number of new genera per million years (origination rate). (Redrawn from G. G. Simpson, *The Meaning of Evolution*, ed. 2, copyright 1967, Yale University Press, New Haven, Conn.; by permission.)

occurs in horses, and is probably limited to the slower part of the tachytelic range. Evolutionary changes taking place not in millions but in thousands or hundreds of years usually leave no fossil record. Yet we have experimental and field evidence in plants of speciational shifts occurring in scores or hundreds of years. Tachytelic rates may sometimes be more rapid than is indicated by the paleontological evidence.

Factors Affecting Evolutionary Rates

Let us list here in taxonomic form the factors that affect, or have been thought to affect, evolutionary rates. We can then discuss the various factors briefly later in this chapter. Discussion of the important factor 4 will be deferred until Chapter 30.

 I. Relation between population and environment
  1. Stability of environment

 II. Inherent characteristics of population
  2. Supply of genetic variability
  3. Length of generation
  4. Population structure
  5. Position of organism in hierarchy of nutrition
  6. Position of organism in hierarchy of reproduction

The Environmental Factor

If the environment is relatively stable, evolution slows down or stops when a population becomes well adapted to that environment. When the population reaches a high adaptive level in a stable environment, any new mutations or recombinations are likely to be selectively disadvantageous. Almost any change in the population is opposed by stabilizing selection.

A new variation may appear occasionally that has a positive selective value in some other available habitat with different ecological conditions. Then a branch phyletic line may arise. But the central group remains constant. This happened in the history of the opossum, which gave rise to four major branch lines of family rank in South America (i.e., the rat-opossums or Caenolestidae, etc.), but remained little changed itself (Simpson, 1953, p. 332).

Bradytelic or low-rate animals occur in and are broadly adapted to habitats that have had great stability through geological time. They are found in tropical and warm-temperate forests (opossum), major tropical rivers (crocodiles), the sea *(Ostrea, Lingula, Limulus)*, and other similarly long-lasting habitats. Bradytelic groups are not found, by contrast, in impermanent lakes, high mountains, volcanic deposits, arctic tundra, etc.

A population in a changing environment has three alternatives: extinction in situ, migration, or evolution. Bradytely is impossible in the area of environmental change. If the population is to persist in the area undergoing environmental change, it must evolve. The rate of evolution now depends first on the rate of environmental change, and second on the ability of the population to keep pace with the external changes.

This ability of the population to sustain horotelic or tachytelic rates depends, in turn, on various biological characteristics, to be considered next.

Intrinsic Biological Factors

The possibility has been considered that the mutation rate or amount of polymorphic variation might be a limiting factor on evolutionary rates in some cases. The available evidence does not support this idea. Morphological evidence points to a normal or high degree of genetic variation in populations of the bradytelic opossum (Simpson, 1944). Electrophoretic evidence likewise points to ample stores of genetic variability in such bradytelic groups as *Limulus* and the fern ally *Lycopodium* (Selander et al., 1970; Levin and Crepet, 1973).

Natural interspecific hybridization is a source of recombinational variation in many groups of plants. Furthermore, the hybrid derivatives often become established in new habitats created by environmental disturbances. A plant group possessing decentralized stores of potential variation – decentralized in two or more species capable of hybridization – probably has an advantage over a strictly monotypic group in responding successfully to rapid environmental change.

Another organismic characteristic that has been considered to be a limiting factor on evolutionary rate is length of generations. Since environmental change takes place on a geological or chronological time scale, whereas natural selection operates on generation time, it is logical to expect that organisms with short life cycles could achieve more rapid evolutionary rates than long-cycle organisms. But actually there is little paleontological evidence to support this logical deduction.

Among mammals, short-cycle opossums are bradytelic, whereas slow-breeding elephants have had rapid evolutionary rates. Short-cycle rodents and long-cycle ungulates have evolved at approximately the same rate in South America since the Pliocene and Pleistocene. Long-cycle carnivores have evolved much faster than short-cycle pelecypods or bivalve molluscs (Simpson, 1944, pp. 20, 25, 63).

The evidence in plants, though mixed, is generally in agreement with that in animals. For example, in the California coastal and foothill regions, annual herbaceous plants have evolved rapidly, but so also have long-lived woody plants such as *Quercus, Ceanothus,* and *Arctostaphylos* in the same areas (Stebbins, 1949, 1950). However, the woody plants in the California chaparral can be subdivided further into two categories on the basis of reproductive method and generation time. The woody chaparral plants that reproduce sexually by seeds have only moderately long generations, whereas those that reproduce vegetatively by crown sprouts have exceedingly long generations; the former category, as exemplified by *Ceanothus* and *Arctostaphylos,* shows a much higher rate of speciation than the latter (Wells, 1969).

There probably are some rapid evolutionary rates—rates measured in historical time rather than geological time—that are possible only in short-cycle organisms. Short-cycle and long-cycle organisms may be equally able to respond evolutionarily to some rates of environmental change. But let the rate of environmental change become ever more rapid. Sooner or later a threshold is reached that separates the long-cycle organism from the short-cycle organism on ability to generate the evolutionary rate needed to keep pace with the environmental change.

Length of generation is probably a factor affecting evolutionary rates, therefore, but it is a factor that comes into play only in the fast end of the rate spectrum, where it is not detected by paleontological methods.

The Hierarchies of Nutrition and Reproduction

Schmalhausen (1949) observed a correlation between the position of an organism in the food pyramid and its evolutionary rate. Organisms standing at the bottom of the pyramid tend to have low rates, and those at the top, high rates.

Schmalhausen divides the food pyramid, or hierarchy of nutrition, as he calls it, into four levels. (1) The lowest position is occupied by organisms whose only defense against predation is rapidity of reproduction (e.g., plankton, bacteria, green algae). (2) The next level consists of organisms with purely passive means of protection (e.g., molluscs, heavily armored animals, higher plants). (3) Above these in the hierarchy are animals that escape from aggressors by rapid locomotion. (4) The top position is occupied by animals that are themselves predators.

Schmalhausen points out that low evolutionary rates are characteristic in organisms belonging to levels 1 and 2 in the nutrition hierarchy; by contrast, rapid rates are often found in organisms occupying levels 3 and 4.

The next problem is to identify the causal factors responsible for the observed correlation. Schmalhausen's (1949) view was that an organism's relation to other organisms, as reflected in its position in the nutrition hierarchy, determines the mode of selection that prevails, and this in turn determines the rate of evolution.

In the case of organisms that stand low in the food chain and are defenseless, elimination by predators is largely non-selective, and under these conditions evolution remains slow and does not follow progressive trends. Predatory animals at the top of the chain (level 4), on the other hand, compete with one another for food. Success or failure in this competition depends to a considerable extent on the native ability of the individual animals. Elimination is highly selective, in other words, and permits rapid evolution of more efficient characteristics. Elimination is also highly selective in animals that depend on their own activity to escape from predators (level 3); such animals also tend to develop adaptive specializations at rapid rates (Schmalhausen, 1949).

The observed correlation between nutritional level and evolutionary rate is open to other interpretations. One can argue, for instance, that selection *has* been operative in organisms of levels 1 and 2, but has promoted high fecundity as a defense against predation, rather than rapid locomotion, these being alternative defense strategies.

Furthermore, the factor of population size and structure is an uncontrolled variable in Schmalhausen's nutrition-evolution correlation. Organisms at levels 1 and 2 frequently exist in very large populations, which are capable of only slow responses to selection, while animals belonging to level 4 often have colonial or island-like populations, which can be changed rapidly by either selection or selection-and-drift. This consideration raises a question as to whether the causal connection is between nutritional level and evolutionary rate, or between population structure and evolutionary rate, or is a mixture of both.

Higher plants all occupy the same low level (level 2) in the hierarchy of nutrition, but differ markedly in method of reproduction, and Stebbins (1949) has arranged these differences in a hierarchy of reproduction parallel to Schmalhausen's nutritional hierarchy. (1) The lowest level contains plants that rely on sheer numbers and passive transport for dispersal of their spores, pollen grains, or seeds (e.g., wind-pollinated conifers, Amentiferae, grasses). (2) The middle level contains plants with spores, pollen grains, seeds, or fruits protected by tough and resistant coats (e.g., hard-seeded pines, oaks, legumes). (3) The highest position is occupied by plants that exploit animals for pollination, seed dispersal, or both (e.g., animal-pollinated legumes, orchids).

Evolutionary rates in the organs involved in pollination or dissemination are correlated in a general way with the foregoing levels (Stebbins, 1949). Thus the wind-pollinated conifers, Amentiferae, and grasses (level 1) show little evolutionary diversification in their pollen-producing organs; but the animal-pollinated legumes and orchids (level 3) have evolved advanced specializations, apparently at rapid rates, in their flowers. Parallel differences are found between the seeds, cones, or fruits of wind-dispersed plants (level 1) and animal-disseminated plants (level 3).

Collateral Readings

Simpson, G. G. 1953. *The Major Features of Evolution.* Columbia University Press, New York. Chapters 1, 2, 10.

Simpson, G. G. 1967. *The Meaning of Evolution.* Ed. 2. Yale University Press, New Haven, Conn. Chapter 8.

Stebbins, G. L. 1950. *Variation and Evolution in Plants.* Columbia University Press, New York. Chapter 14.

29

Molecular Changes

Methods

Several biochemical approaches make it possible to measure the amount of change that has taken place in homologous macromolecules in closely or distantly related species during the course of their divergence. The macromolecules investigated are usually proteins but may be DNA's. The principal methods are: DNA hybridization (for DNA); serological tests (as used for blood types); amino acid sequencing (used for hemoglobin, myoglobin, cytochrome c, etc.); and electrophoresis (used for a variety of enzymes).

The direct results of the comparative biochemical study of the macromolecular species can be expressed quantitatively in various ways. Among the measures of biochemical differentiation between different phylogenetic lines are the following. The relative degree of pairing affinity between DNA strands from different species (in DNA hybridization studies). The number and proportion of amino acid substitutions in homologous polypeptide chains (in protein sequencing studies). The probable number of point mutations involved in producing the observed differences between homologous proteins (an extrapolation from protein sequencing studies). And the proportion of the enzyme loci tested electrophoretically that are different (in electrophoretic assays).

The *indirect* results of the biochemical assay are of particular interest to us here. The quantitative measure of biochemical differentiation between the living species A, B, C, etc., whatever the measure used, can be superimposed on a phylogenetic tree with branches terminating in A, B, C, ... The known or inferred geological time that has elapsed since the divergence of the respective branches gives the denominator. Putting the two measures together, we then have the amount of macromolecular change per unit of time, or the molecular rate of evolution.

This sounds simple enough in theory, but carrying it out in practice is a different matter. Molecular biological approaches to the study of evolutionary change and evolutionary rates are still quite new and in the process of being tested. Such approaches have not yet been able to produce the results that they are potentially capable of yielding. Involved herein is a problem that calls for a digression.

A Critique

A tool in science is only as good as the worker using it. Many of the workers who are currently attempting to apply molecular tools to evolutionary problems are molecular biologists, not organismic biologists, and I would have to say that their knowledge of evolution is usually superficial. Some workers competent in biomathematics but not in biology are also trying to use molecular tools to solve evolutionary problems. The results leave much to be desired.

Serious problems of scientific quality exist currently in *a fraction* of the molecular evolution literature, albeit a substantial fraction. My first preference would be to ignore this literature. But it has achieved much prominence and cannot be ignored. A controversial situation exists and we have to face it. Therefore I will outline the problems below as I see them. I propose to speak in generalities, without citing specific targets for my criticisms, although this could easily be done, in order to keep the criticisms on an impersonal basis. However, anyone can read the molecular evolution literature and see the problems for himself.

One common misconception, implicit in many publications in molecular evolution, is that molecular evidence has a higher degree of reliability or significance for measuring "evolutionary distance" than morphological evidence. This idea can be dismissed as sheer prejudice. Molecular evidence has certain advantages and disadvantages; morphological evidence has others.

Another idea put forward with more fanfare than documentation by several biochemists and biomathematicians is that many of the gene alleles involved in polymorphisms or substitutions are selectively neutral. As regards some gene polymorphisms and substitutions this may well be the case. But the conclusion that selectively neutral genes become fixed on a large scale in populations and genotypes is not supported by adequate direct evidence and is only a speculation at present (see also Chapter 14).

Superimposed on this is still another idea, definitely erroneous in this case, that selective neutrality of polymorphic genes can be properly described as a novel "non-Darwinian" theory of evolution. The term "non-Darwinian evolution" has been used in this sense in the titles of a paper (King and Jukes, 1969) and a symposium (LeCam, Neyman, and Scott, 1972). There are several non-Darwinian theories of evolution, but this is not one of them, for Darwin explicitly recognized the possibility of selectively neutral polymorphisms in nature (Darwin, 1859, Chs. 2, 4). In Darwin's own words (1859, facsimile ed., p. 81),

> Variations neither useful nor injurious would not be affected by natural selection, and would be left a fluctuating element, as perhaps we see in the species called polymorphic.

Next there is the problem, or apparent problem, of the adequacy of the techniques used to sample individual variation within populations. A research paper in molecular evolution often describes the comparative biochemistry of a macromolecule found in several species that would be difficult to sample in nature but easy to obtain from a biological supply house or even a grocery store and meat market. Nothing is said in the paper about the presence or absence of individual variation for the macromolecule in question. And the population-conscious reader of the paper, knowing its author for a strictly laboratory-based worker, is bound to wonder how the materials were collected.

The same standard organisms appear again in phylogenetic trees. The trees have branches ending in "rabbit," "bullfrog," "dogfish," "wheat," "yeast," etc., with numerical indices of the "evolutionary distance" between them. The branches are connected below at definite points of bifurcation arranged according to evolutionary distance. Here again the evolutionist has to pause and wonder. Our knowledge of phylogenies is only rarely complete enough to warrant the drawing of cut-and-dried phylogenetic trees. Conversely, phylogenetic trees with all the branches connected at definite points have a degree of precision usually not justified by the facts, which is why this method of representation is not used much in evolutionary biology any more, except under special circumstances.

This brings us to the question of rates. The amount of molecular differentiation—the numerator for rates—may be known quantitatively. But the denominator—evolutionary time—is taken off the phylogenetic tree or from some similar source, and is not known exactly. Quantitative measures of molecular rates are apt to look more precise than they actually are. Furthermore, the available rate estimates have been used for deducing a general conclusion concerning molecular evolutionary rates, which we will have more to say about later.

Meanwhile another group of workers, consisting of population geneticists who have learned electrophoretic methods, are also trying to study evolution at the biochemical level, and are doing so with much greater success, in my own partisan opinion. The main deficiency here is that population geneticists are not equipped to handle the more sophisticated molecular tools such as amino acid sequencing.

It goes without saying that the population-oriented workers are highly critical of many of the molecular workers. The controversy is widespread and often heated. The opposing viewpoints are represented in the symposium volume mentioned earlier (LeCam, Neyman, and Scott, 1972), as well as in the current journal literature.

Controversy is one way to arrive at the truth; another is collaboration. One worker, or one team of like-minded workers, can get only so far in an essentially interdisciplinary field such as molecular evolution. A collaborative team consisting of both molecular and population biologists could avoid some of the pitfalls of the present and the recent past.

Differences in Amino Acid Sequences

Similarities and differences in the amino acid sequences in polypeptide chains of homologous proteins belonging to different species provide a definite and quantitative measure of the amount of molecular differentiation. A large body of information on molecular homologies has now been assembled for hemoglobin, myoglobin, cytochrome c, immunoglobulin, and other proteins (see the valuable compendia of Dayhoff, 1968, 1969, 1972). Only a few representative examples can be presented here.

The adult human hemoglobin molecule consists of two identical alpha polypeptide chains, two identical beta chains, and their associated heme groups. Each alpha chain contains 141 amino acids and each beta chain, 146 amino acids. In normal human hemoglobin each position in a chain is occupied by a specific type of amino acid. The sequences are known (Figures 29.1A, 29.2A). One gene specifies the sequence in the alpha chains and another, separate gene, the sequence in the beta chains. It is of interest in passing that the alpha and beta chains of human hemoglobin, though different, have similar amino acid sequences and probably arose by divergence from some common ancestral polypeptide chain (Ingram, 1963).

We are more interested here in the amount of differentiation in the hemoglobin chains in different species. Let us take normal adult human hemoglobin as a standard. The number of amino acid differences between man and various other species of mammals is given in Table 29.1. The remarkable fact appears that man and chimpanzee have identical amino acid sequences in both the alpha and beta chains. There are only two amino acid differences between human and gorilla hemoglobins. Man and monkeys are close in their hemoglobin structure (Figures 29.1B, 29.2B). Other orders of mammals show greater differentiation from man in their hemoglobin, with differences in the range of 14–33 per chain (Figures 29.1C, 29.2C; Table 29.1).

TABLE 29.1
Number of amino acid differences between the hemoglobin chains of man and various other mammals. (Data from Dayhoff, 1969.)

Species pair	Number of differences in:	
	Alpha chains	Beta chains
Human–chimpanzee	0	0
Human–gorilla	1	1
Human–Rhesus monkey	4	8
Human–spider monkey	–	6
Human–horse	18	25
Human–cattle	17	25
Human–sheep	21	26–32
Human–goat	20–21	28–33
Human–pig	18	24
Human–llama	–	21
Human–mouse	16–19	25
Human–rabbit	25	14

The hemoglobins of man and carp are differentiated to a greater degree, as would be expected, with 71 amino acid differences between their respective alpha chains (Dayhoff, 1969).

Another respiratory protein, cytochrome c, is located in the mitochondria of eukaryotic organisms, and is well suited for comparative biochemical studies of members of different phyla and kingdoms. Table 29.2 gives some data on molecular differentiation in cytochrome c. Man is taken as the standard for one set of comparisons and *Drosophila* for another. Here again we see a general correlation between amount of molecular differentiation and closeness or remoteness of phylogenetic relationship.

Alpha chain

```
                  5                 10                    15
  1 VAL LEU SER PRO ALA ASP LYS THR ASN VAL LYS ALA ALA TRP GLY
 16 LYS VAL GLY ALA HIS ALA GLY GLU TYR GLY ALA GLU ALA LEU GLU
 31 ARG MET PHE LEU SER PHE PRO THR THR LYS THR TYR PHE PRO HIS
 46 PHE ASP LEU SER HIS GLY SER ALA GLN VAL LYS GLY HIS GLY LYS
 61 LYS VAL ALA ASP ALA LEU THR ASN ALA VAL ALA HIS VAL ASP ASP
 76 MET PRO ASN ALA LEU SER ALA LEU SER ASP LEU HIS ALA HIS LYS
 91 LEU ARG VAL ASP PRO VAL ASN PHE LYS LEU LEU SER HIS CYS LEU
106 LEU VAL THR LEU ALA ALA HIS LEU PRO ALA GLU PHE THR PRO ALA
121 VAL HIS ALA SER LEU ASP LYS PHE LEU ALA SER VAL SER THR VAL
136 LEU THR SER LYS TYR ARG  •
```
Λ

```
                  5                 10                    15
  1 VAL LEU SER PRO ALA ASP LYS SER ASN VAL LYS ALA ALA TRP GLY
 16 LYS VAL GLY GLY HIS ALA GLY GLU TYR GLY ALA GLU ALA LEU GLU
 31 ARG MET PHE LEU SER PHE PRO THR THR LYS THR TYR PHE PRO HIS
 46 PHE ASP LEU SER HIS GLY SER ALA GLN VAL LYS GLY HIS GLY LYS
 61 LYS VAL ALA ASP ALA LEU THR LEU ALA VAL GLY HIS VAL ASP ASP
 76 MET PRO ASN ALA LEU SER ALA LEU SER ASP LEU HIS ALA HIS LYS
 91 LEU ARG VAL ASP PRO VAL ASN PHE LYS LEU LEU SER HIS CYS LEU
106 LEU VAL THR LEU ALA ALA HIS LEU PRO ALA GLU PHE THR PRO ALA
121 VAL HIS ALA SER LEU ASP LYS PHE LEU ALA SER VAL SER THR VAL
136 LEU THR SER LYS TYR ARG  •
```
B

```
                  5                 10                    15
  1 VAL LEU SER PRO ALA ASP LYS.THR ASN ILE LYS.THR ALA TRP GLU
 16 LYS.ILE GLY SER HIS GLY GLY GLU TYR GLY ALA GLU ALA VAL GLU
 31 ARG.MET PHE LEU GLY PHE PRO THR THR LYS.THR TYR PHE PRO HIS
 46 PHE ASP PHE THR HIS GLY SER GLU GLN ILE LYS.ALA HIS GLY LYS.
 61 LYS.VAL SER GLN ALA LEU THR LYS.ALA VAL GLY HIS LEU ASP ASP
 76 LEU PRO GLY ALA LEU SER THR LEU SER ASP LEU HIS ALA HIS LYS.
 91 LEU ARG.VAL ASP PRO VAL ASN PHE LYS.LEU LEU SER HIS CYS LEU
106 LEU VAL THR LEU ALA ASN HIS VAL PRO SER GLU PHE THR PRO ALA
121 VAL HIS ALA SER LEU ASP LYS.PHE LEU ALA ASN VAL SER THR VAL
136 LEU THR SER LYS.TYR ARG  •
```
C

FIGURE 29.1
Amino acid sequences in alpha hemoglobin chains of three species of mammals. Amino acids are identified by the standard abbreviations (see code, opposite page). (A) Human. (B) Rhesus monkey. (C) Rabbit. (Rearranged from Dayhoff, 1969.)

FIGURE 29.2 *(opposite)*
Amino acid sequences in beta hemoglobin chains of same three species of mammals. (A) Human. (B) Rhesus monkey. (C) Rabbit. (Rearranged from Dayhoff, 1969.)

Beta chain

A

```
                    5                  10                 15
1   VAL HIS LEU THR PRO GLU GLU LYS SER ALA VAL THR ALA LEU TRP
16  GLY LYS VAL ASN VAL ASP GLU VAL GLY GLY GLU ALA LEU GLY ARG
31  LEU LEU VAL VAL TYR PRO TRP THR GLN ARG PHE PHE GLU SER PHE
46  GLY ASP LEU SER THR PRO ASP ALA VAL MET GLY ASN PRO LYS VAL
61  LYS ALA HIS GLY LYS LYS VAL LEU GLY ALA PHE SER ASP GLY LEU
76  ALA HIS LEU ASP ASN LEU LYS GLY THR PHE ALA THR LEU SER GLU
91  LEU HIS CYS ASP LYS LEU HIS VAL ASP PRO GLU ASN PHE ARG LEU
106 LEU GLY ASN VAL LEU VAL CYS VAL LEU ALA HIS HIS PHE GLY LYS
121 GLU PHE THR PRO PRO VAL GLN ALA ALA TYR GLN LYS VAL VAL ALA
136 GLY VAL ALA ASN ALA LEU ALA HIS LYS TYR HIS  *
```

B

```
                    5                  10                 15
1   VAL HIS LEU THR PRO GLU GLU LYS ASN ALA VAL THR THR LEU TRP
16  GLY LYS VAL ASN VAL ASP GLU VAL GLY GLY GLU ALA LEU GLY ARG
31  LEU LEU LEU VAL TYR PRO TRP THR GLN ARG PHE PHE GLU SER PHE
46  GLY ASP LEU SER SER PRO ASP ALA VAL MET GLY ASN PRO LYS VAL
61  LYS ALA HIS GLY LYS LYS VAL LEU GLY ALA PHE SER ASP GLY LEU
76  ASN HIS LEU ASP ASN LEU LYS GLY THR PHE ALA GLN LEU SER GLU
91  LEU HIS CYS ASP LYS LEU HIS VAL ASP PRO GLU ASN PHE LYS LEU
106 LEU GLY ASN VAL LEU VAL CYS VAL LEU ALA HIS HIS PHE GLY LYS
121 GLU PHE THR PRO GLN VAL GLN ALA ALA TYR GLN LYS VAL VAL ALA
136 GLY VAL ALA ASN ALA LEU ALA HIS LYS TYR HIS  *
```

C

```
                    5                  10                 15
1   VAL HIS LEU SER SER GLU GLU LYS SER ALA VAL THR ALA LEU TRP
16  GLY LYS VAL ASN VAL GLU GLU VAL GLY GLY GLU ALA LEU GLY ARG
31  LEU LEU VAL VAL TYR PRO TRP THR GLN ARG PHE PHE GLU SER PHE
46  GLY ASP LEU SER SER ALA ASN ALA VAL MET ASN ASN PRO LYS VAL
61  LYS ALA HIS GLY LYS LYS VAL LEU ALA ALA PHE SER GLU GLY LEU
76  SER HIS LEU ASP ASN LEU LYS GLY THR PHE ALA LYS LEU SER GLU
91  LEU HIS CYS ASP LYS LEU HIS VAL ASP PRO GLU ASN PHE ARG LEU
106 LEU GLY ASN VAL LEU VAL ILE VAL LEU SER HIS HIS PHE GLY LYS
121 GLU PHE THR PRO GLN VAL GLN ALA ALA TYR GLN LYS VAL VAL ALA
136 GLY VAL ALA ASN ALA LEU ALA HIS LYS TYR HIS  *
```

Ala	Alanine	Gly	Glycine	Pro	Proline
Arg	Arginine	His	Histidine	Ser	Serine
Asn	Asparagine	Ile	Isoleucine	Thr	Threonine
Asp	Aspartic acid	Leu	Leucine	Trp	Tryptophan
Cys	Cysteine	Lys	Lysine	Tyr	Tyrosine
Gln	Glutamine	Met	Methionine	Val	Valine
Glu	Glutamic acid	Phe	Phenylalanine		

TABLE 29.2
Number of amino acid differences between the
cytochrome c of man and various other organisms.
(Data from Dayhoff, 1969.)

Species pair	Number of amino acid differences
Human—Rhesus monkey	1
Human—horse	12
Human—cattle, sheep	10
Human—dog	11
Human—rabbit	9
Human—chicken, turkey	13
Human—pigeon	12
Human—snapping turtle	15
Human—rattlesnake	14
Human—bullfrog	18
Human—tuna fish	21
Human—dogfish	24
Human—fruit fly	29
Human—screw-worm fly	27
Human—silkworm moth	31
Human—wheat	43
Human—*Neurospora*	48
Fruit fly—screw-worm fly	2
Fruit fly—silkworm moth	15
Fruit fly—tobacco hornworm moth	14
Fruit fly—dogfish	26
Fruit fly—pigeon	25
Fruit fly—wheat	47

Dayhoff (1969) and her co-workers point out that the observed number of
amino acid differences between homologous proteins is not necessarily equal to
the number of actual amino acid substitutions during their evolutionary divergence. Where many amino acid differences occur between two polypeptide
chains, the evolutionary distance becomes greater than the observed differences.
Dayhoff and co-workers have devised a unit of evolutionary distance, called
the PAM unit ("accepted point mutations per 100 links" in the chain), which is
designed to give a corrected estimate of molecular evolutionary divergence.

The relation between the observed number of amino acid differences per
100 links in a chain and the evolutionary distance in PAM units is as follows
(Dayhoff, 1969):

Differences	PAM units
1	1
5	5
10	11
25	31
50	83
75	208
85	370

On this basis, the evolutionary distance between species representing different major groups is somewhat greater than is implied by the numbers given in Tables 29.1 and 29.2.

Other Evidences of Molecular Change

DNA hybridization techniques involve the mixing of single-stranded fragments of the DNA of two species. The proportion of the total DNA in the mixture that reassociates to form double-stranded helices, and the rate of the reassociation, are measures of the degree of genetic affinity between the species.

The application of this technique in *Drosophila* yielded the results shown in Table 29.3. *Drosophila melanogaster* and *D. simulans* belong to the same species group. We see that 80% of their DNA forms pairs in DNA hybridization tests. By contrast, the DNA of *D. funebris*, which belongs in a different subgenus, mostly does not pair with that of *D. melanogaster* or *D. simulans* (Laird and McCarthy, 1968).

TABLE 29.3
Proportion of DNA of *Drosophila* species that pairs in DNA hybridization tests. (Laird and McCarthy, 1968.)

Species combination	Proportion of DNA that pairs
D. melanogaster × *simulans*	80%
D. funebris × *melanogaster* or *simulans*	25%

Hubby and Throckmorton (1965) used electrophoretic methods to determine the similarity or difference in proteins within the *Drosophila virilis* group. Ten species in this group (*D. virilis, D. americana, D. texana*, etc.) were compared with respect to numerous proteins and their underlying gene determinants. They found that 60% of the proteins sampled are common to the whole species group. Another smaller fraction is common to closely related subgroups (e.g., *D. americana* and *D. texana*). The remaining fraction of the proteins tested is unique in each species; this fraction amounts to 2.6% for *D. virilis,* 5.3% for *D. americana,* and has an average value of 14% for all ten species combined.

Another series of studies using gel electrophoresis compared four sibling species of the *Drosophila willistoni* group with respect to 14–28 enzyme loci (Ayala et al., 1970; Ayala and Tracey, 1974). The species turn out to be different at about half of the loci sampled.

A close serological relationship exists between man and apes in the ABO blood groups. (Regarding the ABO blood groups in man, see Chapter 2.) The following blood types are known to occur in four species:

Man	A, B, AB, O
Gibbon	A, B, AB
Orangutan	A, B, AB
Chimpanzee	A, O

The agglutination reactions of chimpanzee blood are indistinguishable from those of human blood, and the same is true of the blood of the gibbon. The gorilla, however, differs from man and the other great apes and resembles monkeys in its ABO blood groups (Wiener and Moor-Jankowski, 1971).

Differences Between Man and Chimpanzee

The molecular relationships of man and the chimpanzee *(Pan troglodytes)* have been extensively explored with a variety of techniques, including all of those mentioned earlier in this chapter. King and Wilson (1975) pooled the extensive molecular data from all sources. They find that humans and chimpanzees are remarkably similar at the macromolecular level, with many identical proteins and many very similar ones.

Amino acid sequencing studies of several homologous proteins in man and chimpanzee reveal that some of these proteins are completely or nearly identical. The following homologous proteins have 0 or 1 amino acid difference, as indicated below:

Hemoglobin	0
Cytochrome c	0
Fibrinopeptide	0
Lysozyme	0?
Delta hemoglobin	1
Myoglobin	1

And other proteins in the two species show only a few amino acid differences: for example, serum albumin, with about 6 such differences at 580 sites (King and Wilson, 1975).

Electrophoretic analysis of 44 homologous proteins points again to close similarity between man and chimpanzee. About half of the proteins assayed are electrophoretically identical in the two species (King and Wilson, 1975).

The amount of molecular differentiation between man and the chimpanzee is on a par with that between sibling species in other groups, such as the *Drosophila willistoni* complex; it is much less than that between non-sibling species in various other genera (King and Wilson, 1975).

Yet *Homo sapiens* and *Pan troglodytes* are undeniably members of different genera on morphological, ecological, and behavioral grounds. Their chromosomes also differ with respect to a number of inversions and translocations. The assignment of the two species to different families is accepted almost universally. A discordance thus exists between the organismic evidence and the molecular evidence concerning the evolutionary distance between humans and chimpanzees (King and Wilson, 1975).

King and Wilson attempt to resolve this paradox by suggesting that the basic genetic differences between humans and chimpanzees are due not so much to point mutations in ordinary or so-called "structural" genes as they are to changes in the regulatory systems of the two organisms. The changes in the regulatory systems would come about partly by mutations in regulatory genes and partly by chromosomal rearrangements that change the gene order and affect gene expression. "A relatively small number of genetic changes in systems controlling the expression of genes may account for the major organismal differences between humans and chimpanzees" (King and Wilson, 1975, p. 115).

I would like to make a few comments on this interpretation. Several different issues are involved here and need to be sorted out.

First, there are certain special features in comparisons of man and chimpanzee that are not found in other species pairs in higher animals. The overall differences between man and chimpanzee are due not only to biological evolution but also to cultural evolution; cultural evolution, conversely, has greatly amplified the differences between the two species. The amplified differences then enter into the taxonomic judgment that places *Homo* and *Pan* in different families. The Hominidae could not be maintained separate from the Pongidae on morphological grounds alone, and, accordingly, we should reduce the two groups in taxonomic rank when we are discussing their strictly biological differences. In short, *Homo* and *Pan are* similar at the organismic level, as zoologists, animal behaviorists, and general naturalists have long recognized.

The gross phenotypic differences are also there. What is their genetic basis? Since other approaches to this question have proven inadequate or misleading, we may perhaps fall back on the extensive body of evidence of plant genetics for suggestive clues. In plants, unlike higher animals, it is often possible to raise large F_2 and F_3 progeny from interspecific and intergeneric crosses, and subject these to factorial analysis. Hybridization studies in many plant groups indicate that species and genera differ with respect to various types of gene systems in which "structural" or Mendelian genes play a large role (for review see Grant, 1975, Chs. 7–12). By extension, we have grounds for expecting numerous and complex gene differences to exist between any two species of higher primates.

Another problem, as King and Wilson realize, concerns the nature of regulatory systems, on which so much stress is placed. Such systems are heterogeneous and poorly understood. Ordinary "structural" or Mendelian genes are among the components of regulatory systems in plants, at least (see again Grant, 1975, Ch. 8), and probably also in animals. I would agree with King and Wilson that regulatory systems of one sort or another probably play an important part in the differentiation of humans and chimpanzees. But it does not necessarily follow from this that the number of ordinary gene differences between these species is small.

The discordance between the protein evidence and the gross phenotypic evidence indicates, to my mind, that different gene systems are involved in the determination of the contrasting sets of characteristics. The genes revealed by molecular techniques appear to be largely different from those revealed by classical genetic techniques. At least the protein-determining genes are not a representative sample of the genotype, contrary to a currently widespread assumption (but see the discussion in the second section of Chapter 14). It follows that, contrary to another common assumption, molecular evidence is only one among several valid measures of genetic and evolutionary distance.

In short, we are not getting close to the adaptively important morphological, ecological, and behavioral differences between humans and chimpanzees by the use of molecular methods alone.

Molecular Rates

The average rate of molecular change in several species of protein molecules has been computed by Dayhoff (1972, p. 50) and is presented here in Table 29.4. The rate of change is expressed in number of PAM units per 100 million years. One PAM unit is one point mutation that brings about an amino acid substitution in 100 links of a polypeptide chain. Use of PAM units thus

TABLE 29.4
Average rate of evolutionary change in various protein macromolecules. (Dayhoff, 1972.)

Type of protein	Number of PAM units per 100 million years
Fibrinopeptide	90
Growth hormones	37
Immunoglobulin	32
Ribonuclease	33
Hemoglobin	14
Myoglobin	13
Gastrin	8
Encephalitogenic proteins	7
Insulin	4
Cytochrome c	3
Glyceraldehyde-3-PO_4 dehydrogenase	2
Histone	0.06

enables us to compare proteins with different molecular weights on a common denominator.

Table 29.4 shows that different species of proteins have different average rates of change. Histones are extremely conservative. Cytochrome c, insulin, and some other proteins are very conservative. By contrast, fibrinopeptides and growth hormones have changed rapidly. And the hemoglobins show an intermediate evolutionary rate.

Several molecular evolutionists have claimed that the rate of molecular change is constant through time for a given type of protein (Ohta and Kimura, 1971; Kimura and Ohta, 1971, 1972; Jukes, 1972). Molecular evolution supposedly proceeds with clock-like regularity. This conclusion is, of course, at variance with what we know about evolutionary rates in gross phenotypic characteristics (see Chapter 28).

However, the conclusion that molecular rates are constant is based on an elementary fallacy, as pointed out by several organismic evolutionists (Stebbins and Lewontin, 1972; Clarke, 1973; Lewontin, 1974, pp. 228–229). Some molecular evolutionists have taken the total amount of molecular change for a long period of geological time (of the order of 10^8 years), computed the average rate of molecular change, and then gone on to present the average rates as constant rates, thus confusing two quite different parameters. By using this procedure one could also reach the erroneous conclusion that gross phenotypic characters evolve at constant rates.

Collateral Readings

Cronquist, A. 1976. The taxonomic significance of the structure of plant proteins: a classical taxonomist's view. *Brittonia* **28:** 1–27.

Dayhoff, M. O., ed. 1968, 1969, 1972. *Atlas of Protein Sequence and Structure.* Vols. 3, 4, 5. National Biomedical Research Foundation, Silver Spring, Md.

Dobzhansky, Th. 1970. *Genetics of the Evolutionary Process.* Columbia University Press, New York. Chapters 8, 12.

Ingram, V. M. 1963. *The Hemoglobins in Genetics and Evolution.* Columbia University Press, New York.

LeCam, L. M., J. Neyman, and E. L. Scott, eds. 1972. *Darwinian, Neo-Darwinian, and Non-Darwinian Evolution.* Berkeley Symposia on Mathematical Statistics and Probability. Proceedings of the Sixth Symposium, Vol. 5. University of California Press, Berkeley.

Lewontin, R. C. 1974. *The Genetic Basis of Evolutionary Change.* Columbia University Press, New York. Chapters 4, 5.

Simpson, G. G. 1974. Recent advances in methods of phylogenetic inference. *Wenner-Gren Cent. Int. Symp. Ser.* **61:** 1–35.

30

Population Structure in Relation to Macroevolution

The Adaptive Landscape

The components of non-trivial evolutionary change have been discussed in previous chapters under such headings as: adaptive characters and character combinations; adaptive gene combinations; broad and narrow specializations; evolutionary trends; and evolutionary rates. The metaphor of the adaptive landscape is a useful way of representing the interactions among these components. This metaphor has been employed by Wright, Dobzhansky, Simpson, Stebbins, and other evolutionists.

Consider a landscape consisting of hills and valleys. The topography symbolizes the distribution of adaptive fields. The hilltops are "adaptive peaks," and the valleys between them constitute a no-man's-land for organisms, or a series of "adaptive valleys." Populations and species occupy the various adaptive peaks by virtue of their adaptive character combinations and the underlying gene combinations. Some hilltops or adaptive peaks are narrow, and others broad, corresponding to the relative breadth of specialization. The hilltops also vary in height; this suggests that some adaptive peaks are easier to climb and occupy than others.

The metaphor can be extended so as to embrace species groups, genera, and larger phyletic assemblages. The peaks on the adaptive landscape are not randomly distributed, but are clustered, to some extent at least, in separate ranges. A range of hills would then represent the adaptive zone of a genus.

So far we have considered only the statics of the situation. What about the dynamics: How do species come to occupy their adaptive peaks, and genera, their mountain ranges? Evolutionary trends are the pathways to the adaptive peaks and ranges. And evolutionary rates correspond to the rate of ascent. Adaptive peaks of different height may call for different combinations of trends and rates.

The adaptive landscape does not necessarily remain static and unchanging. Like a real landscape it is subject to the forces of erosion and mountain-building. These forces may act slowly or rapidly. This represents the role of environmental change in evolution. An adaptive peak can move, at one rate or another, and the population inhabiting that peak must then try to keep pace with it.

Evolutionary Potential of Different Types of Populations

The leading idea concerning the role of population structure in the climbing of adaptive peaks was presented by Wright in his classic paper of 1931 (Wright, 1931; see also 1949, 1960). Wright sets forth the evolutionary characteristics of different types of populations. Three types are especially relevant to this discussion.

The three types of populations are: (1) small isolated populations; (2) large continuous outcrossing populations; and (3) a particular intermediate condition that Wright refers to as the subdivided population. Type 3 is a system of semi-isolated small populations; we have referred to this elsewhere in this book as the colonial or island-like type of population system.

Let us consider the expected performance of each type of population in relation to three modal types of environmental condition: (A) stable environment; (B) slowly changing environment; and (C) rapidly changing environment.

(1) In a small isolated population.   Here most polymorphic genes are fixed quickly by drift or by selection and drift. Consequently little genetic variation remains available in storage with which to meet future challenges (see Chapter 13).

(A) In a stable environment the population quickly reaches the best state of adaptedness with the variation on hand, and then becomes evolutionarily stagnant. (B, C) In a slowly or rapidly changing environment, the population is handicapped by lack of variability, and cannot make an adequate evolutionary response. It may migrate to a refuge, or it is likely to become extinct.

(2) In a large continuous outcrossing population. Here selection is relatively ineffective on recessive alleles at low frequencies, including most new mutations, and the selection-drift combination of forces is totally ineffective. New gene combinations do not persist but are swamped out by wide outcrossing. Furthermore, gene systems promoting homeostatic buffering properties tend to develop, and these homeostatic systems resist change.

(A) In a stable environment the large population maintains the existing adaptations but does not change (e.g., opossum). (B) In a slowly changing environment the population evolves slowly and steadily (e.g., horse). (C) But it cannot keep up with a rapidly changing environment. For the various reasons given in the preceding paragraph, the large continuous population is not able to respond successfully to intense selection. It may break up into small relictual populations (e.g., sequoia) or suffer extinction.

Conditions are thus unfavorable for sustained rapid evolutionary rates in both very small and very large populations.

Now consider the evolutionary potential of the colonial population system. It avoids the disadvantages found in small isolated and large continuous populations, and possesses certain advantages of its own.

(3) In an island-like population system. Here the individual colonies may be monomorphic or nearly so, but the population system as a whole has a decentralized store of genetic variation, distributed among the different colonies; and this variability can spread from colony to colony by occasional interdeme gene flow. Each colony can respond rapidly to local environmental conditions by an interplay among the variations on hand at the time, and selection and drift. Favorable new gene combinations can be fixed in a colony, as well as single gene alleles, because of the protection from wide outcrossing afforded by semi-isolation.

(A) In a stable environment the colonial population system becomes differentiated into local races by selection or selection-drift. (B) In a slowly changing environment some local races are likely to be favored selectively, and will expand, while other local races contract. Some interdeme selection is postulated to occur (see Chapter 11). (C) A rapidly changing environment evokes the same type of response. And other processes may also occur. Expansion of some colonies may bring them into closer proximity, promoting increased rates of interdeme gene flow, and perhaps interracial hybridization followed by the segregation of new types. Rapid evolutionary changes can take place within favored colonies as a result of the operation of selection-drift (see Chapter 13) and the possibility of evading the cost-of-selection restriction on evolutionary rate (see Chapter 14).

The colonial population system is thus the most favorable setup for rapid and drastic evolutionary changes (Wright, 1931, 1949, 1960). Many evolutionists believe, on theoretical considerations, that this type of population structure has been involved in the great historic bursts of evolution that have produced new major groups such as the mammals and the angiosperms.

Quantum Evolution

The next important concept concerning populational aspects of macroevolution is quantum evolution (Simpson, 1944; also 1953). Quantum evolution is a special form of phyletic evolution that occurs when a population is shifting from one adaptive peak to another new or previously unoccupied adaptive peak. The shift entails the crossing of an adaptive valley.

Simpson suggests that the shift to the new adaptive peak usually involves small populations evolving at extraordinarily rapid rates. The transitional period while the population is occupying the new peak is relatively short. During this short period the population is represented by relatively few individuals, in comparison with more normal periods in the history of the lineage. But the amount of phenotypic and genotypic change may be much greater during the period of the shift than it is during vastly longer periods of normal phyletic evolution. The relationship between evolutionary rate and population size is shown diagrammatically in Figure 30.1.

The history of a phyletic group often consists of long periods of horotelic

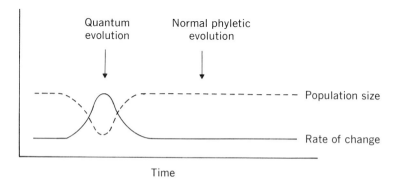

FIGURE 30.1
Comparison of quantum evolution and normal phyletic evolution with respect to evolutionary rate and population size. (Redrawn from Simpson, 1953.)

or bradytelic rates of change interrupted by rare short periods of tachytelic rates and quantum evolutionary changes (Figure 30.1). The term quantum evolution carries the connotation of occasional spurts of major evolutionary activity (Simpson, 1944, p. 206).

There are several reasons for postulating small population size in quantum evolution. While crossing the adaptive valley on its way to the new peak, the population is, by definition, poorly fitted and under stress; and when it reaches the new adaptive peak, it faces strong new selective pressures (Simpson, 1944, 1953). Furthermore, as noted in the preceding section and elsewhere, rapid adaptive evolution can be brought about more readily by the selection-drift combination than by selection alone, and selection-drift requires small population size.

Comparable reasons exist for suggesting a rapid evolutionary rate in quantum evolution. The population must cross the adaptive valley and occupy the new adaptive peak in a short time or else not at all. Horotelic rates do not suffice in this situation.

The product of quantum evolution is a new group of organisms, at any taxonomic level from species to genus and higher categories. Quantum evolution is not necessarily involved in species formation, but may be in certain instances (see Chapter 22). On the other hand, quantum evolution is probably usually involved in the origin of new major groups, such as families and orders (Simpson, 1944, 1953).

The paleontological history of many animal families and orders consists of a long, recorded sequence when horotelic or bradytelic rates prevailed, preceded by a much shorter, unrecorded time span during which the origin of the group must have taken place (see Table 31.2). The fossil record gives evidence of large population size accompanied by horotelic or bradytelic rates during periods of normal phyletic evolution. Conversely, the absence or rarity of fossils during the period of origin and main development of the group is consistent with the postulate of small population size in quantum evolution. The postulated features of quantum evolution provide an explanation for the great rarity of so-called missing links in the fossil record.

Direct paleontological evidence concerning quantum evolution is expected to be scanty, by the nature of the case, and it is. Nevertheless, quantum shifts are documented by fossil evidence in a few favorable cases. One of these is the change from a pad-footed to a springing-foot mechanism in the Equidae, and another is the change from browsing to grazing food habits in the same family (see Chapter 26). Simpson (1953, pp. 389–393) also cites examples in two other mammal orders. One is the origin of the Artiodactyla, as worked out by Schaeffer (1948), and the other is the origin of the subfamily Stylinodontinae in the edentate-like Early Tertiary order Taeniodonta, as studied by Patterson (1949).

Phylogenies inferred from comparative studies of living members of a group can also yield plausible indirect evidence of quantum evolution in certain cases. I have previously (Grant, 1959; 1963, pp. 557–563) interpreted the origin of the two temperate-zone tribes of the plant family Polemoniaceae as instances of quantum evolutionary shifts.

Genetic Revolutions

The next significant concept, in order of appearance, is that of Mayr (1954; see also 1963, Ch. 17) on genetic reorganization in isolated populations. Mayr contrasts the internal genetic environment in large widespread populations of a species with that in small isolates, usually isolated peripheral populations, of the same species.

A local subpopulation in a large continuous population is exposed to a stream of gene flow from neighboring and occasionally from distant subpopulations. The more or less frequent immigration of alien genes into the local subpopulation in question sets up selective pressures favoring genes with good combining ability. The gene pool comes to consist of arrays of alleles that give rise to normal viable products in a wide variety of heterozygous gene combinations. The pool of "good mixer" genes that develops in this genetic environment also, as a corollary, has homeostatic buffering properties.

This gives a certain conservative character to the variation pattern of the large continuous population. It exhibits clinal variation and geographical racial variation (see Chapter 18). But deviant or novel forms of the species cannot and do not persist as racial entities.

Now consider what happens when a few emigrant individuals from the large cross-fertilizing population succeed in founding a small daughter colony on the periphery of the species area. In the first place, the founder population will contain only a small non-random sample of the gene pool of the parent population. But furthermore, it will be isolated from the stream of gene flow. This changes the genetic environment. The selective values of the genes in the isolated daughter gene pool are altered; "good mixer" genes are no longer favored, whereas genes that produce viable types in homozygous condition now have a positive selective value (Mayr, 1954).

The passage from a large cross-fertilizing population to a small isolated daughter colony thus in itself changes the genetic environment and alters or even reverses the selective values of the genes at many loci. The change in internal genetic environment affects many gene loci simultaneously, like the change brought about by hybridization or polyploidy, and it may be the beginning of a "genetic revolution" (Mayr, 1954).

The population may not be able to make the passage through the bottleneck from large to small size. It may not be able to stand a genetic revolution; in this case it will soon become extinct. But in a series of trials, as where numerous independent founder events take place, there is a chance that one or a few founder populations can survive. Survival in this situation depends on the founder population's inheriting a gene pool that can tolerate homozygosity at many loci and that is preadapted to some available new ecological zone.

The genetic revolution in a successful daughter colony results in a more or less radical change in ecological preferences and morphological characters. Novel types arise, adapted to new ecological zones beyond the frontiers of the large conservative ancestral population. Taxonomic evidence agrees with this expectation. Some species and species groups in birds, insects, and annual plants do in fact exhibit novel variations in their isolated peripheral populations (see Chapters 18, 21, and 22).

In short, isolated peripheral populations derived by founder events from a large cross-fertilizing ancestral population may be sites of rapid evolution and of the origin of novel forms (Mayr, 1954).

The divergent population does not necessarily remain small, isolated, and peripheral forever. It may build up in size, spread in area, and eventually reinvade the territory of the ancestral species. In time it too may become conservative in its variation pattern. But then it can give rise to a new generation of isolated daughter colonies, which break out of the ancestral mold once again (Mayr, 1954).

Mayr developed his concept of genetic revolutions largely independently of Wright's views on subdivided populations and Simpson's views on quantum evolution. Nevertheless, good congruence is evident between the three concepts as regards the expected end results in small populations. One area of disagreement concerning the controlling factors is mentioned below.

Mayr rejected genetic drift as a factor in the divergence of the small daughter colonies. Instead he attributes the initial change in the gene pool of the daughter colonies to what he terms the founder effect, that is, to the accidents of sampling that occur when a few emigrant individuals from a large ancestral population found a new daughter colony (Mayr, 1954; 1963, pp. 203ff., 518ff., 527ff.). However, other students have argued that the founder effect is actually a special case of genetic drift (Dobzhansky and Spassky, 1962; Grant, 1963, p. 286). Moreover, the effectiveness or ineffectiveness of drift alone is not the real issue here; the controlling forces during founder events are probably drift and selection, acting jointly.

Carson has published a series of papers in recent years elaborating on the concept of genetic revolutions in peripheral isolates in the light of population studies in *Drosophila* (see Carson, 1959, 1971, 1975; Carson and Heed, 1964). Carson believes that divergent isolated peripheral populations may arise from single founder individuals in many instances. He also introduces some new terminology for various aspects of the original (Mayr, 1954) hypothesis (e.g., homoselection, population flush-crash cycles).

Quantum Speciation

Quantum speciation was described as a mode of speciation in Chapter 22, to which the reader is referred. Our interest in quantum speciation here is in its postulated role in macroevolution. It has been suggested that some evolutionary trends consist of series of consecutive quantum speciational events (Grant, 1963). We will reintroduce quantum speciation here as a prelude to a discussion of such trends.

Let us recall that quantum speciation is the budding off of a daughter species from a large, polymorphic, outcrossing, ancestral population via the intermediate stage of a small local race. The local race is founded by one or a few individuals from the ancestral species population. It is spatially isolated or semi-isolated from the ancestral species; it usually occurs on the periphery of the old species area, and it may occur in a new ecological zone.

The gene pool of the local race is affected by drift and inbreeding at the beginning of its existence, and later by the selection-drift combination and often by genetic revolutions. These factors are capable of bringing about rapid and sometimes drastic changes in ecological preferences, morphological characters, and fertility relationships. Therefore the local race may diverge rapidly to the species level, and furthermore, the daughter species may occupy a new adaptive zone beyond the ecological range of the ancestral species.

The concept of quantum speciation is essentially a synthetic one. It represents a synthesis of selected parts of the previous concepts of Wright (1931), Simpson (1944), and Mayr (1954), presented above, and also of the concepts emerging from Lewis's work on plant evolution (Lewis and Raven, 1958; Lewis, 1962). But no single previous concept fits the description of quantum speciation exactly. One gets selection-drift from Wright (but not from the other authors), quantum shifts (but not speciation) from Simpson, and rapid speciation (but not selection-drift) from Mayr and Lewis; then one puts the selected ingredients together. The synthetic concept of quantum speciation was proposed by Grant (1963; see also 1971) with the thought that the process involved is significant in both speciation and macroevolution.

Phyletic Trends and Speciational Trends

We must now look more closely at the nature of evolutionary trends. In early parts of this book we presented phyletic evolution within lines and evolutionary divergence between lines as separate patterns of evolution. This is a simplification of real cases, in which the two patterns often become intertwined. Then in our discussion of evolutionary trends (in Chapter 26) we presented trends as phenomena of phyletic evolution, but this again is only part of the story.

It will clarify thinking about this question to recognize two broad types of evolutionary trends: (1) phyletic trends and (2) speciational trends.

A phyletic trend is a gradual, progressive, evolutionary change within a phyletic line. It is a phenomenon of phyletic evolution exclusively. The stages in the trend are represented by successional species rather than by contemporaneous biological species (see Chapter 18). Good examples are afforded by various series in horses, such as *Merychippus* in the Miocene and *Pliohippus–Equus* in the Plio-Pleistocene (Chapter 26). Another example is the trend in the structure of molar teeth in the lineage of now extinct Pleistocene European elephants running from *Elephas planifrons* to *E. meridionalis* (Figure 30.2).

A speciational trend is a stepwise progression resulting from a succession of speciational changes. Each new species in the series advances further in the

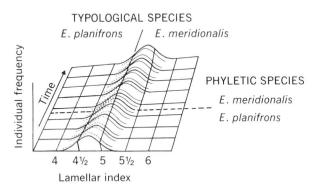

FIGURE 30.2
Phyletic trend in a fossil European elephant lineage (*Elephas planifrons–E. meridionalis*). The lamellar index is a measure of the amount of enamel in the molar teeth. Phyletic species = successional species, and typological species = taxonomic species, in our terminology. (Redrawn from Simpson, 1953.)

direction of the trend. The trend is the line P–Q in Figure 30.3. The stages in the trend are represented by biological, not successional, species (as defined in Chapter 18). We see in Figure 30.3 that the predominating pattern of evolution is speciation rather than phyletic evolution. The trend is the resultant of the successive speciations.

In a phyletic trend there is only one biological species at any given time level. In a speciational trend the conservative ancestral form and the advanced descendant coexist temporarily as two contemporaneous biological species. The line P–Q in Figure 30.3, although it consists of a succession of species in the literal sense, does not contain successional species; the line P–Q, therefore, is not a phyletic trend, but rather a speciational trend, by the criteria proposed here.

Evolutionary changes of substantial magnitude can be and probably are compressed into shorter time periods in speciational trends than they are in phyletic trends. Speciational trends are consequently capable of proceeding at more rapid rates than phyletic trends.

Speciational trends could be fairly common. It is difficult to say how frequent they are, however, owing to the problem of identifying a speciational trend as

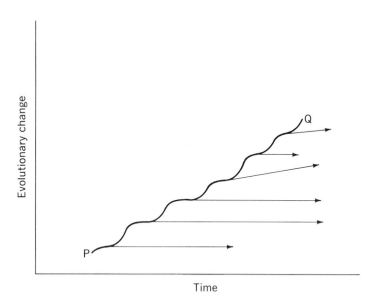

FIGURE 30.3
Speciational trend. The trend (line P–Q) is the resultant of successive speciational changes in a given direction. (From V. Grant, *The Origin of Adaptations*, copyright 1963, Columbia University Press, New York; reproduced by permission.)

such whenever the fossil record is incomplete, as it usually is. A speciational trend can easily be misread as a phyletic trend, and vice versa.

The fossil record of a group, being collected from different deposits, is usually scattered with respect to the real phylogenetic tree. The published phylogenetic tree based on this record—that is, the phylogenetic hypothesis— has gaps and uncertainties in it. In fact, different phylogenies can often be drawn from the same incomplete fossil record, depending on how one connects the branch lines in the unrecorded gaps.

Abel (1929) and Simpson (1953) make a distinction between *Ahnenreihen* and *Stufenreihen* that helps to resolve this problem of interpretation. The two modes are shown diagrammatically in Figure 30.4, where the sequence a–e represents an *Ahnenreihe* and A–E a *Stufenreihe*. The fossil record of a group often lies in the positions A, B, C, D, and E, whereas a, b, c, d, and e are un- known. The first tendency in such a situation might be to connect A, B, C, D, and E in a phylogenetic series. But this would obviously be inaccurate, and, being aware of this pitfall of interpretation, one can be prepared to recognize A, B, C, D, and E as side branches and to infer the existence of the real line of descent, a–e.

A case in which both *Stufenreihe* and *Ahnenreihe* are known is the Equidae. A fossil series in the Old World consisting of *Hyracotherium, Anchitherium, Hypohippus, Hipparion,* and *Equus* was at first believed to be a phylogenetic sequence, but was later found to be a series of side branches, a *Stufenreihe*. The

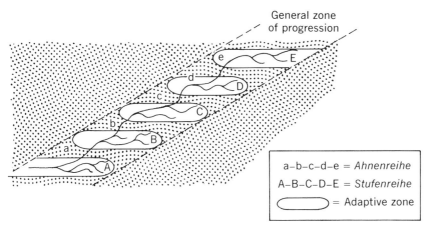

FIGURE 30.4
Ahnenreihe and *Stufenreihe* in the occupation of a series of adaptive zones. (Redrawn from G. G. Simpson, *The Major Features of Evolution*, copyright 1953, Columbia University Press, New York; by permission.)

true phylogenetic lineage, the *Ahnenreihe*, is found in North America, which was the main theater of horse evolution. The series *Hyracotherium, Miohippus, Merychippus, Pliohippus,* and *Equus* in North America constitutes the *Ahnenreihe* in this group (Simpson, 1953, pp. 364–366).

It is evident that the distinction between *Ahnenreihen* and *Stufenreihen* is similar to but different from the distinction between phyletic trends and speciational trends. *Ahnenreihe* is synonymous with either phyletic or speciational trend, depending on the type of trend in the case at hand. But *Stufenreihe* is not synonymous with speciational trend. The populations in a speciational trend occur in a direct line of descent (as in line P–Q in Figure 30.3). No such direct phylogenetic connection exists between the stages in a *Stufenreihe* (e.g., A, B, . . . , E in Figure 30.4).

The *Ahnenreihe-Stufenreihe* distinction of Abel (1929) and Simpson (1953) foreshadowed and paved the way for the distinction between phyletic and speciational trends. In one discussion Simpson (1953, pp. 382–383) comes close to the concept of speciational trends. The first explicit recognition of speciational trends as a special type of evolutionary trend was made by Grant (1963, pp. 566–568, 570; also 1971, pp. 42–43). The term speciational trend was not used in my earlier discussions, in order to economize on terminology, but now seems needed, and is accordingly introduced in the present account.

Views on phylogeny similar to the above have been stated recently by Eldredge and Gould (1972) and Stanley (1975).

Eldredge and Gould (1972) distinguish between "phyletic gradualism" and "punctuated equilibria." These terms apply to approximately the same modes as do the terms phyletic trends and speciational trends, respectively. Eldredge and Gould (1972) do not refer in their paper to relevant earlier work along the same lines by Simpson, Grant, and others. Eldredge and Gould (1972) propose to reinterpret all or nearly all paleontological series as "punctuated equilibria," and they would reduce the role of "phyletic gradualism" to a corresponding degree. This is going too far, in my opinion; as I see it, both speciational trends and phyletic trends have a place in nature.

Two probable examples of speciational trends in plants will be mentioned. In both cases the trend is inferred from the comparative biology of recent forms. This approach has its weaknesses but also some strengths, since the species can be studied in a living condition and their interrelationships determined.

A probable speciational trend in *Clarkia* (Onagraceae) in Pacific North America runs from relatively mesic to xeric forms. The ancestral mesic form was a species close to the present *Clarkia amoena* or *C. rubicunda*. The intermediate stage in the series is represented by the existing *C. biloba*, and the terminal xeric stage by *C. lingulata*. The transition from *C. biloba* to *C. lingulata* was brought about by quantum speciation (Lewis and Lewis, 1955; Lewis and Roberts, 1956).

The probable trend in *Polemonium* (Polemoniaceae) in western North America runs from mild coastal temperate forms to alpine types. The stages in this trend are represented by the following species:

Polemonium carneum	Moist coastal lowlands
P. caeruleum	Coniferous forest zone
P. californicum-delicatum group	Subalpine forest zone
P. pulcherrimum	Alpine zone
P. eximium group	High peaks in alpine zone

This series of species occurs in the zones indicated, on transects from coast to high mountains in western North America. The series does not manifest itself as a continuum on such transects, as is theoretically conceivable, but as a stepwise succession of distinct species, suggesting a speciational trend (Grant, 1959, 1963).

Rapid Evolution in Speciational Trends

The rate of evolution in speciational trends is probably correlated with the mode of speciation involved. Successions of geographical speciations in continuously large populations would be expected to be capable of moderate or normal rates, but not tachytelic rates. Successive quantum speciations, on the other hand, could theoretically generate rapid rates in an evolutionary trend. This latter type of speciational trend is especially interesting.

We have previously seen that single events of quantum speciation are capable of producing new forms adapted to new or different ecological conditions at a rapid rate. Quantum speciation has this capability because: (1) it utilizes the selection-drift combination of forces (see above and Chapter 13); and (2) it can evade the cost-of-selection restriction on rapid multiple-gene substitution (see Chapter 14).

Now if the special advantages of quantum speciation can be exploited once, at one time level, they can be exploited again and again. Quantum speciational shifts can occur repeatedly in cycles. And these shifts can have an overall trend, like the line P–Q in Figure 30.3. A quantum speciational trend of this sort is a means for *sustained* rapid evolution in a given direction. It is perhaps the only means for bringing about an evolutionary trend at a tachytelic rate.

Let us return to the metaphor of the adaptive landscape. Imagine an unoccupied high adaptive peak and a population camped at its base. There are different possible pathways from base to peak. Some pathways ascend gradually and are consequently long and slow. These routes symbolize phyletic trends and geographical speciational trends. They are adequate for the ascent if sufficient time is available.

Other pathways are steep but short and potentially quick. Such routes are analogous to a quantum speciational trend. If the population at the base of the mountain must climb the peak in a short time, i.e., if the time available for the ascent is limited, the population will have to take the short steep route.

Concluding Statement

This chapter reviews a series of theoretical concepts that have developed progressively, step by step, over the past four or five decades. The concepts in question have explanatory value in relation to important but recondite macroevolutionary phenomena. These concepts are consistent with and supported by the factual evidence to a moderately satisfactory extent. However, the factual support cannot be said to be overwhelming; much more evidence needs to be gained on the macroevolutionary processes discussed in this chapter.

Collateral Readings

Grant, V. 1963. *The Origin of Adaptations.* Columbia University Press, New York. Chapters 18, 19.

Mayr, E. 1954. Change of genetic environment and evolution. In *Evolution as a Process,* ed. by J. Huxley, A. C. Hardy, and E. B. Ford. George Allen & Unwin, London.

Mayr, E. 1963. *Animal Species and Evolution.* Harvard University Press, Cambridge, Mass. Chapter 17.

Simpson, G. G. 1944. *Tempo and Mode in Evolution.* Columbia University Press, New York. Chapter 7.

Simpson, G. G. 1953. *The Major Features of Evolution.* Columbia University Press, New York. Chapters 7, 12.

Wright, S. 1931. Evolution in Mendelian populations. *Genetics* **16:** 97–159. Reprinted in *Systems of Mating,* by S. Wright. Iowa State University Press, Ames, 1958.

Wright, S. 1960. Physiological genetics, ecology of populations, and natural selection. In *Evolution After Darwin,* Vol. 1, ed. by S. Tax. University of Chicago Press, Chicago.

31

Evolution of Major Groups

Introduction

One of the big problems in macroevolution concerns the evolution of major groups. What circumstances surround the origin of tribes, families, orders, classes? What evolutionary factors are involved in the origin and development of groups of medium or high taxonomic rank?

The general problem contains a number of subordinate problems, such as the following. Are special novel variations necessary for the origin of major groups? Do major groups make their first appearance by saltation or by gradual divergence from the ancestral stock? Are special modes of evolution involved?

It is difficult to obtain direct evidence concerning these problems. And in the absence of compelling evidence, a variety of speculations have been able to flourish. Nevertheless, some indirect evidence is available, which can serve as a guide to thinking.

It is useful to recognize two main phases in the evolution of a major group: an early formative phase, and a subsequent period of expansion and proliferation. This is useful, since different evolutionary processes are probably involved in the different phases.

Evolutionary Characteristics of Major Groups

In order to discuss the origin of major groups it is necessary to define the essential features of a major group. One could define a major group as a phylogenetically natural group of substantial taxonomic rank. This, however, focuses attention on the wrong end of the problem for our present purposes. Assignment of taxonomic rank is a retrospective decision reached after a group has evolved up to the present time, the era of human taxonomists. We are interested in the beginning stages as well as the eventual flowering of the group.

One could also define a major group in terms of its key character(s). This approach is also unfruitful, and for the same reason. A major group may or may not possess a key character when it is originating.

A more useful criterion of a major group is a state of adaptedness to a particular broad zone on the adaptive landscape. The *origin* of the group coincides with the occupation of a new adaptive zone (Wright, 1949; Simpson, 1953, p. 349). The group *becomes* a major group later by development and proliferation within this zone.

The successful initial occupation of the adaptive zone can come about in two main ways.

(1) A colonizer species may move into a more or less uninhabited territory. The new territory presents no particularly novel environmental conditions other than relative emptiness at the time of colonization. Nor does the original colonizer species necessarily differ much from its ancestral stock. But in the new zone it finds an ecological opportunity for expansion and diversification.

(2) Occasionally a species may acquire a character that opens up a new way of life and preadapts it for a previously unexploited habitat. Thus, in the evolution of vertebrates, the development of lungs and legs opened up the land for colonization, and the development of wings opened up the air. The preadapted colonizer species, in possession of its new pivotal character, invades a new habitat and undergoes further evolution there (Huxley, 1943; Wright, 1949).

In some cases the new character of pivotal importance had evolved earlier in relation to ancestral environmental conditions, but turns out to have potential value in a new type of habitat. Thus the water-breathing lungs and paired fins of some fishes were the forerunners of the air-breathing lungs and crawling legs of primitive land vertebrates.

In both modes 1 and 2, the event of colonization of the new adaptive zone is usually followed by an adaptive radiation in that zone. The original colonizer species usually gives rise to a diverse array of descendant lines occupying different subzones. It is this adaptive radiation that makes the group a recognizable major group. But we must bear in mind that the adaptive radiation represents the successful exploitation, at a later stage, of the ecological opportunity seized by the original colonizer species.

Mode of origin 1, above, is simpler than mode 2; conversely, mode 2 probably accounts for more important evolutionary developments than does mode 1. Mode 1 is a source of endemic groups of medium taxonomic rank. Good examples are the Galápagos finches (Fringillidae, subfamily Geospizinae) on the Galápagos archipelago, and the Hawaiian honeycreepers (family Drepanididae) on the Hawaiian archipelago. Whether higher categories of continental or cosmopolitan distribution originate by mode 1 is open to question. Mode 2 seems to be the main source of truly major and widespread groups, such as the amphibians and the birds.

Saltation vs. Gradualism

Some students of evolution (e.g., Goldschmidt, 1940) have proposed that new groups arise by saltations, that is, by drastic and sudden or nearly sudden mutational changes in pre-existing organisms. Most evolutionists from Darwin on, however, have opposed strict saltatory explanations and have favored some version of gradualism instead.

Saltation in its extreme forms seems to be about as likely to succeed as spontaneous creation. All the evidence of genetics points to the conclusion that the genotype is composed of numerous genes that must interact in a harmonious fashion if the organism is to have normal viability and fertility. Drastic mutations tend to upset the internal genic balance in ways that are selectively disadvantageous. A genotype that has been built up by selection in stages cannot be remodeled successfully overnight; it has to be remodeled in stages.

This consideration does not rule out a reorganization of the genotype in a relatively few generations. In fact there is experimental evidence for rapid reorganization of the genotype in plants (see Grant, 1971, Chs. 13, 14). Such rapid changes have been referred to as saltations by some authors. Toned-down versions of saltation such as these are realistic possibilities in evolution.

Extreme versions of gradualism have other drawbacks as explanations of major evolutionary developments. Gradualism in its extreme form is too slow. Strict gradualism is inconsistent with some paleontological evidence (see Chapter 28). The period of origin of a major group usually represents a relatively small part of that group's total geological time span. This point will be illustrated again later in this chapter.

On a spectrum ranging from saltation to gradualism, the two extremes have to be ruled out, but this leaves a broad middle zone in which it is biologically realistic to look for modes of evolution of major groups. Exceptionally rapid rates are probably called for in the origin of most new groups. But such rapid rates are not inconsistent with gradual stepwise changes. Saltatory elements are probably combined with gradualistic elements in the formation of many new groups. Modern views of quantum evolution and quantum speciational trends (as outlined in Chapter 30) do contain both elements and thus take an intermediate position.

Formation of the Orders of Mammals

An intermediate position, involving gradual change at exceptionally rapid rates, is illustrated by the histories of various orders of mammals.

Recent fissiped carnivores and Recent ungulates (perissodactyls and artiodactyls) represent two very well differentiated major groups. These groups, in the Recent fauna, are differentiated in a series of characters. One thinks of the generally carnivorous diet, clawed feet, and large canine and carnassial teeth of carnivores, in contrast to the herbivorous diet, hoofed feet, and grinding molar teeth of ungulates. These and other character differences are listed in Table 31.1.

Simpson (1953, pp. 342–346) has shown that the ancestral members of the same phyletic groups in the Early Paleocene were not well differentiated but in fact were very similar. The ancestral carnivores (creodonts) and ancestral ungulates (condylarths) shared many characters in common. Thus the creodonts and condylarths had the same mode of walking, the same number of digits (five), the same dental formula, and probably about the same omnivorous diet.

As Simpson (1953, p. 345) notes, the carnivores and ungulates were not different orders or even different families in the Early Paleocene. In fact, the Early Paleocene carnivores and ungulates combined were less varied than some single Recent carnivore families, such as the Mustelidae or Viverridae. One Paleocene genus, *Protogonodon* (Arctocyonidae), contained some species that on balance would be classified as carnivores, and other species that would be classified as ungulates. Later authors differ in the classification of the Early Tertiary family Arctocyonidae; some (e.g., Colbert, 1955) retain it in the Creodonta, while others (e.g., Romer, 1966) place it in the Condylarthra.

The carnivore character combination and the carnivore order thus developed in the historical course of divergence from its Paleocene ancestors; likewise, the ungulate character combination and the two ungulate orders (Perissodactyla and Artiodactyla) developed by divergence from their Paleocene ancestors. It was the divergence that made the major groups (Simpson, 1953).

TABLE 31.1
Character differences between Recent carnivores and ungulates.
(Rearranged from Simpson, 1953.)

Character	Carnivores (Fissipedia)	Ungulates (Perissodactyla and Artiodactyla)
Habitat	Terrestrial, arboreal, or semi-aquatic	Terrestrial
Mode of walking	Digitigrade or plantigrade	Usually unguligrade, sometimes digitigrade
Number of digits	4–5	1–4, rarely 5
Toenails	Well-developed claws, usually hooded	Well-developed hoofs (except hyraxes)
Diet	Generally carnivorous, sometimes omnivorous, rarely herbivorous or insectivorous	Generally herbivorous, sometimes omnivorous
Dental formula	$\dfrac{3.1.4-1.3.1}{3.1.4-1.3.1}$ $\left(\text{Molars exceptionally } \frac{4}{5}\right)$	$3-0.1-0.4-2.3$ $3-0.1-0.4-2.3$
Canine teeth	Large	Usually small or absent, sometimes large
Premolars	Usually simple, not molariform	Generally complex, often molariform
Molars	Usually simple, without grinding pattern	Complex, with grinding pattern
Carnassial teeth	P^4 and M_1 teeth enlarged and carnassial	No carnassials
Crown height	Brachydont	Brachydont to hypsodont

The Perissodactyla and Artiodactyla are treated as one monophyletic group for this comparison. There is a difference of opinion among paleontologists concerning the phylogenetic connections of these two orders. But note that even if the Artiodactyla do not belong in the same natural group with the Perissodactyla, it would still be valid to compare the Perissodactyla with the carnivores. The characters listed in the right-hand column of Table 31.1 would be modified relatively little if the Artiodactyla were removed, and the general conclusions would be unchanged.

Some degree of gradualism is indicated by the carnivore-ungulate case. Other evidence indicates that the evolutionary rates were relatively fast during the early history of these and other mammal orders, when the orders were developing their distinctive features. The evidence, though indirect, is suggestive of a general pattern; it is summarized in Table 31.2 (from Simpson, 1953).

The estimated maximum length of the unrecorded period of origin of an order is given in column 4 of the table. It will be seen that this period usually

TABLE 31.2
Duration of the orders of mammals, in millions of years (MY);
the figures in columns 2 and 4 are estimated maxima.
(Rearranged from Simpson, 1953.)

Order	Total duration, MY	Length of recorded sequence, MY	Length of unknown period of origin, MY	Unknown period as % of total
Extant orders				
Marsupialia	130	80	50	38.5
Insectivora	130	80	50	38.5
Edentata	75	60	15	20.0
Lagomorpha	75	60	15	20.0
Rodentia	75	60	15	20.0
Cetacea	75	45	30	40.0
Carnivora	80	75	5	6.2
Proboscidea	65	40	25	38.5
Sirenia	70	50	20	28.6
Perissodactyla	65	60	5	7.7
Artiodactyla	65	60	5	7.7
Extinct orders				
Taeniodonta	40	35	5	12.5
Condylarthra	40 (55?)	35 (50?)	5	12.5 (9.1?)
Liptoterna	70	60	10	14.3
Notoungulata	75	60	15	20.0
Pantodonta	45	35	10	22.2
Mean	*73.4*	*55.9*	*17.5*	*23.8*

represents a small proportion of the total estimated duration of the order. This proportion is 6.2% for the carnivores and 7.7% for the Perissodactyla and Artiodactyla, and has an average value of 23.8% for all 16 orders combined. These figures probably overestimate the length of the period of origin of the orders.

The data in Table 31.2 are consistent with the hypothesis of rapid evolution and quantum evolution during the origin of the orders of mammals (see also Chapters 28 and 30).

The Role of Quantum Evolution

The process of quantum evolution was described in Chapter 30 and does not require repetition here. We do, however, wish to recall Simpson's (1944, 1953) conclusion, based on the convergence of various lines of evidence, that quantum evolution is the normal mode of evolution involved in the origin of major groups.

We can extend the conclusion. It is probable that the quantum evolutionary change often or usually takes the form of speciation, and thus becomes an event of quantum speciation. Furthermore, a number of quantum speciational events may be linked together in a time series to form a speciational trend. Such a quantum speciational trend may well be involved in the formative period of evolution of a major group (see Chapter 30).

A further deduction can be drawn as regards the type of terrain in which a major group is most likely to originate. Quantum evolution and quantum speciation require island-like population systems. Such population systems are not found commonly in plains country. They do occur commonly in country with broken terrain, such as mountains and archipelagos in the case of terrestrial plants and animals.

Plant evolution does seem to be rapid and diversified in mountainous country in the modern world. On the basis of this and other evidence, various plant evolutionists have concluded that mountains are probable sites of major evolutionary developments in the plant kingdom (e.g., Kerner, 1894–1895; Axelrod, 1970, 1974; Stebbins, 1974).

Adaptive Radiation

When a species succeeds in establishing itself in a new territory or new habitat, it gains an ecological opportunity for expansion and diversification. The original species may respond to this opportunity by giving rise to an array of daughter species adapted to different major niches within the territory or habitat. These daughter species become the ancestors of a series of branch lines when they, in turn, produce new daughter species. The group enters its second phase of development, the phase of proliferation.

Adaptive radiation is the pattern of evolution in this phase of proliferation. And speciation is the dominant mode of evolution in adaptive radiation. The first generation of speciational events gives rise to the primary branches in the now growing phylogenetic tree. These primary branches correspond to different primary ecological niches. Second, third, and later generations of speciation in each primary branch then parcel out the available ecological niches in the territory or habitat into a larger number of more highly specialized species.

The new territory or habitat fills up with species. And as it does, the group becomes a major group. And the major group comes to have a taxonomic structure organized along adaptive lines.

At one stage of development the group has the taxonomic rank of a family.

Its primary subdivisions are then tribes and genera with different primary adaptive modes; each generic line consists of species groups, representing secondary subdivisions of the adaptive zone; and each species group consists of species representing tertiary subdivisions of the adaptive zone. At a more advanced stage of development the group may be an order. Now the primary divisions, still along adaptive lines, are families, and the second-, third-, and fourth-order subdivisions appear as genera, species groups, and species, respectively.

The concept of adaptive radiation was formulated by Osborn (1910) in relation to mammal evolution. Osborn recognized five main lines of specialization in the limbs and feet of mammals, fitting their possessors for diverse habitats and modes of locomotion, as follows. (1) Swift running in terrestrial habitats; (2) digging habits for underground life; (3) swimming in amphibious and aquatic mammals; (4) climbing in arboreal mammals; and (5) parachuting and flying in aerial groups. Other lines of specialization were recognized in the teeth and diet: omnivorous, carnivorous, herbivorous, etc. The orders of mammals, according to Osborn (1910), represent different combinations of these sets of characteristics and hence different adaptive modes.

Adaptive Radiation in the Hawaiian Honeycreepers

An excellent modern example of adaptive radiation is furnished by the bird family Drepanididae in the Hawaiian Islands. This family has been studied by a series of workers over the years. The monographs of Amadon (1950) and Baldwin (1953) are the basic works; supplementary information and interpretations are given by Bock (1970) and Carlquist (1974).

The family consists of nine genera grouped into two subfamilies, as shown in Table 31.3. The family is monophyletic and is endemic to the Hawaiian Islands. It should be noted that, unfortunately, a number of the species have become extinct or very rare in modern times. The Drepanididae belong to the suborder Passeres, or songbirds, of the order Passeriformes; within the Passeres they are related to the true honeycreepers (Coerebidae) and/or the tanagers (Thraupidae) (Amadon, 1950; Baldwin, 1953).

Since the family Drepanididae is endemic to a geologically young oceanic island chain, and has affinities to mainland passerine groups, its phylogenetic roots must lie in the latter, but the identity of the ultimate mainland ancestor remains uncertain. The ultimate ancestor, whatever it was, colonized the Hawaiian Islands and developed into the immediate ancestor of the family. This immediate ancestor is inferred to have been a short-billed bird with a diet consisting of insects and nectar (Baldwin, 1953). This diet, bill type, and the

TABLE 31.3
System of classification of the Hawaiian
honeycreepers (family Drepanididae).
(Amadon, 1950.)

Classification	Number of species
Subfamily Psittirostrinae (yellow or green plumage)	
Loxops	5
Hemignathus	4
Pseudonestor	1
Psittirostra	6
Subfamily Drepaniinae (black and red plumage)	
Himatione	1
Palmeria	1
Ciridops	1
Vestiaria	1
Drepanis	2

corresponding type of tongue are found in the more primitive living members
of both subfamilies.

A great diversity of feeding habits and bill types has evolved in the Hawaiian
drepanids. This diversity is much greater than is normal for a passerine bird
family. There are warbler-like, creeper-like, finch-like, fruit-eating, flower-
feeding, and other types within the limits of a single family.

The probable phylogeny of the family down to the genus level is shown in
Figure 31.1 (from Amadon, 1950; a different phylogeny is suggested by Bock,
1970). The two subfamilies, Psittirostrinae and Drepaniinae, represent the
primary branches in the family tree, and the genera, the secondary branches.
Third-order branchings produced the species in the non-monotypic genera
such as *Loxops* and *Hemignathus*.

The branchings constitute an adaptive radiation. The adaptive radiation in
the Hawaiian honeycreepers is related mainly to feeding habits and is manifested
in bill and tongue characters as well as in other features. The main adaptive
zones that have been occupied are insect-eating, nectar-eating, fruit-eating,
and seed-eating. Some of these zones were further subdivided. Thus, among
insect-eating drepanids there are short-billed foliage gleaners and long-billed
bark probers (Baldwin, 1953).

The subfamily Drepaniinae consists mainly of flower-feeding nectar eaters.
Some are short-billed nectar eaters, e.g., *Himatione* (Figure 31.2A); others are

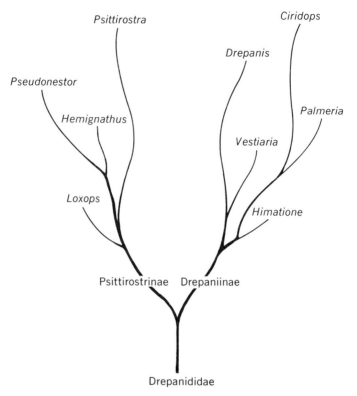

FIGURE 31.1
Suggested phylogeny of the Hawaiian honeycreepers (Drepanididae).
(Amadon, 1950.)

long-billed nectar eaters, e.g., *Vestiaria* and *Drepanis* (Figure 31.2B). *Ciridops* in this subfamily is a fruit eater (Figure 31.2C). Some of the bill types in the flower-feeding drepanids are correlated with various flower forms in the Hawaiian lobelioids (Campanulaceae, subfamily Lobelioideae).

The subfamily Psittirostrinae consists of insect, fruit, and seed eaters. The genus *Loxops* contains five species of insectivorous birds with small or medium-sized bills. The small-billed species glean insects from leaves (e.g., *L. coccinea*, Figure 31.2D), while the species with medium bills probe for insects in bark crevices. *Hemignathus lucidus* (Figure 31.2E) is a sickle-billed insect eater that obtains insects from wood more or less in the fashion of a creeper. Other sickle-billed species of *Hemignathus* have been reported to feed on flowers.

Pseudonestor in the same subfamily Psittirostrinae has a parrot-like bill that

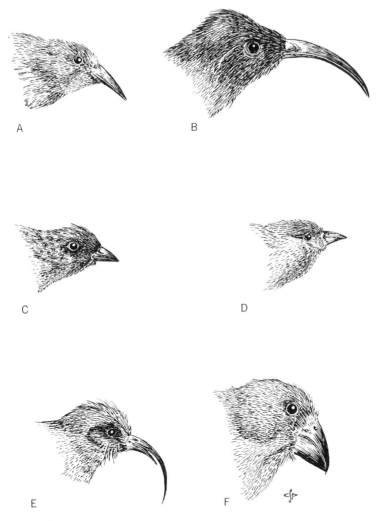

FIGURE 31.2
Types of bills in the Hawaiian honeycreepers. (A) *Himatione sanguinea* (nectar feeder).
(B) *Drepanis pacifica* (nectar feeder). (C) *Ciridops anna* (fruit eater). (D) *Loxops coccinea*
(insect eater). (E) *Hemignathus lucidus* (insect eater). (F) *Psittirostra palmeri* (fruit and
seed eater). (Drawn from photographs in Amadon, 1950; drawings by Charles Papp.)

is used for crushing twigs and prying loose bark to obtain wood-inhabiting insects and grubs. The related genus *Psittirostra* (Figure 31.2F) also possesses powerful bills, used in this case for crushing fruits and cracking seeds.

A parallel example of adaptive radiation in another insular bird group is that of the Galápagos finches (Fringillidae, subfamily Geospizinae) in the Galápagos Islands (see monograph of Lack, 1947). An example of a different sort in a plant group, in which the adaptive radiation is related to mode of pollination and is manifested in a diversity of flower types, is found in the family Polemoniaceae (see monograph of Grant and Grant, 1965).

The pattern of adaptive radiation is efficient for the group concerned in the sense that it represents a maximum exploitation, by that group, of the ecological opportunities presented by a new territory or habitat.

32

Extinction

Introduction

The species living in one geological period are the lineal descendants of only a small fraction of the species of an earlier period. Extinction is the fate of the vast majority of species. Extinction is likewise the fate of most phyletic groups of low or medium taxonomic rank, such as genera, families, and orders, although supraspecific groups generally have a longer duration than species. Furthermore, there are periods in earth history when extinction occurs on a wide scale and affects many species belonging to a number of different groups.

It is useful to distinguish between three levels of extinction: extinction of particular species (e.g., the Irish elk); extinction of supraspecific groups (e.g., the ammonites in the Cretaceous); and widespread or mass extinction (as in the Late Cretaceous).

A further distinction can be made with regard to extinction in the case of supraspecific groups. Some such groups become completely extinct, leaving no phylogenetic descendants, while other groups become extinct themselves, but give rise to descendants that are very different and form a separate supraspecific group.

In this chapter we will examine some examples of extinction at each of the levels mentioned above. We will see that the precise causes of extinction of species and of groups are often unknown or very uncertain. Episodes of wide-spread extinction, however, can often be correlated with geological revolutions, suggesting general causes of extinction.

The Irish Elk in Ireland

One of the famous cases of species extinction is that of the Irish elk, *Megaloceros giganteus,* a giant deer of Pleistocene and early Recent time. The species ranged from Ireland across Europe to Siberia and China and south to northern Africa. It flourished in Ireland during a phase following the retreat of ice at the end of the last glaciation. It became extinct in Ireland about 11,000 years ago. It may have survived into early Recent time in parts of continental Eurasia (Gould, 1974).

Most of the specimens of the Irish elk have been obtained from lake deposits and peat beds in Ireland. The specimens are those of a gigantic deer, standing about 2 meters tall at the shoulders. The stags bore huge antlers weighing up to 50–60 pounds (ca. 25 kg) and spanning a width of about 3 meters (Figure 32.1). These are maximum dimensions. Detailed measurements are given by Gould (1974). The antlers are the biggest known in the deer family. Various authors have noted that the large and heavy antlers of the Irish elk, which were shed and regrown each year, must have been both a physical burden and a physiological drain on the animals.

The older explanations of the extinction of the Irish elk focused attention on the probable disadvantages of the huge antlers. Depending on their philosophical preferences, some older authors couched their explanations of the extinction in terms of orthogenesis and others in terms of natural selection. In either case the antlers were assumed to be non-adaptive. The antlers of the Irish elk may well have been a contributing factor in the animal's extinction, but, as Simpson (1949, 1967) pointed out, there is no reason to believe that the antlers were inadaptive when they developed and while the species was flourishing.

Simpson (1949, 1967) pointed out further that antler size in the Irish elk should not be considered as an isolated character in itself, but rather as a product of relative growth rates in different parts of the body during phylogeny. In the members of the deer family, the phylogenetic growth rate of the antlers is much greater than that of the body, so that the largest species of deer have the *proportionately* largest antlers. The large-bodied Irish elk has antlers of about the expected size for a very large deer.

It may be, therefore, that selection promoted large body size in the Irish

FIGURE 32.1
Antlers of the Irish elk (*Megaloceros giganteus*). (A) Display of palm when stag
is looking straight ahead. (B) Exposure of tines when head is bowed. (Redrawn
from Gould, 1974, after Millais and Cuvier.)

elk, and large antler size was a side effect (Simpson, 1949, 1967). Increase in
body size is a common trend in mammals, and possesses various selective advantages, as discussed in Chapter 26. An alternative possibility is that sexual
selection favored increased antler size in the Irish elk because of the advantage of large-antlered males in mating, and large body size was the side effect
(Gould, 1974).

The causes of extinction of the Irish elk, in Ireland at least, and perhaps elsewhere, can be sought in other aspects of its life. The habitat of the Irish elk was apparently open grassy country with scattered woods. This type of vegetation flourished during one phase in the warming trend at the end of the last ice age, and the Irish elk flourished with it. In a later phase of the same climatic trend, however, the grassland was replaced by forest, and the Irish elk, which was not a forest deer, probably could not survive in dense forest. Climatic and vegetational changes in late Glacial time, unfavorable to the Irish elk, are the probable cause of its extinction (Gould, 1973, 1974).

The Horse in North America

The horse, *Equus,* became extinct in North America at the beginning of the Recent epoch, about 8,000 to 10,000 years ago. North America, as will be recalled from Chapter 26, had been the main theater of evolution of *Equus,* and the source of migrant stocks of *Equus* on three other continents. Vast herds of horses, belonging probably to several species of *Equus,* had existed in North America during the Pleistocene. Then *Equus* became completely extinct in North America (Simpson, 1951, and personal communication).

Millennia later, in 1519, the European domestic horse, *Equus caballus,* was introduced in North America by the Spaniards. The introduced horse was very successful and established wild herds in the ancient homeland of its forebears.

The reasons for the extinction of *Equus* in North America are unknown. And the mystery is compounded by the fact that *Equus* was able to reestablish itself successfully in the wild in North America several millennia after its extinction here.

The various suggested explanations of the extinction are all more or less unconvincing. Was the native North American horse a victim of the Ice Age? *Equus* had survived through the Pleistocene, and, moreover, it occurred in areas beyond the limits of the glaciation. Was the horse hunted to extermination by the early American Indians, with whom it overlapped in time? This is a possibility. But the Indians hunted the bison and did not exterminate it. Did the horse fall prey to a disease epidemic? This is possible too. But there is no evidence of a disease epidemic in the fossil record (Simpson, 1951).

A parallel series of events took place in South America. There also, *Equus* flourished during the Pleistocene, became extinct in the Early Recent epoch, but was introduced from Europe in historic times and subsequently ran wild. The reasons for the extinction in South America are also unknown (Simpson, 1951).

The Sabertooth Cats

An often-cited case of extinction of a supraspecific group is that of the saber-tooth cats. The sabertooths were one of the main branches of the Felidae, consisting of such genera as *Hoplophoneus, Machairodus,* and *Smilodon.* The group differed from the modern feline cats in the great size of the upper canine teeth and in other related features of the jaws and teeth (see Figure 32.2).

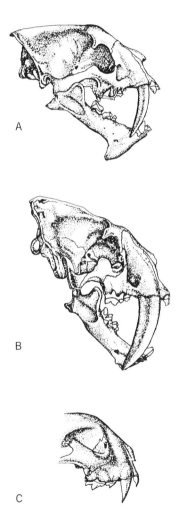

FIGURE 32.2
Skulls of sabertooth cats. Not shown to same scale. (A) *Hoplophoneus,* Oligocene, skull 6.25 inches long. (B) *Smilodon,* Pleistocene, skull 12 inches long. (C) *Pseudaelurus,* Pliocene, a feline cat, showing canine teeth for comparison. (Redrawn with modifications from Romer, 1966, after Matthew and Zdansky.)

Sabertooth cats occurred in North America and Eurasia throughout most of the Tertiary from the Early Oligocene on, and reached South America in the Pleistocene. *Hoplophoneus* was a characteristic Oligocene member of the group in North America, and *Smilodon,* a larger Pleistocene member in North and South America (Figure 32.2A, B). The group became extinct in the Pleistocene (Romer, 1966).

It was commonplace at one time to attribute the increase in canine size and body size of sabertooth cats through the Tertiary to orthogenesis. The extinction was then explained as a result of the orthogenetic trend's overstepping the bounds of functional usefulness. The large canine teeth in *Smilodon* supposedly interfered with biting, whereupon the phyletic group became extinct. If so, as Simpson (1953, p. 269) notes, the animals took 40 million years to starve to death.

As a matter of fact, the large canine teeth did not interfere with biting. The jaw was capable of opening wide and closing tight.

The sabertooth cats were probably specialized for preying on large, slow-moving, thick-hided herbivores, such as mastodons and giant sloths. Their large canine teeth, with the jaw in an open position, would have served as effective stabbing and tearing organs, backed up by a strong neck and shoulders.

During the Pleistocene, mass extinction thinned the ranks of the category of thick-hided herbivores. The reduction in numbers and eventual extinction of this type of prey animal shrank or closed the adaptive zone of the sabertooth cats. Meanwhile the alternative method of ambush-and-pursuit hunting of fast-moving prey animals had been perfected by the feline branch of the cat family. With their own adaptive zone closing up in the Pleistocene, and alternative adaptive zones preempted, the sabertooth cats had come to the end of their string (Simpson, 1953, p. 222).

The Dinosaurs

One of the most extensive group extinctions of all time was that of the dinosaurs in the Late Cretaceous. The dinosaurs, constituting the two reptile orders Saurischia and Ornithischia, had been the dominant land animals through the Mesozoic, and the Mesozoic has been called the Age of Dinosaurs.

A great diversity of forms had developed during the course of the Mesozoic, as indicated by the classification of the dinosaurs into two orders, 25 families, and over 200 genera (see Colbert, 1961, pp. 265–274; Colbert lists 218 genera). These included such giants as *Brontosaurus* and *Tyrannosaurus,* as well as various medium-sized and small forms. The group as a whole was worldwide in distribution and had been dominant on the land for over 100 million years. Then in the Late Cretaceous, in the space of several million years, the entire group became extinct.

The causes of this extinction are not known. This is not for lack of attempts to find an explanation. In fact the literature abounds in simplistic and unsubstantiated speculations concerning dinosaur extinction. The dinosaurs became too cold in the climate of Late Cretaceous time. They became too hot. The primitive mammals ate their eggs. The herbivorous dinosaurs destroyed the Mesozoic vegetation by overgrazing. The dinosaurs were killed off by bursts of radiation from an exploding nova in the heavens. A reversal of the earth's magnetic field permitted lethal doses of cosmic rays to reach the land surface.

The last two suggestions can probably be ruled out. Unique astronomical or geophysical events that yield bursts of lethal radiation are essentially one-shot or two-shot factors, and would be expected to produce immediate mass killings, whereas the dinosaur extinctions were spread out over thousands and millions of years (Simpson, 1968). Gradual climatic changes in the Late Cretaceous correlate better with the actual course of dinosaur extinction. Adverse climatic changes have been the preferred factor in the minds of many students, and are implicit in the first two suggestions mentioned above, although these are probably too simplistic as they stand.

Axelrod and Bailey (1968) place the emphasis on the change from an equable to an inequable climate in the latter part of the Cretaceous. Equable warm climates had prevailed during most of the Mesozoic, and the dinosaurs had been evolving for about 130 million years under such climatic conditions. The latter part of the Cretaceous saw a change to inequable continental climates with greater extremes of temperature. Other phyletic groups of reptiles in the Cretaceous, such as lizards, snakes, and turtles, could escape very cold or hot periods by burrowing, hibernating, or submerging, and have survived to the present. But the terrestrial dinosaurs did not have these means of escape from temperature extremes and became extinct (Axelrod and Bailey, 1968; see also Spotila et al., 1973).

Body thermal stress in a continental climate could explain much, perhaps most, but not quite all of dinosaur extinction. Among the reptile groups that survived through the Cretaceous and to the Recent are the tropical and sub-tropical crocodilians and the New Zealand rhynchocephalian *Sphenodon*. Why did the loss of equable climates in the Late Cretaceous spare these two orders but not the two orders of dinosaurs?

A further complication arises. The various climatic hypotheses of dinosaur extinction are predicated on the assumption that the dinosaurs, being reptiles, were cold-blooded. However, some indirect paleontological evidence, such

as that of bone histology, has recently been brought forward to indicate that
the dinosaurs were not cold-blooded, as supposed, but rather warm-blooded
or endothermic (Bakker, 1975). Whether this new conclusion will stand up,
and if so, whether it can be reconciled with the inequable-climate hypothesis
of dinosaur extinction, remains to be seen. Meanwhile the mystery is still
unsolved.

Widespread Extinction in the Late Cretaceous

Dinosaur extinction is part of a more extensive pattern of mass extinction
in the Late Cretaceous. Other reptile orders that became extinct at this time
were the marine ichthyosaurs, plesiosaurs, and mosasaurs, and the flying ptero-
saurs. Among marine invertebrates, the previously successful mollusc group
of ammonites died out in the Late Cretaceous.

The disappearance of several marine groups in the Late Cretaceous is difficult
to explain on radiation theories of Cretaceous extinction, since marine animals
would presumably be shielded from the postulated radiations. These same facts
are not inconsistent with climatic theories of Cretaceous extinction, as applied
to land reptiles, but they do nothing to strengthen such theories either, for
marine reptiles would have been more or less insulated from the new inequable
climates, but died out anyway.

We are witnessing a very widespread turnover in the fauna of the land, sea,
and air. This suggests the operation not of single specific ecological factors,
but of general and pervasive ones with numerous branching and interacting
effects.

Pleistocene Extinctions in Mammals

Another episode of widespread extinction occurred in the Pleistocene. Mammals had been the dominant land animals throughout the Tertiary, and the
Tertiary is often referred to as the Age of Mammals. But this age came to an
end, or to the beginning of the end, in the Late Pleistocene, with widespread
extinction in the mammalian fauna, and the process has continued in the Recent.

The mammalian extinctions involved groups of various taxonomic rank,
from single species (e.g., the Irish elk) to supraspecific groups (e.g., sabertooth
cats, a subfamily). Some species that became extinct are the Irish elk, cave bear,
cave lion, and giant beaver. Extinctions of supraspecific groups include the
sabertooth cats, ground sloths, giant armadillos, glyptodons, mastodons, mam-
moths, woolly rhinoceros, and North American horses.

The extinctions fell most heavily on species and supraspecific groups of large mammals. Small mammals survived through the Late Pleistocene and Early Recent to a greater extent than did large mammals.

There are two main current hypotheses as to the cause of Pleistocene extinction of large mammals. One of these attributes the extinctions to big-game hunting by early man (Martin, 1973), the other to glacial climate (Axelrod, 1967). There are difficulties with both hypotheses.

Martin's hypothesis applies mainly, though not exclusively, to the large mammals of North America. Martin argues that early man, in the hunter stage of culture, on entering North America from Siberia about 11,500 years ago, found a rich new food resource in the large mammals then existing in North America. These mammals, as exemplified by the mastodon, mammoth, horse, camel, and bison in North America, had not previously been exposed to predation by hominid hunters. The early human population expanded rapidly and migrated rapidly through North America and to South America, on the basis of the big-game food resource, driving most of their game animals to extinction through excessive hunting. Some of these species did apparently become extinct at about the same time as the postulated early human population explosion, i.e., about 10,000–11,000 years ago (Martin, 1973; see also Martin and Wright, 1967).

This hypothesis hinges on the widely held view that human entry into North America occurred in relatively recent time. Some evidence indicates that man has been present in North America for a much longer time than the 11,000–12,000 years postulated by Martin and others. Dating of Paleo-Indian bones from southern California by one of the newer methods (the aspartic acid racemization method) gives ages of up to 48,000 years for these bones (Bada, Schroeder, and Carter, 1974). The whole question of dating remains open, at this writing, and it remains to be seen whether man is relatively recent or relatively old in North America.

Axelrod (1967) seeks the cause of mammalian extinctions in climatic changes that culminated in the Pleistocene. Tertiary climates had been generally equable, and the mammals were adapted to such equable climatic conditions. The new climates of the Pleistocene were highly inequable, with severe cold seasons or drought seasons. Severe cold brings about the death of individual mammals, by physiological effects or by starvation, and these individual deaths could accumulate to the level of species extinction. New cold (or dry) climates would also have adverse indirect effects on many mammals by causing changes in the vegetation and in their habitats.

The body-size distribution of the extinctions conforms to the climatic hypothesis. The small mammals did not suffer widespread extinction. Most small mammals can escape a cold season by retreating into burrows, dens, or nests.

Furthermore, small mammals tend to have short reproductive cycles that can be geared to a seasonal climate. In many small mammals the young can grow to maturity during a short favorable season. Large mammals mostly do not have these ways of fitting into a highly seasonal cold (or dry) climate (Axelrod, 1967).

One fact not explained by the climatic hypothesis is that there was relatively little extinction of mammals in the Early Pleistocene, when the actual change to glacial climate occurred. Another difficulty with the climatic hypothesis is that many large mammals became extinct not only in northern continents but also in Africa, which escaped the full force of the Ice Age (Simpson, personal communication).

Episodes of Extinction

It has long been recognized that the geological history of life is punctuated by episodes of mass extinction separated by long periods of gradual evolutionary change. The fossil record points to a condition somewhat intermediate between the old catastrophism (of Cuvier) and the strict uniformitarianism (of Lyell), a condition of episodic evolution. There are periodical episodes of mass extinction, when many old groups die out, followed by the development and adaptive radiation of new groups (Newell, 1967).

Indeed, the boundaries between the geological eras and periods are based partly on the episodes of mass extinction, particularly in animals. The defining characteristic of animal episodic evolution is embedded in the terminology of the eras (Paleozoic, etc.).

A significant feature of episodes of extinction is the occurrence of high rates of extinction in quite independent groups at about the same time. Thus both the ammonoids and the reptiles suffered high rates of extinction in the Late Permian, Late Triassic, and Late Cretaceous (Newell, 1967).

Newell (1967) has collated and analyzed the data on the first and last appearances in the fossil record of 2250 animal families belonging to all major groups. A summary of his findings is presented in Figure 32.3. Figure 32.3A shows the rate of extinction of animal families, as indicated by their last known occurrence in the record, through geological time. We see that the curve for extinction rate is very uneven. Major peaks, representing episodes of widespread extinction, occur in the Late Cambrian, Late Devonian, Late Permian, Late Triassic, and Late Cretaceous; minor peaks occur at the ends of some of the intervening periods.

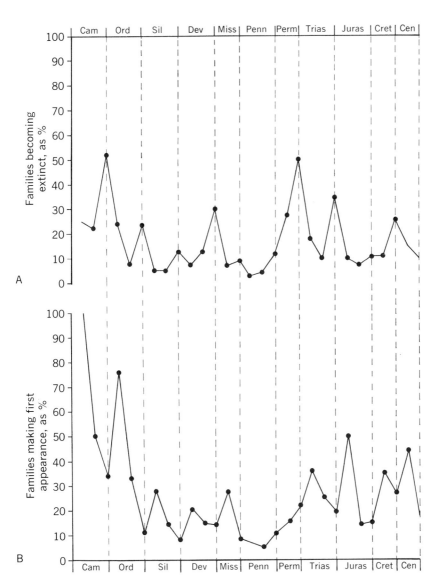

FIGURE 32.3
Episodes of extinction and replacement in animals. (A) Rate of extinction of families.
(B) Rate of first appearances of new families in the record. Further explanation in text.
(Redrawn from Newell, 1967.)

Most of these episodes of extinction can be correlated with times of continental emergence and continental climates. This correlation suggests that environmental disturbances were the start of complex chain reactions ending in the widescale extinctions (Newell, 1967; Axelrod, 1974).

Figure 32.3B shows that the first appearance of new animal families in the record is also very uneven through time. There are peaks in the origination curve as well as in the extinction curve. Furthermore, the peaks in the origination curve tend to lag behind those in the extinction curve. Episodes of mass extinction are followed by the formation and development of new groups (Newell, 1967).

This timing in the appearance of new groups is consistent with the conclusion expressed in the preceding chapter. An ecological opportunity is a necessary condition for the formation of a new group. Mass extinction provides such ecological opportunities.

Collateral Readings

Axelrod, D. I. 1967. Quaternary extinctions of large mammals. *Univ. Calif. Publ. Geol. Sci.* **74:** 1–42.

Bakker, R. T. 1975. Dinosaur renaissance. *Sci. Amer.* **232**(4): 58–78.

Colbert, E. H. 1961. *Dinosaurs: Their Discovery and Their World.* E. P. Dutton & Co., New York.

Martin, P. S., and H. E. Wright, Jr., eds. 1967. *Pleistocene Extinctions: The Search for a Cause.* Yale University Press, New Haven, Conn.

Simpson, G. G. 1953. *The Major Features of Evolution.* Columbia University Press, New York. Chapters 7, 9.

Simpson, G. G. 1967. *The Meaning of Evolution.* Ed. 2. Yale University Press, New Haven, Conn. Chapters 11, 13.

Simpson, G. G., and W. S. Beck. 1965. *Life: An Introduction to Biology.* Ed. 2. Harcourt, Brace & World, New York. Chapters 30, 31.

Special Topics in Macroevolution

Space limitations prevent our discussing various other interesting topics in macroevolution. We can, however, give honorable mention, as it were, to five such topics, and cite a few bibliographical references for each of them, for the benefit of readers wishing to pursue them further.

1. ONTOGENY AND MORPHOGENESIS IN RELATION TO EVOLUTION

De Beer, G. R. 1951. *Embryos and Ancestors.* Ed. 2. Oxford University Press, Oxford.

Huxley, J. S. 1932. *Problems of Relative Growth.* Methuen, London.

Rensch, B. 1960. *Evolution Above the Species Level.* Translation. Columbia University Press, New York. Chapter 6.

2. EVOLUTION OF POLLINATION SYSTEMS

Faegri, K., and L. van der Pijl. 1971. *The Principles of Pollination Ecology.* Ed. 2. Pergamon Press, Oxford.

Grant, K. A., and V. Grant. 1968. *Hummingbirds and Their Flowers.* Columbia University Press, New York.

Proctor, M., and P. Yeo. 1973. *The Pollination of Flowers.* Collins, London.

3. COEVOLUTION OF PHYTOPHAGOUS INSECTS AND PLANTS WITH REFERENCE TO CHEMICAL SUBSTANCES

Brower, L. P. 1970. Plant poisons in a terrestrial food chain and implications for mimicry theory. In *Biochemical Coevolution.* Oregon State University Press, Corvallis.

Brower, L. P., and J. V. Z. Brower. 1964. Birds, butterflies, and plant poisons: a study in ecological chemistry. *Zoologica (New York)* **49:** 137–159.

Ehrlich, P. R., and P. H. Raven. 1964. Butterflies and plants: a study in coevolution. *Evolution* **18:** 586–608.

Levin, D. A. 1971. Plant phenolics: an ecological perspective. *Amer. Nat.* **105:** 157–181.

Rothschild, M. 1970. Toxic Lepidoptera. *Toxicon* **8:** 293–299.

Rothschild, M. 1975. Remarks on carotenoids in the evolution of signals. In *Coevolution of Animals and Plants,* ed. by L. E. Gilbert and P. H. Raven. University of Texas Press, Austin.

4. HISTORICAL BIOGEOGRAPHY

Axelrod, D. I. 1972. Ocean-floor spreading in relation to ecosystematic problems. In *A Symposium on Ecosystematics,* ed. by R. T. Allen and F. C. James. University of Arkansas Museum, Fayetteville, Occasional Paper No. 4.

Gill, E. D. 1975. Evolution of Australia's unique flora and fauna in relation to the plate tectonics theory. *Proc. Roy. Soc. Victoria* **87:** 215–234.

Hallam, A., ed. 1973. *Atlas of Paleobiogeography.* American Elsevier, New York.

Keast, A., F. C. Erk, and B. Glass, eds. 1972. *Evolution, Mammals, and Southern Continents.* State University of New York Press, Albany.

Lillegraven, J. A. 1974. Biogeographic considerations of the marsupial-placental dichotomy. *Ann. Rev. Ecol. Syst.* **5:** 263–283.

Raven, P. H., and D. I. Axelrod. 1974. Angiosperm biogeography and past continental movements. *Ann. Missouri Bot. Gard.* **61:** 539–673.

5. EVOLUTION ON ISLANDS

Carlquist, S. 1974. *Island Biology.* Columbia University Press, New York.

Lack, D. 1947. *Darwin's Finches.* Cambridge University Press, Cambridge.

VII

FROM ORGANIC EVOLUTION TO HUMAN EVOLUTION

33

Progressive Evolution

The Concept of Progress in Evolution

The concept of progress is basically a subjective one, and in common usage it relates primarily to human affairs. A trend in human history is said to exhibit progress when a later stage is better than an earlier stage in some characteristic singled out for emphasis by an observer of the trend. The viewpoint and value system of the observer play an essential role in the identification of progress. We can readily think of trends in modern times that a land developer calls progress and that a conservationist considers to be a lamentable step in the wrong direction.

The questions before us here are: first, whether the concept of progress can be transplanted satisfactorily from human affairs to evolutionary biology; and second, assuming that it can be transplanted, whether it can be converted from a subjective to an objective concept, that is, whether objective criteria can be found for progress in organic evolution.

In a series of thoughtful and provocative essays, Huxley (1942, Ch. 10; 1954, 1958) has argued in the affirmative on both questions. He argues for the objective reality of a special type of evolutionary trend to be known as progressive evolution. In fact, Huxley transplants progress from human affairs to organic evolution and then brings it back to human affairs again, making mankind the culminating development of progressive evolution.

Huxley's thesis has been discussed by a number of evolutionists, usually with something less than full agreement (e.g., Simpson, 1949, 1967, 1974*a*; Thoday, 1958; Rensch, 1960; Stebbins, 1969; Ayala, 1974). My own position is also somewhat ambivalent, as will be seen from the following presentation. Nevertheless, portions of Huxley's thesis survive the criticism, as will also be seen.

There is no particular difficulty with regard to the first question mentioned above. One can make an a priori decision to equate progressive evolution with human evolution if one wishes, and one is equally free to reject such a choice of definitions, as I am inclined to do. The important issue concerns the second question: whether progressive evolution is an objectively real process in nature. This is a biological question.

Progress vs. Specialization

Huxley (1942) makes a distinction between progress and specialization in evolution:

"Specialization . . . is an improvement in efficiency of adaptation for a particular mode of life . . ." (p. 562). But "Specialization . . . always involves the sacrificing of certain organs or functions for the greater efficiency of others" (p. 567). Consequently the improvements incorporated into a specialization for a particular adaptive zone tend to limit the possibilities of future change to a continuation in the same adaptive zone. Specializations are self-limiting. Most of organic evolution is the development of specializations. And evolution is largely a series of blind alleys (Huxley, 1942).

Progress in evolution, on the other hand, is improvement in the *general* efficiency of the machinery of life; it is "all-round biological improvement" (Huxley, 1942, pp. 562, 567). Progressive evolution is consequently a trend in which past changes and present characteristics do not limit future possibilities. It is open-ended (Huxley, 1942).

A survey of the organic world indicates that, as of now, the criteria of progressive evolution are realized only in the human species, with its intelligence, culture, and flexible behavior. "Only along one single line is progress and its future possibility being continued—the line of man" (Huxley, 1942, p. 571).

Succession of Dominant Types

The problem can be approached from another angle by surveying the paleontological history of animal life. The fossil record reveals a succession of dominant types through the course of time. Thus the following groups were (are) dominant in the eras or periods listed:

Marine molluscs, trilobites, and other marine invertebrates	Ordovician
Fishes	Late Paleozoic
Dinosaurs and other reptiles	Mesozoic
Mammals	Tertiary
Man	Quaternary

Running through this succession of dominant forms is an improvement in the general efficiency of the organism. The succession can be seen as a series of ever more efficient structural and functional organizations for carrying out major life processes. The successive developments of a bony skeleton, lungs, warm blood, and an intelligent brain are milestones in this series (Huxley, 1942).

The milestones themselves—the main stages in the series—are called grades (Huxley, 1958). A grade is a level of improvement in the structural and functional organization of the organism. Thus in the foregoing series, bony fishes represent one grade, amphibians another, and terrestrial reptiles another.

The series of grades exhibits the definitive features of progressive evolution (Huxley, 1942, 1958).

The actual succession of dominant types has been a good deal more complex than is indicated above, however, as Simpson (1967, 1974) has pointed out. There may be two or more different dominant types at the same time in the same broad adaptive zone (e.g., insects and terrestrial vertebrates), or in different adaptive zones (fish and terrestrial vertebrates). Consequently there is not just one series, but several or many.

Criteria of Progressive Evolution

The evolutionary series from the lower marine animals through the vertebrates to man entails changes in a number of characteristics. If the series constitutes a progressive evolutionary trend, the changing characteristics give us a set of criteria of progressive evolution, arrived at more or less independently of subjective considerations.

Among the changes in the series from lower animals to higher vertebrates are the following (Huxley, 1942; Simpson, 1949, 1967; Rensch, 1960). (1) Increase in energy level of life processes, that is, increase in metabolic rate; (2) increase in efficiency of reproduction, including increase in care of eggs and/or young; (3) improvements in perception of signals from environment and in ability to react to environmental stimuli; and (4) increase in control over and independence of the environment.

Huxley (1942) emphasizes the last characteristic—control and hence relative independence of the environment—as the most significant distinguishing criterion of progressive evolution. On this criterion, mammals rank high; certain mammals, like the beaver, very high; and man, highest of all.

Huxley's criterion of independence and control of the environment comes close to the mark. But in Huxley's treatment, the term "environment" is construed broadly, and "control" narrowly, in such a way as to slant the application of the criterion toward mankind. One would prefer a criterion that is less anthropocentric and more biological.

An alternative criterion can be suggested. Progressive evolution consists of adaptations to environments successively further removed from the original and ancestral environment of life. This criterion will be discussed in the next section.

Discussion

Like various other evolutionists, I have certain reservations concerning Huxley's thesis. I think that Huxley's idea of progressive evolution contains a valuable kernel of truth that should be preserved; but at the same time I also believe that Huxley, in equating progressive evolution mainly with human evolution, has construed progressive evolution in too narrow a sense.

Organic evolution is nothing if it is not diverse. Human evolution, or, more precisely, the biological part of human evolution, is undoubtedly an important development of organic evolution; but important developments have occurred in other phyletic lines too. There are progressive developments, for example, in other mammals, in birds, in social insects, and in the plant kingdom, that tend to be ignored in discussions of evolutionary progress.

Furthermore, the age of man will not last forever. Organic evolution will probably be making new advances in future ages, when the human species is gone from the scene. The nature of future dominant types of organisms is unpredictable at present.

Let us restate the concepts of progress and specialization in terms of types of evolutionary trends. We can recognize three modal types of evolutionary trend as classified on this basis. (1) Short-term specialization trends, as illustrated by a branch line in the Hawaiian honeycreepers (Chapter 31). (2) Long-term specialization trends, as in the evolution of the horse's hoof (Chapter 26). These two types of trends are channelled and self-limiting. (3) Progressive evolutionary trends, which entail general improvements in biological organization. The series—land reptile, warm-blooded mammal, intelligent mammal, intelligent hominid—is only one example of a progressive trend; another is the series—pteridophyte, seed plant, woody angiosperm, annual herbaceous angiosperm.

We can think of progressive trends as ultra-long-term specialization trends, thus eliminating the dichotomy between progress and specialization. Progressive trends are specialization trends in which the organisms involved become specialized for adaptive zones that are common on earth but far removed from the ancestral nutrient soup.

Life began in nutrient-rich waters. The exhaustion of nutrients in the seas stimulated the development of photosynthetic aquatic forms, and of grazers on these, and of predators on the grazers. The filling up of the seas then stimulated the colonization of the land, by several phyletic groups, and of the air, also by several groups. Terrestrial organisms progressed from warm moist land areas to dry and cold lands. Man is the culmination of one progressive trend; but rodents, birds, winged insects, desert shrubs, and desert annual plants are the culminations, up to this point in earth history, of other such trends.

The stepwise conquest of environments that are widespread on earth but basically inhospitable to life, such as nutrient-poor water, dry land, and air, is the common theme running through those long-term adaptive trends that we call progressive.

Collateral Readings

Huxley, J. S. 1942. *Evolution: The Modern Synthesis.* George Allen & Unwin, London. Chapter 10.

Huxley, J. S. 1958. Evolutionary processes and taxonomy with special reference to grades. In *Systematics of Today,* ed. by O. Hedberg. Uppsala Universitets Årsskrift, Uppsala.

Rensch, B. 1960. *Evolution Above the Species Level.* Translation. Columbia University Press, New York. Chapter 7.

Simpson, G. G. 1967. *The Meaning of Evolution.* Ed. 2. Yale University Press, New Haven, Conn. Chapter 15.

Simpson, G. G. 1974. The concept of progress in organic evolution. *Social Res.* **41:** 28–51.

34

Hominid Phylogeny

Classification of the Primates

The order of Primates is remarkable not only for its diversity, but more especially for the range of this diversity. The range between such extreme types as a tree shrew or lemur on the one hand, and an ape or man on the other, is enormous. The diversity in other large mammalian orders is pretty much on one grade, but that in the primates presents itself in a series of grades.

Table 34.1, taken at face value only, indicates the total amount of diversity in the primates. There are 4 suborders, 11 families, and 60 genera. But this is not the whole story. Read Table 34.1 in conjunction with mental images of the animals involved, as recalled from a trip to the zoo or a perusal of a book on the natural history of mammals. Such an exercise conveys a notion of the range as well as the amount of the diversity.

The diversity in the primates ranges through about five grades. The grades would be: (1) the tree shrews, which are segregated from the order in some classifications; (2) the prosimians proper (lemurs, lorises, tarsiers, etc.); (3) the monkeys; (4) the apes; and (5) men.

At the lower end of the spectrum of grades, the primates approach the somewhat more primitive order Insectivora. The tree shrews are intermediate between the two orders. The primates are probably derived from the insectivores. The recorded fossil history of the insectivores goes back to the Cretaceous, and that of the primates, to the Paleocene.

TABLE 34.1
System of classification of Recent Primates. (Compiled from Colbert, 1969, and Walker, 1964.)

Major group	Genera and species	Geographical distribution
Suborder Lemuroidea		
Superfamily Tupaioidea		
Family Tupaiidae; tree shrews	5 genera (*Tupaia*, etc.); ca. 15 species	Tropical Asia and Malaysia
Superfamily Lemuroidea		
Family Lemuridae; lemurs	6 genera (*Lemur*, etc.); ca. 16 species	Madagascar
Family Indridae; woolly lemurs	3 genera; 4 species	Madagascar
Superfamily Daubentonioidea		
Family Daubentoniidae; aye-aye	Monotypic; *Daubentonia* *madagascariensis*	Madagascar
Superfamily Lorisoidea		
Family Lorisidae; lorises, galagos	6 genera (*Loris, Galago,* etc.); ca. 12 species	Old World tropics
Suborder Tarsioidea		
Superfamily Tarsioidea		
Family Tarsiidae; tarsiers	Single genus *(Tarsius);* 3 species	Malaysia
Suborder Platyrrhini		
Superfamily Ceboidea		
Family Cebidae; New World monkeys	12 genera (*Cebus, Alouatta,* etc.); ca. 37 species	New World tropics
Family Callithricidae; marmosets	4 genera (*Callithrix*, etc.); ca. 33 species	Tropical South America and Panama
Suborder Catarrhini		
Superfamily Cercopithecoidea		

Family Cercopithecidae;
      Old World monkeys,
      baboons
	16 genera (*Macaca,*
	Chaeropithecus, Papio,
	etc.); ca. 60 species
		Old World tropics
		and subtropics

Superfamily Hominoidea		

Family Pongidae;
      apes
	5 genera; ca. 10 species.
	Hylobates (gibbon),
	Symphalangus
	  (Siamang gibbon),
	Pongo (orangutan),
	Pan (chimpanzee),
	Gorilla (gorilla)
		Old World tropics

Family Hominidae; man	Monotypic; *Homo sapiens*	Cosmopolitan

The fossil record indicates that the first primates were prosimians, that they were widespread in distribution, and that they underwent an adaptive radiation in the Paleocene and Eocene. Some of the branch lines died out in the Eocene and Oligocene. Other branch lines persisted to the Recent. The surviving major branch lines correspond to the various superfamilies listed in Table 34.1 (Simpson, 1967, pp. 87–93).

The Hominoids

One of the main branches of the primates, having a separate existence since the Eocene, is the hominoid line. The Hominoidea consist of three families: the two Recent families (Pongidae and Hominidae) and one extinct family of apes (Oreopithecidae, including the genus *Oreopithecus*). These families correspond to three secondary branches in the hominoid line.

The Oligocene and Miocene fossil records of these families are fragmentary. As a result many critical points of phylogeny are uncertain.

The *Oreopithecus* lineage appeared in the Miocene and died out in the Pliocene; the lineage may well have existed prior to the Miocene. The true apes (Pongidae) are represented by *Aegyptopithecus* in the Oligocene and by *Dryopithecus* in the Miocene and Pliocene. (We are taking a conservative view of the genera here; specialists often recognize various segregate genera.)

The oldest known Hominidae is *Ramapithecus,* of Miocene and Early Pliocene age. Not much is known about *Ramapithecus.* It looks as though the ancestors of the Hominidae diverged from the dryopithecine assemblage of apes. The Hominidae were evidently separate from the Pongidae by Miocene time (see Romer, 1966, pp. 382–383; Simpson, 1967, pp. 92–93; Colbert, 1969, pp. 311–313).

The Hominidae contain three genera, again adopting a conservative genus concept. These are: *Ramapithecus* (early man-ape), *Australopithecus* (australopithecines), and *Homo* (man). Their ages are given in Table 34.2 (following Campbell, 1974). The age estimates are in an almost constant state of flux as new fossil finds are made and as methods of dating are improved.

The Plio-Pleistocene australopithecines are known mainly from South Africa and East Africa. They were ape-like in skull, face, and brain size (Figure 34.1), but human-like in dentition, erect posture, and bipedal locomotion. There were at least two contemporaneous species differing in size (*A. africanus* and *A. boisei*), and perhaps three species. [The fossil forms known as *A. robustus* and *Homo habilis* are placed in *A. boisei* by Campbell (1974, p. 95), but are regarded as separate entities by some other authors.] The australopithecines died out before the end of the Pleistocene.

TABLE 34.2
Known and probable ages of the genera and species of Hominidae.
(Data from Campbell, 1974, p. 98.)

Genus and species	Epoch	Age in years (approximate)
Ramapithecus	Miocene to Pliocene	Probably ca. 18,000,000 to 6,000,000 B.P.
R. wickeri		Known 14,000,000 B.P.
R. punjabicus		Known 9,000,000 B.P.
Australopithecus	Pliocene to Pleistocene	Probably ca. 6,000,000 to 1,000,000 B.P.
A. africanus		Known 5,500,000 to 1,300,000 B.P.
A. boisei		Known 3,000,000 to 1,200,000 B.P.
Homo	Pleistocene to Recent	Probably since ca. 1,300,000 B.P.
H. erectus		Known 1,300,000 to 400,000 B.P.
H. sapiens		Known 250,000 B.P. to present

NOTE ADDED IN PROOF: Recent provisional reports indicate older ages than those given above for the genus *Homo* and for *H. erectus*.

The large-brained genus *Homo* represents another lineage that, though later, overlapped in time with *Australopithecus*. Two species of *Homo* are recognized: *Homo erectus* (early man) and *H. sapiens* (including *H. neanderthalensis*; modern man) (Figure 34.1). Both existed in Africa and elsewhere in the Old World in the Pleistocene. *Homo erectus* and *H. sapiens* formed a pair of successional species, with the former grading into the latter through time. Only one species of *Homo* has existed at any one time level, but *H. erectus* or its immediate ancestor apparently coexisted with *Australopithecus*.

Homo is close to and probably derived from *Australopithecus*. The exact structure of the phylogenetic tree is uncertain, however, and different anthropologists favor different phylogenetic hypotheses (see Brace et al., 1971; Pilbeam, 1972, pp. 13, 144, 156; Pilbeam and Gould, 1974). One likely possibility is that *Homo erectus* was derived from *Australopithecus africanus*. Future fossil finds may clarify the phylogeny.

The Arboreal Heritage

The primates are predominantly animals of warm regions (see Table 34.1); they are predominantly arboreal, the ground-living forms like baboons and man being exceptional; and they are predominantly omnivorous. Life in an

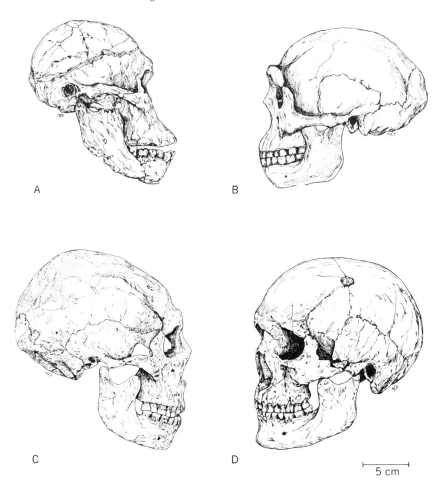

FIGURE 34.1
Skulls of fossil hominids. All drawings to same scale. (A) *Australopithecus africanus,* Early
Pleistocene, South Africa. (B) *Homo erectus,* Middle Pleistocene, Peking. (C) *Homo sapiens,*
Neanderthal man, Late Pleistocene, Iraq. (D) *Homo sapiens,* Late Pleistocene, Peking.
(Rearranged from Brace, Nelson, and Korn, 1971; drawings by Mrs. M. L. Brace from
specimens in various museum collections.)

arboreal habitat resulted in the development of some adaptations that were
later to be exploited by man.

An arboreal mammal must be able to judge distances accurately, focus on
objects, hold on to branches, and hold onto food objects. The corresponding
adaptations in the primates are a well-developed visual sense, grasping hands
with an opposable thumb, and a balancing tail.

The visual sense in primates is a composite of several features. The eyes are well developed. They are placed in the front of the face, which permits focusing on objects and judgment of distance by parallax. A spot in the retina of the eye permits fine focusing. Finally, a color sense, which is lacking in many groups of mammals, is developed in the primates, where it probably serves to give a better sense of perspective.

The primate visual sense is an important component, in turn, of primate intelligence. Primate intelligence is another composite trait. There is, to begin with, the brain, which is well developed in all the higher primates. But what that brain is able to learn about the external world depends largely upon the *combination* of primate eyes and primate hands. Other sense organs, particularly hearing, play supplementary roles.

The primate brand of intelligence is based mainly on the combination of focusing eyes and grasping or feeling hands. This is not the only possible brand. Canine intelligence, for example, depends mainly on the combination of focusing eyes, sense of smell in the nose, and sensitive ears; bee intelligence depends on the combination of compound eyes, the insect type of color vision, and flexible, odor-detecting antennae.

The eye-hand combination was and is the basis of concentration on concrete objects in the pre-hominid primates. The same combination was to become the basis of tool making, tool use, experimentation, and problem solving in the hominids.

The Terrestrial Heritage

The apes are forest animals and are arboreal, or at least partially so. The gibbons spend most of their lives in the trees and move about by brachiating and leaping. Chimpanzees and gorillas are semi-terrestrial, and move on the ground by knuckle-walking.

Australopithecus and *Homo,* by contrast, were and are animals of open savannas and plains. Their mode of locomotion is walking and trotting on two legs. Bipedalism and its corollary, erect posture, are the most distinctive morphological characteristics of the Hominidae.

Forest habitats were diminishing in extent and savanna country was expanding during the Middle and Late Tertiary. Climatic and vegetational changes opened up new habitats. Invasion of this new habitat probably marked the divergence of the primitive hominids from the ancestral forest-dwelling apes. Bipedal locomotion and erect posture were part of the adaptive character combination fitting the hominids for life on savannas and plains.

Much of the later recorded evolution of the hominids in the Pleistocene consists of a trend toward a more fully erect posture, with neck and head thrown back, and with the body modified for an improved walking gait.

Bipedal locomotion had the important consequence of freeing the hands for the making and use of tools and weapons. *Australopithecus* used primitive stone tools, and *Homo erectus,* more advanced ones (Figure 34.2). The handling of objects is an important component of intelligence in arboreal primates, as noted previously. And now with the change from an arboreal to a terrestrial habitat, and the release of the hands from the function of locomotion, the manipulation of things could be developed to its full potential. Primitive technology was an early result.

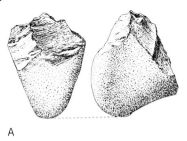

A

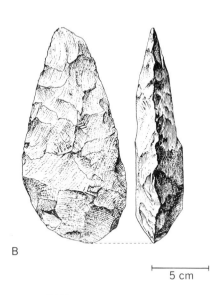

B

|—————| 5 cm

FIGURE 34.2
Lower Paleolithic stone tools from East Africa. Made from lava rocks. Shown to same scale. (A) Early Pleistocene. Pebble tool, Oldowan stage, Olduvai Gorge, Tanzania. (B) Middle Pleistocene. Hand axe, Acheulian stage, Kenya. (Oakley, 1949.)

Intelligence is an adaptively valuable trait in the primates, and is undoubtedly favored by natural selection. An evolutionary trend toward increased intelligence can be discerned in the non-hominid primates. This trend reaches a very high level in *Pan* (chimpanzee) and *Pongo* (orangutan). It continues to still higher levels in the hominids.

In the Hominidae, judging by cranial capacities, *Homo erectus* was considerably more intelligent than *Australopithecus* (See Figure 34.1). Cranial capacities for the two groups show the following ranges and means (from Campbell, 1974, p. 272):

	Range, cm^3	Mean, cm^3
A. africanus	435–815	588
Homo erectus	775–1225	950

It is quite likely that the superior intelligence and superior tools and weapons of *Homo erectus* or its immediate ancestor gave it an advantage over *Australopithecus* in direct and indirect competiton between the two groups during the Pleistocene. At any rate, *Australopithecus* died out, while *Homo erectus* survived and later developed into *H. sapiens*.

Racial Variation

The modern historical age opens with *Homo sapiens* worldwide in distribution and differentiated into a series of geographical races as well as innumerable local races. The geographical races recognized by Garn (1961) are the following:

> Asiatic mongoloids
> American Indians
> Indians (of India, Pakistan, etc.)
> European whites
> African blacks
> Australian blacks
> Melanesians
> Micronesians
> Polynesians

These are typical geographical races, as in other species of mammals. Some of the racial characters in body size, body proportions, pigmentation, etc., are correlated with climate (see Coon, 1955; Garn, 1961). It is well known, for instance, that racial variation in skin color is correlated with the sunniness of the climate.

Skin color regulates vitamin D synthesis in man. Vitamin D in the right amount is necessary for proper growth, too little vitamin D being a cause of rickets, and too much, a cause of bone calcification. Vitamin D is synthesized in the skin. And its synthesis in the skin is controlled by sunlight: the more sunlight, the more vitamin D.

Dark-skinned races live in sunny areas. Here the dark skin pigments filter out sunlight and prevent the synthesis of too much vitamin D. Conversely, the white race occurs naturally in northern Europe, where it could not synthesize enough vitamin D without a pale skin, which lets in maximum sunlight. But white skin absorbs too much sunlight in sunny climates, and white persons transplanted to such climates become suntanned as a phenotypic modification to prevent this excessive light absorption. On the other hand, black races would probably die of rickets under natural conditions in northern Europe. Racial differences in skin color are thus adaptive in relation to the sunniness of the climate (Loomis, 1967).

Facts such as these bespeak the role of natural selection in the development of at least some of the racial characteristics of *Homo sapiens*. Local racial variation has a haphazard pattern, in some cases, suggesting that drift or the selection-drift combination also plays a role (see Chapter 13). In still other cases, the features of human racial variation point to a controlling influence of migration and gene flow (see Chapter 18).

Conclusions

The aspects of hominid evolution discussed in this chapter are extensions of non-hominid primate evolution. We see evolutionary trends in hand use, intelligence, posture, etc.; we see signs of a quantum shift to a new mode of locomotion in a new type of habitat; there is interspecific selection between hominid lines; and there is geographical race formation in the surviving widespread hominid species.

These aspects of hominid evolution are a part of the larger picture of organic evolution. They can be explained as results of the operation of the evolutionary factors and forces set forth in earlier chapters of this book.

In this connection it is worthwhile to note that 600 generations or more can occur in man in a period of 10,000 years. Substantial evolutionary changes could be brought about by selection or selection-drift in this short time period. And longer periods, which are available according to the fossil record, would of course allow for more far-reaching evolutionary changes.

Collateral Readings

Campbell, B. 1974. *Human Evolution.* Ed. 2. Aldine-Atherton, Chicago and New York.

Colbert, E. H. 1969. *Evolution of the Vertebrates.* Ed. 2. John Wiley & Sons, New York. Chapter 21.

Le Gros Clark, W. E. 1960. *The Antecedents of Man.* Quadrangle Books, Chicago.

Pilbeam, D. 1972. *The Ascent of Man.* Macmillan, New York.

Simpson, G. G. 1967. *The Meaning of Evolution.* Ed. 2. Yale University Press, New Haven, Conn. Chapter 7.

Primate Social Evolution

Social Groups in Non-Human Primates

Primates are social animals. Individuals in primates are usually members of family groups and of larger troops.

Social grouping in the higher primates above the prosimian grade is promoted by two conditions. The infant and juvenile stages are relatively long in monkeys and apes, as compared with most mammals, and this leads to a prolonged association between mother and offspring. And the absence of a clearly demarcated breeding season in higher primates and in many prosimians as well leads to a continuous or at least prolonged association of the sexes (for more on this see DeVore, 1965, Chs. 14, 18). The combination of female-offspring bonds and female-male bonds results in family groups and troops.

There is much species-to-species variation in social structure in the primates. A detailed account of primate social organization is beyond the scope of this chapter (but see DeVore, 1965; Eisenberg et al., 1972; Wilson, 1975, Ch. 26). Eisenberg and co-workers (1972) give a useful classification of the modes of social organization, which is presented here with minor modifications.

(1) Maternal family consisting of mother and offspring. Adult females and males have separate centers of activity. Found in some species of lemur and loris, and the aye-aye. (Refer to Table 34.1 for system of classification of primates.)

(2) Biparental family. The family group consists of a female, a male, and young. Found in a species of woolly lemur, and in some marmosets, New World monkeys, and gibbons.

(3) One-male troop. The troop consists of several maternal families and one adult male in contact with them all. The male has a strong intolerance of other mature or maturing males. Found in some New World monkeys (including the howler monkey) and Old World monkeys (including Hamadryas baboons and Gelada baboons).

(4) Age-graded multiple-male troop. A cohesive group consisting of several females, several males, and young. There is an intermediate degree of male tolerance. This degree of tolerance permits the coexistence of several males of varying ages, mostly young males, with the dominance order corresponding to the age of the males. Found in some New World monkeys (including, again, the howler monkey) and Old World monkeys (including macaques), and the gorilla.

(5) True multiple-male troop. As in mode 4, but with a high degree of male tolerance, permitting the coexistence of several adult males. These are codominant and cooperative and maintain a flexible oligarchy in the troop. Found in a species of lemur, a species of woolly lemur, in some Old World monkeys (including baboons and macaques), and in the chimpanzee.

The above series of types of social organization probably corresponds approximately to an evolutionary trend. It is probable that mode 1 is primitive, mode 3 intermediate, and modes 4 and 5 derived, in the primates. Going along with the changes in social structure is a trend toward larger-sized groups (Eisenberg et al., 1972).

The types of social organization also correlate roughly with habitat. One-male troops (mode 3) are found mainly though not exclusively in species with an arboreal foraging habitat. Conversely, true multiple-male troops (mode 5) tend to occur preponderantly among species with semi-terrestrial foraging habits (Eisenberg et al., 1972).

Adaptive Value of Social Grouping

The social group serves several useful functions in the higher non-human primates, on which we will focus our attention for now. The members of a group band together in holding a territory and in defense against predators. Furthermore, and most importantly, the troop is the medium for the sharing of experiences and teaching of the young about a variety of vital matters—matters such as the recognition of edible fruits and leaves and of poisonous plants, the whereabouts of enemies, and the like.

Some of the functions of social groups in monkeys and apes will be discussed briefly in the following two sections.

Language

The requirements for communication and coordination of individual activities in troops of monkeys and apes are met by body languages involving facial expressions and by vocal animal languages. Both modes of expression are well developed in monkeys and apes (see DeVore, 1965). Monkeys are noted for their incessant and noisy chatter.

Walker (1964, pp. 394, 426) collected and deciphered the vocal sounds of a douroucoulis monkey (*Aotus trivirgatus,* Cebidae). He found that this monkey has about 50 vocal sounds. By means of these vocalizations the monkey can express observations and feelings such as the following:

Suspicion of danger	"book"
Warning of danger	"wook"
Curiosity without suspicion	"uuhh"
Curiosity with suspicion	"huh," "wheu"
Friendly greeting	"chrrr"
Enthusiasm about something	bird-like notes
Annoyance	"ack," "kack"
Objection to being bothered	"uk," repeated
Give it to me	"ehe"
I want to go with you	"eh, eh, eh"
I love you	"ooohh, ooohh"

Washburn relates the following anecdote concerning baboons in Nairobi Park, Africa. These baboons are protected and had become accustomed to people in cars. A parasitologist once shot two of the baboons from a car. Some baboons were presumably eye-witnesses of the event, but others undoubtedly were not. News of the event quickly spread throughout the baboon community, with the result that it became impossible to get close to baboons while in a car, and remained so for at least eight months (Washburn, in DeVore, 1965).

Intelligence and Learning

A language with any substantial repertoire requires a social group above a critical size for its maintenance and perpetuation. It also requires a certain degree of intelligence and learning ability on the part of the members of the group. Complex interactions and feedback exist between the social group and the various facets of intelligence.

The higher primates are noted for their intelligence. The size of the brain in proportion to body size is far greater in monkeys and apes than it is in other orders of mammals. A crude measure of intelligence in higher non-human primates is given by the cranial capacity. Some mean cranial capacities are listed below (from Campbell, 1974, p. 272):

	Mean capacity, cm^3
Macaque *(Macaca)*	100
Baboon *(Papio)*	200
Gibbon	103
Chimpanzee	383
Orangutan	405
Gorilla	505

The trend continues in the hominids, as follows:

	Mean capacity, cm^3
Australopithecus africanus	588
Homo erectus	950
Homo sapiens	1330

Many examples have been recorded of the problem-solving ability of monkeys and apes. The chimpanzee is a favorite subject in this regard. Chimpanzees not only use tools for obtaining food, as do a few other mammals and birds, but actually make crude tools out of sticks for accomplishing a definite objective (Goodall, in DeVore, 1965).

The learning of languages and of complex skills requires time as well as native intelligence. Prolongation of the immature stages of individual development is another well-known trend in the primates, as the figures in Table 35.1 show. It has been remarked that the great apes are probably smart enough to

TABLE 35.1
Duration of stages of life in various primates. (Wilson, 1975.)

Species	Gestation period, days	Infantile phase, years	Juvenile phase, years	Total post-natal immature stage, years	Total life span, years
Lemur	126	0.75	1.75	2.5	14
Rhesus macaque	168	1.5	6.0	7.5	27–28
Gibbon	210	2.0 (?)	6.5	8.5	30+
Chimpanzee	225	3.0	7.0	10.0	40
Gorilla	265	3.0+	7.0+	10.0+	35 (?)
Orangutan	275	3.5	7.0	10.5	30+
Man	266	6.0	14.0	20.0	70–75

learn a basic human language, but develop to maturity too fast to be able to do so in practice.

The Hominids

The various adaptive features of social grouping listed above for the higher non-human primates apply with equal force in the hominids. In addition, certain other adaptive advantages of social grouping have even greater force in the hominids.

The hominids diverged from the ancestral forest-dwelling stock of primates as terrestrial inhabitants of savanna and plains country, as noted in Chapter 34. Primitive terrestrial hominids living in open country would have been exposed and vulnerable to large predators. A group could have put up an effective defense against such predators where one or a few individuals could not. A social organization must have had a high selective value, in relation to defense against large predators, in the newly occupied open habitat.

Furthermore, groups of early hominids were probably more successful in hunting and other modes of food-getting than solitary individuals.

There is every reason to expect, therefore, that the australopithecines were social, and there is some indirect evidence to confirm this expectation. Australopithecines hunted baboons. Baboons are social, and are well defended against solitary hominid hunters armed with rocks and clubs. The obvious inference is that the australopithecines must have hunted in organized groups (Bartholomew and Birdsell, 1953).

The most primitive type of social group known in *Homo sapiens* is the hunting-and-gathering band. Generalized food-gathering peoples have existed in various parts of the world up to the recent past and in some cases into the present. Examples include the Shoshone and Algonquin Indians of North America, the Bushmen and Negritos of Africa, the aborigines of Australia and Tasmania, and the jungle tribes and Andaman Islanders of India (Birdsell, 1958).

A band typically consists of about 20 to 100 individuals and is made up of a series of biparental families. The band claims and defends a territory. Food-gathering within the territory is opportunistic, depending on the food resources available, and varies with the seasons. Both women and men participate in foraging. There is no accumulation of food reserves or of other forms of wealth. The society is basically egalitarian; there is no ruling class and no inferior sex (Birdsell, 1958).

The !Kung people of the Kalahari Desert of South Africa furnish a modern example, still extant and recently described (Kolata, 1974), of a primitive foraging society. The nomadic !Kung people have lived in their area as hunters and gatherers for at least 11,000 years. The situation is changing now, however, as the !Kung are being absorbed by the surrounding civilization (Kolata, 1974).

Hunting-and-gathering bands in primitive peoples appear to have a social organization very similar to that of multiple-male troops in higher non-human primates.

Human societies at a pre-civilized stage of culture are usually more complex in economic basis and social structure and larger in population size than the foraging band. One thinks of big-game-hunting tribes, pastoral tribes, and primitive agricultural village communities. Size and complexity continue to increase as we enter the stage of civilized societies. And here we can think of the wide range of socio-political units, from ancient agricultural civilizations to modern industrial nation-states and post-industrial super-states.

Discussion

The trends in primate social evolution thus continue in *Homo sapiens*. Through the various grades of the primates, from prosimian to modern man, we see an overall increase in the size of the social group, the complexity of social organization, the power of technology, the sophistication of language, intelligence, and the length of the immature stage of individual development.

These are not separate, but interrelated and interacting trends; and they are not linear, but rather overall trends.

It is customary in discussions of behavioral and social phenomena to differentiate between man and beast. Mankind is given credit for controlling man's own destiny, but blind forces shape the fate of other mammals, including the non-human primates. Competition for and defense of feeding and breeding grounds in the vertebrates are called "territoriality" and are studied by zoologists; homologous interactions in mankind are referred to as power struggles, or in the extreme case as wars, and are studied by historians. Natural selection can operate in its inexorable fashion in the non-human primates to produce certain social and psychological results, while homologous results in man are ascribed to "cultural evolution."

But what *is* the difference between territoriality in vertebrates and that in mankind? How does natural selection differ between man and other primates? Does man really control his own fate, as he likes to think, and if so, why hasn't he done a better job of it—or is he deluding himself?

In tracing out social evolutionary trends in primates, we find ourselves forced to cross disciplinary lines time and again. We cross from animal behavior to primatology to paleoanthropology to anthropology to archeology to history to current affairs—with no apparent place in which to draw a natural boundary line separating man from other higher primates.

This chapter ends with some questions about the evolution and nature of man that are very difficult to answer. Attempts to find answers, or at least partial answers, will be made in the next chapter.

Collateral Readings

DeVore, I., ed. 1965. *Primate Behavior: Field Studies of Monkeys and Apes.* Holt, Rinehart & Winston, New York.

Eisenberg, J. F., N. A. Muckenhirn, and R. Rudran. 1972. The relation between ecology and social structure in primates. *Science* **176:** 863–874.

Washburn, S. L., and R. Moore. 1974. *Ape Into Man.* Little, Brown & Co., Boston.

Wilson, E. O. 1975. *Sociobiology: The New Synthesis.* Harvard University Press, Cambridge, Mass. Especially Chapters 26 and 27, but also many earlier chapters.

CHAPTER

36

Cultural Evolution

The Cultural Heritage

The cultural or traditional heritage is the accumulated store of knowledge, understanding, arts, customs, and technological ability available to a human social group at any given time in its history. This body of knowledge and traditions is a product of the discoveries and inventions of preceding generations. It has been and will be transmitted from generation to generation by education in the broad sense of the word. New additions to the cultural heritage can be made in any generation, and these too can be passed on to succeeding generations by the same process of education.

The progressive development and accumulation of the cultural heritage constitute cultural evolution. The inheritance of acquired characters is a real process in cultural evolution.

The differences between twentieth-century man and Stone Age man in morphological characters, including cranial capacity, are relatively slight. But the cultural differences between them are enormous. The changes in *Homo sapiens* from the Paleolithic to the modern stage of development have been brought about mainly by cultural evolution.

The forces of organic evolution can account for the origin and early evolution of the genus *Homo*. But organic evolution alone will not account for the change from early man to modern man. The process of cultural evolution enters the picture here and provides an explanation for the changes that cannot be explained by organic evolutionary forces.

Cultural evolution has a momentum of its own that is different from that of organic evolution. And cultural evolution can be considered as a process in itself. In practice, however, cultural evolution interacts with organic evolution. Although it is useful to isolate cultural evolution for purposes of special study, it is more correct, in any study of mankind, to view modern man as a product of the joint action of organic and cultural evolution.

A useful summary of the timing of the main evolutionary changes, both organic and cultural, in the genus *Homo* is given by Bodmer and Cavalli-Sforza (1976), and is reproduced here as Figure 36.1.

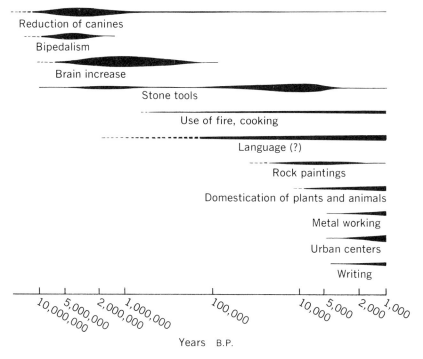

Reduction of canines

Bipedalism

Brain increase

Stone tools

Use of fire, cooking

Language (?)

Rock paintings

Domestication of plants and animals

Metal working

Urban centers

Writing

10,000,000 5,000,000 2,000,000 1,000,000 100,000 10,000 5,000 2,000 1,000

Years   B.P.

FIGURE 36.1
Approximate timing of various organismic and cultural changes in the genus *Homo*.
(Redrawn from W. F. Bodmer and L. L. Cavalli-Sforza, *Genetics, Evolution, and Man*, copyright 1976, W. H. Freeman and Co., San Francisco.)

Education

Man, considered as one of the mammals and one of the higher primates, is a product of organic evolution; but man as a human is a product of both organic and cultural evolution, as noted above. Moreover, the truly distinctive features of the human state, those that distinguish man from animal, are the features that have been produced by cultural evolution.

Individual organisms of *Homo sapiens* have, at birth, an animal and a primate constitution, but not a cultural endowment. The latter is acquired during individual development by education in the broad sense. During the early years of ontogeny each individual develops from a human animal condition (at birth) through a barbarian stage (as a juvenile) to the human condition, in any real or diagnostic sense of the word human. And this individual mental development takes place under the molding influence of education.

Furthermore, the individual acquires the culture that he (or she) is born and raised in, except in those rare instances where an individual develops the ability to think for himself and goes beyond his native cultural horizons—and to a certain extent even then.

The essential difference between man and animal lies, then, in the realm of the human mind, not in the human body, and the distinctively human aspects of the human mind are a product of education (Briffault, 1927).

Conceptual Thought

One of the distinctive features of human mental life is the capacity for conceptual thought—the ability to think in terms of concepts and to formulate ideas.

Now conceptual thought depends on language. It depends, moreover, on a language structure more sophisticated than that of animal languages, in which specific emotions and directives are expressed by particular signals. Human language, as distinguished from animal language, contains words for general classes of items—thus, words for classes of objects or for abstract ideas. Furthermore, human language consists of varied combinations of words in phrases and sentences; these word combinations enable a speaker to express relationships among ideas and to formulate complex concepts. A language with these properties can serve as a foundation for conceptual thought.

Such a type of language requires a social group above a critical size for its maintenance and transmittal, and it requires a prolonged learning period in each new generation for its acquisition. A social group above a critical size, composed of educable individuals, is essential for the development and maintenance of conceptual thought. And conceptual thought is, in turn, an essential ingredient in the higher forms of human culture.

The Evolutionary Nature of Man

The age-old question—What is man?—has been answered in a variety of ways. Man is a political animal; man is a tool-making animal; a religious animal; and so on. Evolutionary biology provides an approach to this question that makes use of historical depth rather than aphoristic phrases.

As seen in the light of evolutionary biology, man is a mammal, more particularly a primate, more particularly a hominoid, and a hominid of an advanced type. These successive layers or grades of organic evolutionary development are built into the human organism.

But man is more than an animal and more than a product of organic evolution alone. He is also a product of cultural evolution. Cultural evolution adds another layer, or series of layers, if you will, to man's nature. A dualistic constitution, partly biological and partly cultural, is built into mankind by the course of his evolutionary development.

The top layer in the stratified human make-up, the layer implanted by cultural evolution, is the definitive one for distinguishing the human condition from the animal state. We can add to our list of aphorisms: Man is the cultural animal.

The next question concerns the placement of the boundary line between the animal state and the human state in evolutionary development. When, in the course of phylogeny, did man enter into the human state? When did *Homo* or some earlier hominid become human in the strict sense of the word as defined above?

We must recall that the elements that make up the human condition are also found in the non-human primates: elements such as social grouping, prolonged immaturity, intelligence, language, and tool-making ability (see Chapter 35). These elements developed during the primate animal stage of evolution; they were available to the early hominids; and they were developed further to new heights in human evolution.

The transition from the animal to the truly human condition was a gradual one during a moderately long phylogenetic history. A similar gradual transition occurs in the ontogeny of every individual person. The boundary line can be drawn only in an arbitrary way in individual ontogeny, as is done in legal definitions of maturity, and it is doubtful whether any non-arbitrary boundary can be found in phylogeny. The evolutionary change of state was historical and gradual.

Collateral Readings

Coon, C. S. 1962. *The Story of Man.* Ed. 2. Alfred A. Knopf, New York.

Farb, P. 1968. *Man's Rise to Civilization as Shown by the Indians of North America from Primeval Times to the Coming of the Industrial State.* E. P. Dutton & Co., New York.

37

Determinants of
Human Evolution

The Unfinished Quest for Explanations

Modern man is the culmination of a series of evolutionary trends that can be traced back into the non-human primates. The more important trends in human evolution were discussed in a general way in the preceding chapters (34–36). These and other trends are described more fully in works on anthropology and human evolution (e.g., Campbell, 1974). What are the driving forces in these trends?

There is wide agreement on the thesis that human evolution involves a mixture of two processes: organic evolution and cultural evolution. Trends in organic evolution are propelled by long-continued selection, either orthoselection (see Chapter 26) or repeated events of selection-drift (Chapter 30). Cultural evolution has a momentum of its own that can explain the gradual accumulation of the cultural heritage.

But what is the relation between natural selection and organic evolutionary trends, on the one hand, and cultural evolution, on the other?

The simplest and historically oldest answer to this question is the thesis that organic evolution brought the human phyletic lineage up to a certain threshold, to a level where cultural evolution was possible. At this point in human phylogeny, cultural evolution took over and brought about all subsequent developments. Organic and cultural evolution are held to be separate in time and independent in action. This view is implicit in the writings of the older cultural anthropologists (e.g., Briffault, 1927) and is still being expressed in some modern discussions (e.g., Kraus, 1973).

The great development of both evolutionary biology and anthropology during the last two or three decades, and the increased communication between the two formerly isolated fields, have led to a more modern view. This view maintains that natural selection has continued to operate and organic evolution has continued to occur during the era of cultural evolution (e.g., Darlington, 1969; Bajema, 1971). The results of cultural evolution are simply more conspicuous than those of continuing organic evolution.

A further refinement is the concept of continual interaction between organic and cultural evolution. An important effect of such interaction would be selection for educability and for the ability to assimilate culture. Organic and cultural evolution would thus often proceed hand in hand. This idea is stressed by Dobzhansky (1962).

Although much progress has been made, up to now, in our understanding of the determinants of human evolution, much more progress still remains to be made. Existing explanations are sometimes too general, too lacking in specificity, and perhaps insufficient in some cases.

I have before me several recent works on human evolution written from the standpoint of anthropology. These works all describe the evolutionary trends. They identify the significant factors: habitat, hand use, intelligence, etc. Their discussions of the driving force behind the trends, however, amount to little more than a general reference to "natural selection." This is no doubt right as far as it goes, but it doesn't go very far.

Another work on human evolution written by an organic evolutionist (Dobzhansky, 1962) discusses the role of natural selection in much greater detail. But Dobzhansky discusses the various modes of individual selection, as they may apply in human populations, and he does not consider group selection. It is fair to ask whether the genetically determined traits of modern man can all be accounted for by individual selection, or whether, alternatively, some traits may be products of group selection.

Some human traits are themselves very complex. Intelligence is a case in point. Will one single mode of selection explain the development of human intelligence? We should consider the possibility that some aspects of this complex trait have been developed by individual selection and other aspects by group selection.

Turning to the other member of the dual set of forces, cultural evolution, our concepts regarding this also seem to be very general and in need of analytical study.

The objective of the present chapter is to bring some of these problems into better focus.

Group Selection

The conditions exist in the social primates, including man, for the effective operation of group selection. Interdeme selection and group selection were introduced in Chapter 11. It will be recalled that we are using the term group selection for a special form of interdeme selection in which the competing populations are social groups.

The troop or band is a more or less cohesive unit of organization in the primates, including man. It is a unit of maintenance and of reproduction.

Different troops or bands must often come into competition for a foraging territory or some other necessary resource. The success or failure of a troop or band in competitive encounters and territorial disputes with rival groups is determined, among other factors, by the intelligence of the members and/or leaders of the group, its technology and communications, and its strength in numbers. Group selection can come into play when the competing troops or bands differ genetically with respect to these characteristics.

It is biologically reasonable to assume that group-to-group differences in intelligence and in its products, technological ability and communications ability, are determined in part by genetic differences in both the non-human and the human primates. It is ideologically unfashionable at present (in the United States) to seriously consider this premise in regard to human groups, especially if the term race is mentioned in the same context. This constraint does not apply to non-human primates, however. Eschewing such discriminatory attitudes toward monkeys and apes, we will accept the biologically realistic premise of genetic differences in mental and behavioral ability between some groups in human and non-human primates alike.

In a succession of group-selection events in primates, including hominids, the social groups with the best intelligence, coordination, and technology can be expected to prevail, on the average, and their genetically determined capacities will tend to spread within the species. The process of group selection, long continued, would bring about a gradual increase in intelligence, language, and technological ability in each surviving lineage. Group selection can explain the origin of a number of features of social life in primates.

One such feature is language. Note that we are using the term language in the broad sense here to include animal language (see discussions in Chapters 35 and 36). Communication is an adaptively valuable characteristic of the troop or band. It is also a group characteristic. It seems very probable that language ability in primates, including man, has been developed by group selection.

Intelligence is adaptively useful for both the individual animal and the social group. Individual selection is probably an important determinant here, but this does not rule out supplementary effects of group selection. A troop consisting mainly of smart monkeys is likely to be able to cope more successfully with the problems in the world around it than a neighboring competitor troop of stupid monkeys.

Even group size, which is a factor in group success, may come under the control of group selection to some extent. The formation of larger-sized troops in semi-terrestrial and terrestrial primates is correlated with the reduction of male intolerance (see Chapter 35). Large group size is adaptively advantageous in semi-terrestrial and terrestrial primates for defense, competition, and cooperative hunting. If male intolerance stood in the way of expansion of pre-existing small troops, and if some troops had a higher degree of genetically determined male tolerance than others, these latter troops and their genes for male tolerance would tend to prevail.

Group selection is not discussed in seven recent books (1962 to 1974) on human evolution that I have examined. Among recent authors, Mayr (1963, pp. 651–652) and Wilson (1975, pp. 562, 567, 573) do mention group selection in connection with human evolution, but they mention it only in passing, and do not develop the subject to any extent. On the other hand, Darwin, in *The Descent of Man* (1883, Chs. 4–7), attached considerable importance to what we would now call group selection.

In a study begun in the post-Darwin era but written and published much later, Keith (1948) developed the concept that the social group, especially the small group in primitive man, is the basic evolutionary unit in human evolution. Human evolution was supposed to result from competitive interactions between small inbreeding social groups. Keith set this idea in a framework of evolutionary theory that was out of date in 1948 and is even more so today. Keith's thesis has not been accepted by modern anthropologists, and his 1948 book is not cited in most recent books on human evolution. This, however, may be a case of throwing the baby out with the bathwater.

Intelligence

Intelligence is a complex of several different kinds of mental ability, as is widely recognized. There is problem-solving ability, mechanical ability, learning ability, communication skill, and so on.

Some components of intelligence confer an advantage primarily on the individual primate and his or her offspring. This would be the case for ability to learn from experience, memory, ability to predict results of actions, and facility in tool use. The evolutionary development of such components of intelligence can be attributed largely to individual selection.

Other components of intelligence confer their advantage in a social context. This is the case with language skill in group coordination and in education of the young. Such types of mental ability could well be produced largely by group selection.

A logical and reasonable hypothesis, therefore, is that primate intelligence has been built up by a combination of individual and group selection.

Learning ability in primates is a function of the length of the immature stages of individual development. Comparative data on length of immature stages, in living primates representing different grades from prosimian to human, show clearly that the period available for learning has increased from the lower to the higher primates (see Table 35.1). Rate of individual development is a genetically determined trait. Natural selection is thus implicated as a driving force in this evolutionary trend. Furthermore, since primate learning is a social process, group selection must have played a large role in the adjustment of development rates for greater learning ability.

Group selection would tend to increase the level of intelligence in all lineages of social primates. Yet the trend toward greater intelligence has obviously proceeded further in human than in non-human primates. It is probable that the special biological and social factors described in the preceding chapters (34–36) —bipedal locomotion, hand use, etc.—released a potential for the development of superior intelligence in hominids that was not released in arboreal, quadrupedal, non-human primates.

Interspecific Selection

Another mode of supra-individual selection has evidently played a role in hominid evolution. This is interspecific selection leading to species replacement (see Chapter 20).

Homo and *Australopithecus* overlapped in time and were sympatric. It is probable that *Homo erectus* or its immediate ancestor and the *Australopithecus* species were in direct or at least indirect competition for food or foraging territories. *Australopithecus* died out, while *Homo erectus* survived and rose to dominance in the hominid adaptive zone.

Homo erectus had a larger brain and better stone tools than *Australopithecus*. These are probably two of the factors determining the course of the interspecific selection process.

Interspecific selection in the hominids differs from that in most animals in being linked to group selection. The contest was between social groups of *Homo erectus* or "pre-*erectus*" and competing groups of *Australopithecus*. Since the effective units of reproduction in the competing species were social groups, the interspecific selection process takes on a special form, that of interspecific group selection.

Mechanisms of Cultural Evolution

The standard view of cultural evolution holds that this process proceeds by the gradual build-up of the cultural heritage. Useful knowledge is first discovered, then incorporated into the general body of knowledge of the social group, and then transmitted to succeeding generations by education. Each succeeding generation adds to this store of knowledge, by its own discoveries, and these are also transmitted in the same way. Thus the cultural heritage is cumulative.

Undoubtedly cultural evolution does occur in this fashion to some extent. But does it occur only this way—does it always proceed so straightforwardly? A reading of history suggests that it does not. Current views of cultural evolution are as general, and as vague, as current views concerning natural selection in human evolution, and the former, like the latter, stand in need of critical reevaluation.

Let us again consider group size. Group selection acting on genetic group differences in male tolerance may have brought the group size up to the level of the multiple-male troop or band. This level is attained in some advanced non-human primates and in primitive hominids (see Chapter 35).

But the trend toward increasing size and complexity of the social group obviously continues into human prehistory and history. Bands become merged into tribes, and tribes into chiefdoms and states. States increase in size from city-states to small nation-states to large nation-states.

Degree of male tolerance has nothing to do with these latter phases of the group-size trend. Male tolerance is not noticeably different between the Netherlands and the United States. Intertribal and international wars, however, have had much to do with the size trend. Up to a point, bigger tribes or nations are superior to smaller tribes or nations in the test of war, on the average, where other factors are more or less equal, as they often are. In a succession of wars the smaller social-political units tend to disappear as independent groups, surviving in some dependent form—as a colony, a subordinate state in a federation, or a satellite country—while the larger groups emerge as the effective independent political powers. Group-size trends are determined, up to a point, by the survival of the biggest.

I say "up to a point" because very big political units, as exemplified by empires, do not have long-term durability, but always break up eventually into smaller fragments. There are reversals in the group-size trend, especially at the upper level. The size-increase trend is displayed more clearly below this level.

Here we have a social trend determined by non-genetic differences between competing groups. The non-genetic difference in this model happens to concern group size. But one can readily think of differences between groups in other cultural features—differences that, in a succession of inter-group competitive encounters, could produce a cultural evolutionary trend.

Some technological trends, for example, may be a resultant of successive competitive struggles between groups with different technological capabilities. Ideological trends could be produced in the same way. One can predict the average result of numerous encounters between groups that hold a belief in magic and opposing groups that take a pragmatic approach to problems.

We have let the pendulum swing, in this section, from the orthodox view of general cultural evolution on a broad front, which we hold to be too simplistic, to cultural evolution mediated by cultural differences between competing social groups. Of course this latter picture is also too simple. We are forced by lack of space to ignore the numerous complications found in real interactions between social groups. It is perhaps sufficient for the present to call attention to alternative, indirect, group-centered modes of cultural evolution.

Interactions

Interactions between selection for anatomical and morphogenetic characters on the one hand and selection for social traits on the other are an important aspect of human evolution.

The evolution of learning ability entailed some interesting complexities. Increase in brain size beyond a certain point was limited by the size of the birth canal through which the infant head had to pass at birth. Selection for intelligence came into opposition with selection for female viability in childbirth. The impasse was resolved by two compromises.

The first compromise was to widen the female pelvis. This entailed loss of running speed. The human female is a slower runner than the human male. A certain amount of running ability was sacrificed in the human female for the ability to bear larger-brained children.

The second compromise was to slow down the rate of development of the brain. The human brain is incompletely formed at birth, and the newborn human infant is quite helpless; the brain must undergo most of its development during infancy and childhood. Humans differ markedly in these conditions from apes and monkeys, which are more self-sufficient at birth. This second compromise had the secondary effect of prolonging the immature stages of life (Briffault, 1927; Washburn, 1960).

Both compromise solutions reduced the independence of the human female. She was saddled with her young for a longer period of time. And the loss of running ability made her less able to hunt for herself and more dependent on the men for game.

These third-order effects produced fourth-order effects within the primitive human social group. The band had to pick up some of the responsibility for the women and the young. The social group became a more cooperative and communal organization. The cause-effect chain went full circle. Selection for learning ability, a socially useful trait in a social milieu, brought about various morphological and developmental changes that, as a corollary, called forth a still more close-knit social organization.

Concluding Statement

We have enumerated several evolutionary forces that are probably involved in human evolution. Among the organic evolutionary forces are individual selection, intraspecific group selection, interspecific group selection, and the selection-drift combination. Among cultural evolutionary forces are the general accumulation of a cultural heritage, and cultural trends resulting from competition between social groups that differ culturally (but not genetically).

Some of these forces are recognized in current discussions of human evolution and some are not. We have devoted a bit of extra attention to the latter; this attention is for exposition, not for emphasis. I would not know how to assess the relative importance of the various forces mentioned above. To be more specific, I do suggest that group selection is important in human organic evolution, and that differential success of social groups is important in human cultural evolution, but I don't profess to know how important, relative to other forces, these group-centered processes are.

A possible determinant of human evolution that has received some prominence in current literature discussions is the fixation of selectively neutral mutant genes by genetic drift. Here I will venture an opinion. This process could well have occurred occasionally. But I would not expect any important characteristics in human evolution to have resulted from it.

The various evolutionary forces interact with one another, and this interaction greatly complicates the analysis of the process of human evolution.

It is probably safe to say that hominid phylogeny is now much better known than it was a generation ago, although important lacunae remain, and that the evolutionary *factors* in this phylogeny have been identified for the most part; but our understanding of the evolutionary *forces* involved in human evolution is still very incomplete.

Collateral Readings

Dobzhansky, Th. 1962. *Mankind Evolving.* Yale University Press, New Haven, Conn.

Mayr, E. 1970. *Populations, Species, and Evolution.* Harvard University Press, Cambridge, Mass. Chapter 20.

Wilson, E. O. 1975. *Sociobiology: The New Synthesis.* Harvard University Press, Cambridge, Mass. Chapter 27.

VIII

SOCIAL IMPLICATIONS

38

The Evolution Controversy

The Creation

The story of the divine creation of the world is told in Chapters 1 and 2 of the Book of Genesis, ascribed to Moses.

In Chapter 1 of Genesis it is recorded that God caused the earth to bring forth plant life on the third day (or stage) of creation, animal life on the fifth day, and man on the sixth day of creation. God intended that man should "have dominion over the fish of the sea, and over the fowl of the air, and over the cattle, and over all the earth . . ." (Genesis 1:26).

God "formed man of the dust of the ground, and breathed into his nostrils the breath of life . . ." (Genesis 2:7). He next placed the newly formed man in the Garden of Eden and provided him with food-bearing trees, water, and "every beast of the field, and every fowl of the air" (Genesis 2:8–20). Finally, while the man was asleep, God took one of his ribs and made it into a woman (Genesis 2:21–22).

The man was Adam and the woman Eve. Adam and Eve were the parents of mankind; "Eve . . . was the mother of all living" (Genesis 3:20). The year was 4004 B.C., according to the chronology of Ussher, which was based on Biblical records.

The Controversy

For centuries in Western civilization the Mosaic account of creation held sway as the standard explanation of life and of man's origins. Its exclusive position in the realm of thought was maintained by orthodox religious practice and religious teaching.

This monolithic situation came to an end with the publication of Darwin's *Origin of Species* in 1859. The traditional orthodox view now faced a serious rival viewpoint. No longer did one doctrine enjoy exclusive rights in the ideological field; instead, the field became an arena for competition between two alternative doctrines, the one a religious dogma and the other a scientific theory.

The general idea of organic evolution was not new in 1859. That idea had been put forward earlier by Lamarck in the *Philosophie Zoologique* (1809), by Chambers in *Vestiges of Creation* (1844), and by others. Lamarck's work was unconvincing as to both factual evidence and theory, and so were the treatments of the other early-nineteenth-century authors. These treatments did, however, prepare the way for the reception of Darwin's *Origin of Species*.

Some of the milestones in Darwin's career are listed in Table 38.1. See also de Beer (1964).

In *The Origin of Species,* Darwin marshaled an overwhelming body of evidence, derived from several independent lines of investigation, pointing to the past and continuing occurrence of organic evolution. And he developed the theory of natural selection to provide a motive force for the evolutionary changes. Darwin's treatment of the subject, unlike those of previous authors, was convincing to zoologists, botanists, and geologists. The years and decades after 1859 saw a gradual and widespread acceptance of the evolution theory throughout the scientific world.

There was, to be sure, some controversy after 1859 in scientific circles. But the main controversy took place in lay circles and involved attempts by theologians and religious groups to defend the act-of-creation doctrine. The controversy between creationism and evolutionism during the late nineteenth century in England, Europe, and America forms an interesting but often bitter chapter in the history of thought (see, inter alia, F. Darwin, 1958; Fothergill, 1952; Eiseley, 1958; Glick, 1974).

The creationism-evolutionism controversy is still going on today in the United States. In fact, it has flared up during the 1960s and 1970s. Undoubtedly it is an anachronism today.

The ranks of creationists have dwindled since the nineteenth century in modern Western nations. Secularization of a part of the population is one reason; liberalization of views within many religious sects is another. Many religious groups and individuals take an allegorical interpretation of Genesis that permits

TABLE 38.1
Key events and dates in the life of Charles Darwin.

1809	Born in Shrewsbury, England.
1827–1831	Student at Cambridge University.
1831–1836	Naturalist aboard H.M.S. *Beagle* on voyage around the world.
1839	Married to Emma Wedgwood. Publication of *Journal of Researches* (= *Voyage of the Beagle*). (Ed. 2 in 1845.)
1842	*Structure and Distribution of Coral Reefs.* (Ed. 2 in 1874.)
1851	Monograph on Cirripedes.
1858	Joint papers by Darwin and Wallace on natural selection and evolution communicated to the Linnean Society, and published in the *Proceedings of the Linnean Society.*
1859	Publication of *The Origin of Species.* (Ed. 1; ed. 6 appeared in 1872.)
1868	*The Variation of Animals and Plants Under Domestication.* (Ed. 2 in 1875.)
1871	*The Descent of Man and Selection in Relation to Sex.* (Ed. 2 in 1874.)
1872	*The Expression of the Emotions in Man and Animals.*
1875–1881	Publication of various botanical works, including *The Effects of Cross and Self Fertilisation in the Vegetable Kingdom* (1876).
1882	Died at home in Down, buried in Westminster Abbey.

them to accept the conclusions of evolutionary biology. Fundamentalist sects in the United States, however, still insist today on a strictly literal interpretation of Genesis. These are the groups that keep the anachronistic controversy alive.

The modern controversy in the United States focuses on the teaching of biology in the public schools. Creationists want to see an even-handed presentation of the Mosaic account and the Darwinian account in biology textbooks. Furthermore, they want evolution played down and presented as "a theory rather than a fact." Fundamentalist groups advocating creationism have influenced school boards to establish the above propositions as guidelines in several states, including California and Texas. The questions of educational principle thus raised have been debated in board meetings and courtrooms.

Critique

The creationist's position in the contemporary scene is wrong on several counts. In the first place, the public schools are not a proper medium for the promulgation of particular sectarian religious doctrines.

The demand for equal time in textbook and classroom presentations of creationism and evolution is based on the premise that the two subjects are commensurate—that they warrant equal consideration—and this premise is false. Creationism and evolutionism are not in the same league. Creationism is not a scientific theory that can be weighed against the evolution theory. It is a religious dogma. There is no independent evidence to support its account of the origin of plant, animal, and human life. The acceptance of the creationist story depends on faith rather than reason.

Let us consider the point, stated so often by creationists, that evolution is a theory, not a fact. We will accept for the moment the naive assumption that a scientific subject can be categorized as either theory or fact. And we will refrain from pointing out that creationism is neither a theory nor a fact.

Evolution was indeed a theory at one time, but a bit of water has flowed over the dam since 1859. Much of evolution is now in the category of verified fact. Evolutionary changes at the levels of microevolution and speciation have been observed by competent biologists and can now be regarded as proven facts.

Evolutionary changes at the level of macroevolution are in a somewhat different position, in that they are historical phenomena and consequently were not observed directly by any human observer. Lack of eye-witness testimony is an inherent problem in any attempt to reconstruct past events. The Mosaic account of creation, which was written long after the event, suffers from this same difficulty.

The simple dichotomy of theory vs. fact does not cover the situation in macroevolution. Evolutionary history cannot be observed directly, but can be inferred from the fossil record. The fossils themselves are facts. Macroevolution is a necessary inference from these facts.

Evidence for Macroevolution

It may be worthwhile to summarize briefly the various lines of evidence that point to the reality of macroevolution. Such a summary is desirable in view of the continuing opposition to evolution in some quarters. Most of the lines of evidence listed below were brought to bear on the question by Darwin in *The Origin of Species*. The amount of evidence in each line is much greater today than it was in Darwin's time.

(1) Economy of hypotheses. Microevolutionary changes and some modes of speciation are established by direct observational and/or experimental evidence. There is a continuous series of levels from the local population and

geographical race through the species to the species group, subgenus, and genus. To anyone who has studied living nature and has become imbued with a sense of the unity of nature, it is inconceivable that organic diversity should be produced by evolution at the lower levels but by some other means at supraspecific levels.

(2) Fossil record. Many groups of organisms with good fossil records show a succession of forms through geological time (see Chapter 26). In some cases transitional forms between two different major groups are preserved as fossils (see Chapters 31 and 34).

(3) Taxonomic pattern of relationships among living species. Species are clustered naturally into genera, genera into families, families into orders, and so on. This "natural subordination of organic beings in groups under groups," as Darwin put it, is a result of a branching phylogeny. A natural genus Y is composed of species with a common ancestry; a related genus Z contains another set of species; the genera Y and Z are two branches on an older branch, the family; and so on through the taxonomic hierarchy. The hierarchical nature of taxonomic relationships was not invented by evolutionary biologists; in fact it was discovered by pre-evolutionary taxonomists who accepted creationism, but it was later correctly interpreted by evolutionists. Living species would not be expected to cluster in groups within groups if they were products of separate acts of creation.

(4) Geographical distribution. Many genera, tribes, families, and other groups of medium taxonomic rank are confined to one geographical region— a particular archipelago, subcontinent, or continent, etc. The group has its center of distribution in this region. The logical inference is that the living species belonging to the group, or many of them, evolved in the region of their present abundance and diversity (see Chapter 31, latter part). The doctrine of creationism provides no explanation for the observed patterns of geographical distribution of supraspecific groups.

(5) Homologies. When the members of a major group are compared, they are found to possess a common general plan of structural organization, but to differ with respect to certain homologous parts in the body. One can think of the different forms taken by the homologous forelimbs in different orders of mammals, of the homologous hindlimbs in different mammalian orders, of the limbs in different classes of terrestrial vertebrates, of the corolla parts in the flowers of different angiosperm families and orders, and so on.

Let Darwin discuss the significance of homologous organs in his own words (1872, Ch. 14):

> What can be more curious than that the hand of a man, formed for grasping, that of a mole for digging, the leg of the horse, the paddle of the porpoise, and the wing of the bat, should all be constructed on the same pattern, and should include similar bones, in the same relative positions? . . .
>
> Nothing can be more hopeless than to attempt to explain this similarity of pattern in members of the same class, by utility or by the doctrine of final causes. The hopelessness of the attempt has been expressly admitted by Owen in his most interesting work on the 'Nature of Limbs'. On the ordinary view of the independent creation of each being, we can only say that so it is;— that it has pleased the Creator to construct all the animals and plants in each great class on a uniform plan; but this is not a scientific explanation.
>
> The explanation is to a large extent simple on the theory of the selection of successive slight modifications,—each modification being profitable in some way to the modified form, but often affecting by correlation other parts of the organisation. In changes of this nature, there will be little or no tendency to alter the original pattern, or to transpose the parts. The bones of a limb might be shortened and flattened to any extent, becoming at the same time enveloped in thick membrane, so as to serve as a fin; or a webbed hand might have all its bones, or certain bones, lengthened to any extent, with the membrane connecting them increased, so as to serve as a wing; yet all these modifications would not tend to alter the framework of the bones or the relative connection of the parts. If we suppose that an early progenitor—the archetype as it may be called—of all mammals, birds, and reptiles, had its limbs constructed on the existing general pattern, for whatever purpose they served, we can at once perceive the plain signification of the homologous construction of the limbs throughout the class.

(6) Vestigial organs. Some members of a major group often possess an organ that is atrophied and non-functional. This rudimentary organ is homologous with a well-developed functional organ in other members of the same group. Thus flightless birds have rudimentary wings, some whales have rudimentary pelvis bones, and some snakes, including the python, have rudimentary hindlimbs. These structures are interpreted as reduced vestiges of their well-developed homologues in other members of the same major group. The subgroup possessing the rudimentary organ entered into a habitat or way of life in which the formerly functional organ was no longer useful, and it was greatly reduced by selection, but vestiges of it persist as phylogenetic remnants. There is no good explanation for the existence of useless rudimentary organs in the doctrine of creationism.

(7) Biochemical similarities. A modern line of evidence, not available in Darwin's time, is the close similarity in the biochemical composition and molec-

ular structure of homologous proteins in members of different related families or orders. Good examples are the homologous forms of hemoglobin and of cytochrome c in humans, apes, and monkeys (see Chapter 29).

It should be noted, finally, that the case for the reality of macroevolution rests not on one line of evidence alone, but on the concurrent testimony of several independent classes of facts, as Darwin showed in *The Origin of Species*.

Collateral Readings

Anonymous. 1967. *Did Man Get Here by Evolution or by Creation?* Watchtower Bible and Tract Society, New York.

Darwin, C. 1872. *On the Origin of Species*. Ed. 6. John Murray, London. Various modern reprints of this edition are available.

De Beer, G. 1964. *Charles Darwin*. Doubleday & Co., Garden City, N.Y.

Eiseley, L. 1958. *Darwin's Century*. Doubleday & Co., Garden City, N.Y.

Glick, T. F., ed. 1974. *The Comparative Reception of Darwinism*. University of Texas Press, Austin.

39

Relevance of Evolutionary Biology

Science bears two kinds of fruits: understanding of nature and control of natural processes. Science nourishes both philosophy and technology. These two main aspects of science are found in evolutionary biology.

Evolutionary biology, broadly construed, has many valuable practical applications to its credit in agriculture and medicine. The improvement of cultivated plants and domesticated animals is, in essence, applied organismic evolution. A knowledge of evolutionary biology is a necessary ingredient in any successful program of chemical or biological control of disease organisms in medicine, or of insect pests in agriculture.

But the most important contribution of evolutionary biology, in the opinion of many evolutionists, including myself, lies in the philosophical and intellectual sphere. Evolutionary biology fills a basic human need: the desire to know how life developed on earth and the desire to understand the nature and origin of man. The emancipation of human thought from traditional dogmas concerning these questions is an achievement of evolutionary biology in line of succession with the earlier scientific revolutions in geology and astronomy.

Evolutionary biology provides perspective on man's place in nature. Many specific issues that are troublesome in non-biological reference frames can be seen more clearly in the light of this perspective. Two examples, one ancient and the other modern, will be mentioned.

The question of pain was discussed extensively in a theological framework in the Middle Ages. Why would a good and perfect God let a bad quality such as pain into the world? Conversely, does the existence of pain cast doubt upon the perfection of God? Needless to say, the question was not resolved conclusively in the theological framework. In the framework of evolutionary biology, however, pain falls into place as an adaptive trait. The ordinary sensations of pain are signals warning of potential harm to the animal body.

Today the question of legalized abortion is being discussed in a legal and political framework. Anti-abortionists cite the rights of the fetus, pro-abortionists the rights of women. Again we arrive at an impasse. And again evolutionary biology provides a perspective in which to view the question. A long-term evolutionary trend in the vertebrates, reaching an advanced stage in the mammals and primates, concerns fecundity. The trend is toward fewer offspring and greater care of the young, that is, toward the sacrifice of quantity for the sake of quality in the younger generation. The implications of this trend are not supportive of the anti-abortionist position.

The human condition in the world today is not good. Overpopulation, environmental degradation, poverty, famine, aggressiveness, Frankenstein technology, and bad government stalk the earth; and overhead always hangs the threat of nuclear annihilation. Thoughtful world citizens in all walks of life are now concerned about the human condition. But the older evolutionary biologists can claim to be among the first to see the crisis coming and sound the alarm; if their warnings in the first half of this century had been heeded, much present misery could have been spared.

The situation of the mid-1970s is before us at this writing. And evolutionary biology is again available for help. Viewing man as a part of nature, and more particularly as a dominant animal species in nature, evolutionary biology can see the basic problem — a growing disharmony between man and his environment — and can identify the main cause of the problem: runaway population growth. Evolutionary biology can point out the inadequacy of halfway measures (such as increasing food production alone), and of cosmetic solutions (food conferences), while pointing to the need for real solutions dealing with causes rather than with symptoms.

The future of human civilization, and perhaps of the human species itself, may depend on the adoption of measures to correct the present disequilibrium between man and his earthly environment. Such measures will have to include the stabilization and eventually the reduction of human population size through birth control. But this is not the place to discuss specific remedies. The point to be stressed here is that the insights of evolutionary biology are needed in the search for proper solutions.

We noted earlier that evolutionary biology, like other branches of science, has both a practical and a theoretical side. The distinction between the two aspects is usually clear enough in the short term. But philosophical perspectives are sometimes more practical than practical applications themselves, in the long term. Perspectives drawn from theoretical evolutionary biology that would help to prevent the decline of mankind would be practical applications of the first order.

The knowledge, implications, and warnings of evolutionary biology have been available for years, but have been ignored by the people who run governments, as well as by large segments of the citizenry. Meanwhile the human condition has not been improving, and the big problems are not being solved, under the leadership of the old traditional governments. Obviously evolutionary biology must get into public affairs. For the lessons of evolutionary biology are of concern to everybody.

Collateral Readings

Ehrlich, P. R., and R. L. Harriman. 1971. *How to Be a Survivor.* Ballantine Books, New York.
Hardin, G. 1972. *Exploring New Ethics for Survival.* Viking Press, New York.

Key to Technical Terms

The following list contains terms used in this book and some synonymous terms used in other works. The terms are either defined and exemplified, or sometimes briefly characterized, or placed in synonymy, on the pages indicated. This is not an independent glossary, therefore, but a guide tied in with the text. For an extremely useful glossary see Rieger, Michaelis, and Green (1976).

Bibliography

Abel, O. 1929. *Paläobiologie und Stammesgeschichte.* Gustav Fischer Verlag, Jena.

Alexander, R. D. 1974. The evolution of social behavior. *Ann. Rev. Ecol. Syst.* **5:** 325–383.

Amadon, D. 1950. The Hawaiian honeycreepers (Aves, Drepaniidae). *Bull. Amer. Mus. Nat. Hist.* **95**(4): 151–262.

Anderson, E., and W. L. Brown. 1952. Origin of corn belt maize and its genetic significance. In *Heterosis,* ed. by J. W. Gowen. Iowa State College Press, Ames.

Anderson, W., Th. Dobzhansky, O. Pavlovsky, J. Powell, and D. Yardley. 1975. Genetics of natural populations. XLII. Three decades of genetic change in *Drosophila pseudoobscura. Evolution* **29:** 24–36.

Andrewartha, H. G., and L. C. Birch. 1954. *The Distribution and Abundance of Animals.* University of Chicago Press, Chicago.

Antonovics, J. 1971. The effects of a heterogeneous environment on the genetics of natural populations. *Amer. Sci.* **59:** 593–599.

Avery, O. T., C. M. Macleod, and M. McCarty. 1944. Studies on the chemical nature of the substance inducing transformation of pneumococcal types. *J. Exp. Med.* **79:** 137–158.

Axelrod, D. I. 1959. Late Cenozoic evolution of the Sierran bigtree forest. *Evolution* **13:** 9–23.

———. 1967. Quaternary extinctions of large mammals. *Univ. Calif. Publ. Geol. Sci.* **74:** 1–42.

———. 1970. Mesozoic paleogeography and early angiosperm history. *Bot. Rev.* **36:** 277–319.

———. 1972. Ocean-floor spreading in relation to ecosystematic problems. In *A Symposium on Ecosystematics,* ed. by R. T. Allen and F. C. James. University of Arkansas Museum, Fayetteville, Occasional Paper No. 4.

———. 1974. Revolutions in the plant world. *Geophytology* **4:** 1–6.

Axelrod, D. I., and H. P. Bailey. 1968. Cretaceous dinosaur extinction. *Evolution* **22:** 595–611.

Ayala, F. J. 1969. Experimental invalidation of the principle of competitive exclusion. *Nature* **224:** 1076–1079.

———. 1974. The concept of biological progress. In *Studies in the Philosophy of Biology,* ed. by F. J. Ayala and Th. Dobzhansky. Macmillan, New York and London.

Ayala, F. J., and W. W. Anderson, 1973. Evidence of natural selection in molecular evolution. *Nature* **241:** 274–276.

Ayala, F. J., and M. E. Gilpin. 1974. Gene frequency comparisons between taxa: support for the natural selection of protein polymorphisms. *Proc. Nat. Acad. Sci. USA* **71:** 4847–4849.

Ayala, F. J., C. A. Mourao, S. Perez-Salas, R. Richmond, and Th. Dobzhansky. 1970. Enzyme variability in the *Drosophila willistoni* group. I. Genetic differentiation among sibling species. *Proc. Nat. Acad. Sci. USA* **67:** 225–232.

Ayala, F. J., and M. L. Tracey. 1974. Genetic differentiation within and between species of the *Drosophila willistoni* group. *Proc. Nat. Acad. Sci. USA* **71:** 999–1003.

Babcock, E. B., and G. L. Stebbins. 1938. The American species of *Crepis:* their inter-relationships and distribution as affected by polyploidy and apomixis. Carnegie Institution of Washington, Washington, D.C., Publ. 504.

Bada, J. L., R. A. Schroeder, and G. F. Carter. 1974. New evidence for the antiquity of man in North America deduced from aspartic acid racemization. *Science* **184:** 791–793.

Bajema, C. J. 1971. *Natural Selection in Human Populations.* John Wiley & Sons, New York.

Bakker, R. T. 1975. Dinosaur renaissance. *Sci. Amer.* **232**(4): 58–78.

Baldwin, P. H. 1953. Annual cycle, environment and evolution in the Hawaiian honey-creepers (Aves: Drepaniidae). *Univ. Calif. Publ. Zool.* **52:** 285–398.

Barghoorn, E. S. 1971. The oldest fossils. *Sci. Amer.* **224**(5): 30–42.

Bartholomew, G. A., and J. B. Birdsell. 1953. Ecology and the protohominids. *Amer. Anthropol.* **55:** 481–498.

Bateman, A. J. 1950. Is gene dispersion normal? *Heredity* **4:** 353–363.

Bateman, K. G. 1959. The genetic assimilation of four venation phenocopies. *J. Genet.* **56:** 443–474.

Beardmore, J. A., Th. Dobzhansky, and O. Pavlovsky. 1960. An attempt to compare the fitness of polymorphic and monomorphic experimental populations of *Drosophila pseudoobscura. Heredity* **14:** 19–33.

Beaudry, J. R. 1960. The species concept: its evolution and present status. *Rev. Canad. Biol.* **19:** 219–240.

Bergson, H. 1911. *Creative Evolution.* Translation. Henry Holt & Co., New York. Various later reprints.

Birdsell, J. B. 1950. Some implications of the genetical concept of race in terms of spatial analysis. *Cold Spring Harbor Symp. Quant. Biol.* **15:** 259–314.

———. 1958. On population structure in generalized hunting and collecting populations. *Evolution* **12:** 189–205.

Blair, W. F. 1955. Mating call and stage of speciation in the *Microhyla olivacea–M. carolinensis* complex. *Evolution* **9:** 469–480.

———. 1960. *The Rusty Lizard: A Population Study.* University of Texas Press, Austin.

Bock, W. J. 1970. Microevolutionary sequences as a fundamental concept in macro-evolutionary models. *Evolution* **24:** 704–722.

Bodmer, W. F., and L. L. Cavalli-Sforza. 1976. *Genetics, Evolution, and Man.* W. H. Freeman and Co., San Francisco.

Bonnett, O. T. 1954. The inflorescences of maize. *Science* **120:** 77–87.

Brace, C. L., H. Nelson, and N. Korn. 1971. *Atlas of Fossil Man.* Holt, Rinehart & Winston, New York.

Briffault, R. 1927. *The Mothers.* 3 vols. Macmillan, London.

Brower, L. P. 1970. Plant poisons in a terrestrial food chain and implications for mimicry theory. In *Biochemical Coevolution.* Oregon State University Press, Corvallis.

Brower, L. P., and J. V. Z. Brower. 1964. Birds, butterflies, and plant poisons: a study in ecological chemistry. *Zoologica (New York)* **49:** 137–159.

Brown, W. L., and E. O. Wilson. 1956. Character displacement. *Syst. Zool.* **5:** 49–64.

Brücher, H. 1943. Experimentelle Untersuchungen über den Selektionswert künstlich erzeugter Mutanten von *Antirrhinum majus. Z. Botan.* **39:** 1–47.

Brues, A. M. 1969. Genetic load and its varieties. *Science* **164:** 1130–1136.

Brussard, P. F., and P. R. Ehrlich. 1970. The population structure of *Erebia epipsodea* (Lepidoptera: Satyrinae). *Ecology* **51:** 119–129.

Buffon, G. L. L. 1770. *Histoire Naturelle des Oiseaux.* Paris.

———. 1808. *Natural History of Birds, Fish, Insects, and Reptiles.* Translation, 6 vols. Symonds, London.

Bush, G. L. 1969*a*. Sympatric host race formation and speciation in frugivorous flies of the genus *Rhagoletis* (Diptera: Tephritidae). *Evolution* **23:** 237–251.

———. 1969*b*. Mating behavior, host specificity, and the ecological significance of sibling species in frugivorous flies of the genus *Rhagoletis* (Diptera–Tephritidae). *Amer. Nat.* **103:** 669–672.

Cain, A. J., and P. M. Sheppard. 1954. Natural selection in *Cepaea. Genetics* **39:** 89–116.

Campbell, B. 1966, 1974. *Human Evolution.* Eds. 1 and 2. Aldine-Atherton, Chicago and New York.

———, ed. 1972. *Sexual Selection and the Descent of Man 1871–1971.* Aldine-Atherton, Chicago.

Candela, P. B. 1942. The introduction of blood-group B into Europe. *Human Biol.* **14:** 413–443.

Carlquist, S. 1974. *Island Biology.* Columbia University Press, New York.

Carson, H. L. 1957. The species as a field for gene recombination. In *The Species Problem,* ed. by E. Mayr. American Association for the Advancement of Science, Washington, D.C.

———. 1959. Genetic conditions which promote or retard the formation of species. *Cold Spring Harbor Symp. Quant. Biol.* **24:** 87–105.

———. 1970. Chromosome tracers of the origin of species. *Science* **168:** 1414–1418.

———. 1971. Speciation and the founder principle. *Stadler Genet. Symp.* **3:** 51–70.

———. 1975. The genetics of speciation at the diploid level. *Amer. Nat.* **109:** 83–92.

Carson, H. L., and W. B. Heed. 1964. Structural homozygosity in marginal populations of nearctic and neotropical species of *Drosophila* in Florida. *Proc. Nat. Acad. Sci. USA* **52:** 427–430.

Cavalli-Sforza, L. L., and W. F. Bodmer. 1971. *The Genetics of Human Populations.* W. H. Freeman and Co., San Francisco.

Chambers, R. 1844. *Vestiges of the Natural History of Creation.* London.

Clarke, B. 1973. Neutralists vs. selectionists. *Science* **180:** 600–601.

Clausen, J. 1951. *Stages in the Evolution of Plant Species.* Cornell University Press, Ithaca, N.Y.

————. 1965. Population studies of alpine and subalpine races of conifers and willows in the California high Sierra Nevada. *Evolution* **19:** 56–68.

Clausen, J., and W. M. Hiesey. 1958. Experimental studies on the nature of species. IV. Genetic structure of ecological races. Carnegie Institution of Washington, Washington, D.C., Publ. 615.

Clausen, J., D. D. Keck, and W. M. Hiesey. 1940. Experimental studies on the nature of species. I. Effect of varied environments on western North American plants. Carnegie Institution of Washington, Washington, D.C., Publ. 520.

————, ————, and ————. 1948. Experimental studies on the nature of species. III. Environmental responses of climatic races of *Achillea.* Carnegie Institution of Washington, Washington, D.C., Publ. 581.

Clausen, R. E. 1941. Polyploidy in *Nicotiana. Amer. Nat.* **75:** 291–306.

Cloud, P. 1974. Evolution of ecosystems. *Amer. Sci.* **62:** 54–66.

Colbert, E. H. 1955, 1969. *Evolution of the Vertebrates.* Eds. 1 and 2. John Wiley & Sons, New York.

————. 1961. *Dinosaurs: Their Discovery and Their World.* E. P. Dutton & Co., New York.

Colwell, R. N. 1951. The use of radioactive isotopes in determining spore distribution patterns. *Amer. J. Bot.* **38:** 511–523.

Connell, J. H., and E. Orias. 1964. The ecological regulation of species diversity. *Amer. Nat.* **98:** 399–414.

Coon, C. S. 1955. Some problems of human variability and natural selection in climate and culture. *Amer. Nat.* **89:** 257–279.

————. 1962. *The Story of Man.* Ed. 2. Alfred A. Knopf, New York.

Cope, E. D. 1896. *The Primary Factors of Organic Evolution.* Open Court, Chicago.

Cronquist, A. 1951. Orthogenesis in evolution. *Res. Stud. State Coll. Wash.* **19:** 3–18.

————. 1968. *The Evolution and Classification of Flowering Plants.* Houghton Mifflin, Boston.

————. 1976. The taxonomic significance of the structure of plant proteins: a classical taxonomist's view. *Brittonia* **28:** 1–27.

Crosby, J. L. 1963. The evolution and nature of dominance. *J. Theor. Biol.* **5:** 35–51.

Crow, J. F. 1970. Genetic loads and the cost of natural selection. In *Mathematical Topics in Population Genetics,* ed. by K. Kojima. Springer Verlag, New York.

Crow, J. F., and M. Kimura. 1970. *An Introduction to Population Genetics Theory.* Harper & Row, New York.

Crumpacker, D. W., and J. S. Williams. 1973. Density, dispersion, and population structure in *Drosophila pseudoobscura. Ecol. Monogr.* **43:** 499–538.

Darlington, C. D. 1939, 1958. *The Evolution of Genetic Systems.* Eds. 1 and 2. Ed. 1, Cambridge University Press, Cambridge. Ed. 2, Basic Books, New York.

————. 1969. *The Evolution of Man and Society.* Simon & Schuster, New York.

Darwin, C. 1859, 1872. *On the Origin of Species by Means of Natural Selection.* Eds. 1 and 6. John Murray, London.

————. 1871, 1874. *The Descent of Man and Selection in Relation to Sex.* Eds. 1 and 2. John Murray, London.

————. 1875. *The Variation of Animals and Plants Under Domestication.* Ed. 2, 2 vols. John Murray, London.

Darwin, F., ed. 1958. *The Autobiography of Charles Darwin and Selected Letters.* Reprint. Dover Publications, New York.

Dayhoff, M. O., ed. 1968. *Atlas of Protein Sequence and Structure.* Vol. 3, 1967–1968. National Biomedical Research Foundation, Silver Spring, Md.

———, ed. 1969. *Atlas of Protein Sequence and Structure.* Vol. 4, 1969. National Biomedical Research Foundation, Silver Spring, Md.

———, ed. 1972. *Atlas of Protein Sequence and Structure.* Vol. 5, 1972. National Biomedical Research Foundation, Silver Spring, Md.

De Beer, G. R. 1951. *Embryos and Ancestors.* Ed. 2. Oxford University Press, Oxford.

———. 1964. *Charles Darwin.* Doubleday & Co., Garden City, N.Y.

DeFries, J. C., and G. E. McClearn. 1972. Behavioral genetics and the fine structure of mouse populations: a study in microevolution. *Evol. Biol.* **5:** 279–291.

DeVore, I., ed. 1965. *Primate Behavior: Field Studies of Monkeys and Apes.* Holt, Rinehart & Winston, New York.

De Wet, J. M. J., and J. R. Harlan. 1972. Origin of maize: the tripartite hypothesis. *Euphytica* **21:** 271–279.

De Wet, J. M. J., J. R. Harlan, and C. A. Grant. 1971. Origin and evolution of teosinte [*Zea mexicana* (Schrad.) Kuntze]. *Euphytica* **20:** 255–265.

Dobzhansky, Th. 1937, 1951a. *Genetics and the Origin of Species.* Eds. 1 and 3. Columbia University Press, New York.

———. 1943. Genetics of natural populations. IX. Temporal changes in the composition of populations of *Drosophila pseudoobscura. Genetics* **28:** 162–186.

———. 1947a. Genetics of natural populations. XIV. A response of certain gene arrangements in the third chromosome of *Drosophila pseudoobscura* to natural selection. *Genetics* **32:** 142–160.

———. 1947b. A directional change in the genetic constitution of a natural population of *Drosophila pseudoobscura. Heredity* **1:** 53–64.

———. 1948. Genetics of natural populations. XVI. Altitudinal and seasonal changes produced by natural selection in certain populations of *Drosophila pseudoobscura* and *Drosophila persimilis. Genetics* **33:** 158–176.

———. 1950. Mendelian populations and their evolution. *Amer. Nat.* **84:** 401–418.

———. 1951a. (See Dobzhansky, 1937, 1951a, above.)

———. 1951b. Experiments on sexual isolation in *Drosophila.* X. Reproductive isolation between *Drosophila pseudoobscura* and *Drosophila persimilis* under natural and under laboratory conditions. *Proc. Nat. Acad. Sci. USA* **37:** 792–796.

———. 1955. The genetic basis of systematic categories. In *Biological Systematics.* Biology Colloquium, Oregon State College, Corvallis.

———. 1956. Genetics of natural populations. XXV. Genetic changes in populations of *Drosophila pseudoobscura* and *Drosophila persimilis* in some localities in California. *Evolution* **10:** 82–92.

———. 1958. Genetics of natural populations. XXVII. The genetic changes in populations of *Drosophila pseudoobscura* in the American southwest. *Evolution* **12:** 385–401.

———. 1962. *Mankind Evolving.* Yale University Press, New Haven, Conn.

———. 1970. *Genetics of the Evolutionary Process.* Columbia University Press, New York.

———. 1971. Evolutionary oscillations in *Drosophila pseudoobscura.* In *Ecological Genetics and Evolution,* ed. by R. Creed. Blackwell Scientific Publications, Oxford.

Dobzhansky, Th., L. Ehrman, O. Pavlovsky, and B. Spassky. 1964. The superspecies *Drosophila paulistorum. Proc. Nat. Acad. Sci. USA* **51**: 3–9.

Dobzhansky, Th., and C. Epling. 1944. Contributions to the genetics, taxonomy, and ecology of *Drosophila pseudoobscura* and its relatives. Carnegie Institution of Washington, Washington, D.C., Publ. 554.

Dobzhansky, Th., and H. Levene. 1951. Development of heterosis through natural selection in experimental populations of *Drosophila pseudoobscura. Amer. Nat.* **85**: 247–264.

Dobzhansky, Th., and O. Pavlovsky. 1957. An experimental study of interaction between genetic drift and natural selection. *Evolution* **11**: 311–319.

Dobzhansky, Th., and J. R. Powell. 1974. Rates of dispersal of *Drosophila pseudoobscura* and its relatives. *Proc. Roy. Soc. London, B* **187**: 281–298.

Dobzhansky, Th., and B. Spassky. 1947. Evolutionary changes in laboratory cultures of *Drosophila pseudoobscura. Evolution* **1**: 191–216.

Dobzhansky, Th., B. Spassky, and N. Spassky. 1952. A comparative study of mutation rates in two ecologically diverse species of *Drosophila. Genetics* **37**: 650–664.

Dobzhansky, Th., and N. P. Spassky. 1962. Genetic drift and natural selection in experimental populations of *Drosophila pseudoobscura. Proc. Nat. Acad. Sci. USA* **48**: 148–156.

Dobzhansky, Th., and S. Wright. 1943. Genetics of natural populations. X. Dispersion rates in *Drosophila pseudoobscura. Genetics* **28**: 304–340.

—— and ——. 1947. Genetics of natural populations. XV. Rate of diffusion of a mutant gene through a population of *Drosophila pseudoobscura. Genetics* **32**: 303–324.

Dott, R. H., and R. L. Batten. 1971. *Evolution of the Earth.* McGraw-Hill, New York.

Downs, T. 1961. A study of variation and evolution in Miocene *Merychippus.* Contributions in Science, Los Angeles County Museum, Los Angeles, No. 45.

Dunn, L. C., and S. P. Dunn. 1957. The Jewish community of Rome. *Sci. Amer.* **196**(3): 118–128.

Durrant, A. 1962*a.* The environmental induction of heritable change in *Linum. Heredity* **17**: 27–61.

——. 1962*b.* Induction, reversion and epitrophism of flax genotrophs. *Nature* **196**: 1302–1304.

Ehrlich, P. R. 1961. Intrinsic barriers to dispersal in checkerspot butterfly. *Science* **134**: 108–109.

Ehrlich, P. R., and R. L. Harriman. 1971. *How to Be a Survivor.* Ballantine Books, New York.

Ehrlich, P. R., and P. H. Raven. 1964. Butterflies and plants: a study in coevolution. *Evolution* **18**: 586–608.

Ehrlich, P. R., R. R. White, M. C. Singer, S. W. McKechnie, and L. E. Gilbert. 1975. Checkerspot butterflies: a historical perspective. *Science* **188**: 221–228.

Ehrman, L. 1965. Direct observation of sexual isolation between allopatric and between sympatric strains of the different *Drosophila paulistorum* races. *Evolution* **19**: 459–464.

Ehrman, L., and R. P. Kernaghan. 1971. Microorganismal basis of infectious hybrid male sterility in *Drosophila paulistorum. J. Hered.* **62**: 66–71.

Ehrman, L., and E. B. Spiess. 1969. Rare-type mating advantage in *Drosophila. Amer. Nat.* **103**: 675–680.

Ehrman, L., and D. L. Williamson. 1965. Transmission by injection of hybrid sterility to nonhybrid males in *Drosophila paulistorum:* preliminary report. *Proc. Nat. Acad. Sci. USA* **54:** 481–483.

Eiseley, L. 1958. *Darwin's Century.* Doubleday & Co., Garden City, N. Y.

Eisenberg, J. F., N. A. Muckenhirn, and R. Rudran. 1972. The relation between ecology and social structure in primates. *Science* **176:** 863–874.

Eklund, M. W., F. T. Poysky, J. A. Meyers, and G. A. Pelroy. 1974. Interspecies conversion of *Clostridium botulinum* type C to *Clostridium novyi* type A by bacteriophage. *Science* **186:** 456–458.

Eldredge, N., and S. J. Gould. 1972. Punctuated equilibria: an alternative to phyletic gradualism. In *Models in Paleobiology,* ed. by T. J. M. Schopf. Freeman, Cooper & Co., San Francisco.

Emlen, J. M. 1973. *Ecology: An Evolutionary Approach.* Addison-Wesley, Reading, Mass.

Epling, C., H. Lewis, and F. M. Ball. 1960. The breeding group and seed storage: a study in population dynamics. *Evolution* **14:** 238–255.

Epling, C., D. F. Mitchell, and R. H. T. Mattoni. 1953. On the role of inversions in wild populations of *Drosophila pseudoobscura. Evolution* **7:** 342–365.

Evans, G. M., A. Durrant, and H. Rees. 1966. Associated nuclear changes in the induction of flax genotrophs. *Nature* **212:** 697–699.

Ewens, W. J. 1965. Further notes on the evolution of dominance. *Heredity* **20:** 443–450.

Faegri, K., and L. van der Pijl. 1971. *The Principles of Pollination Ecology.* Ed. 2. Pergamon Press, Oxford.

Falconer, D. S. 1960. *Introduction to Quantitative Genetics.* Ronald Press, New York.

Farb, P. 1968. *Man's Rise to Civilization as Shown by the Indians of North America from Primeval Times to the Coming of the Industrial State.* E. P. Dutton & Co., New York.

Felsenstein, J., and S. Yokoyama. 1976. The evolutionary advantage of recombination. II. Individual selection for recombination. *Genetics* **83:** 845–859.

Fisher, R. A. 1930, 1958. *The Genetical Theory of Natural Selection.* Eds. 1 and 2. Ed. 1, Clarendon Press, Oxford. Ed. 2, Dover Publications, New York.

Fisher, R. A., and E. B. Ford. 1947. The spread of a gene in natural conditions in a colony of the moth *Panaxia dominula* L. *Heredity* **1:** 143–174.

Flake, R. H., and V. Grant. 1974. An analysis of the cost-of-selection concept. *Proc. Nat. Acad. Sci. USA* **71:** 3716–3720.

Ford, E. B. 1955. *Moths.* Collins, London.

———. 1964, 1971. *Ecological Genetics.* Eds. 1 and 3. Ed. 1, Methuen, London. Ed. 3, Chapman & Hall, London.

———. 1965. *Genetic Polymorphism.* Faber & Faber, London.

Fothergill, P. G. 1952. *Historical Aspects of Organic Evolution.* Hollis & Carter, London.

Fox, A. S, W. F. Duggleby, W. M. Gelbart, and S. B. Yoon. 1970. DNA-induced transformation in *Drosophila*: evidence for transmission without integration. *Proc. Nat. Acad. Sci. USA* **67:** 1834–1838.

Fox, A. S., and S. B. Yoon. 1966. Specific genetic effects of DNA in *Drosophila melanogaster. Genetics* **53:** 897–911.

——— and ———. 1970. DNA-induced transformation in *Drosophila:* locus-specificity and the establishment of transformed stocks. *Proc. Nat. Acad. Sci. USA* **67:** 1608–1615.

Fox, A. S., S. B. Yoon, and W. M. Gelbart. 1971. DNA-induced transformation in *Drosophila:* genetic analysis of transformed stocks. *Proc. Nat. Acad. Sci. USA* **68:** 342–346.

Galinat, W. C. 1970. The cupule and its role in the origin and evolution of maize. Agricultural Experiment Station, University of Massachusetts, Amherst, Bull. 585.

———. 1971*a*. The origin of maize. *Ann. Rev. Genet.* **5:** 447–478.

———. 1971*b*. The evolution of sweet corn. Agricultural Experiment Station, University of Massachusetts, Amherst, Bull. 591.

Garn, S. M. 1961. *Human Races.* Charles C. Thomas, Springfield, Ill.

Gause, G. F. 1934. *The Struggle for Existence.* Williams & Wilkins, Baltimore.

———. 1935. Experimental demonstration of Volterra's periodic oscillations in the numbers of animals. *J. Exp. Biol.* **12:** 44–48.

Geist, V. 1971. *Mountain Sheep: A Study in Behavior and Evolution.* University of Chicago Press, Chicago.

Gill, E. D. 1975. Evolution of Australia's unique flora and fauna in relation to the plate tectonics theory. *Proc. Roy. Soc. Victoria* **87:** 215–234.

Glass, B., M. S. Sacks, E. F. Jahn, and C. Hess. 1952. Genetic drift in a religious isolate: an analysis of the causes of variation in blood groups and other gene frequencies in a small population. *Amer. Nat.* **86:** 145–159.

Glick, T. F., ed. 1974. *The Comparative Reception of Darwinism.* University of Texas Press, Austin.

Goldschmidt, R. B. 1940. *The Material Basis of Evolution.* Yale University Press, New Haven, Conn.

———. 1952. Homoeotic mutants and evolution. *Acta Biotheor.* **10:** 87–104.

———. 1953. Experiments with a homoeotic mutant, bearing on evolution. *J. Exp. Zool.* **123:** 79–114.

———. 1955. *Theoretical Genetics.* University of California Press, Berkeley.

Gottlieb, L. D. 1971. Gel electrophoresis: new approach to the study of evolution. *Bioscience* **21:** 939–943.

———. 1974. Genetic confirmation of the origin of *Clarkia lingulata. Evolution* **28:** 244–250.

Gould, S. J. 1973. The misnamed, mistreated, and misunderstood Irish elk. *Nat. Hist.* 82(3): 10–19.

———. 1974. The origin and function of "bizarre" structures: antler size and skull size in the "Irish elk," *Megaloceros giganteus. Evolution* **28:** 191–220.

Grant, A., and V. Grant. 1956. Genetic and taxonomic studies in *Gilia.* VIII. The Cobwebby gilias. *Aliso* **3:** 203–287.

Grant, K. A., and V. Grant. 1964. Mechanical isolation of *Salvia apiana* and *Salvia mellifera* (Labiatae). *Evolution* **18:** 196–212.

——— and ———. 1968. *Hummingbirds and Their Flowers.* Columbia University Press, New York.

Grant, V. 1954. Genetic and taxonomic studies in *Gilia.* IV. *Gilia achilleaefolia. Aliso* **3:** 1–18.

———. 1958. The regulation of recombination in plants. *Cold Spring Harbor Symp. Quant. Biol.* **23:** 337–363.

———. 1959. *Natural History of the Phlox Family. Systematic Botany.* M. Nijhoff, The Hague.

———. 1960. Genetic and taxonomic studies in *Gilia*. XI. Fertility relationships of the diploid Cobwebby gilias. *Aliso* **4:** 435–481.

———. 1963. *The Origin of Adaptations*. Columbia University Press, New York.

———. 1966*a*. Selection for vigor and fertility in the progeny of a highly sterile species hybrid in *Gilia*. *Genetics* **53:** 757–775.

———. 1966*b*. The selective origin of incompatibility barriers in the plant genus *Gilia*. *Amer. Nat.* **100:** 99–118.

———. 1971. *Plant Speciation*. Columbia University Press, New York.

———. 1975. *Genetics of Flowering Plants*. Columbia University Press, New York.

Grant, V., and R. H. Flake. 1974*a*. Population structure in relation to cost of selection. *Proc. Nat. Acad. Sci. USA* **71:** 1670–1671.

——— and ———. 1974*b*. Solutions to the cost-of-selection dilemma. *Proc. Nat. Acad. Sci. USA* **71:** 3863–3865.

Grant, V., and K. A. Grant. 1965. *Flower Pollination in the Phlox Family*. Columbia University Press, New York.

Gray, A. P. 1954. *Mammalian Hybrids*. Commonwealth Agricultural Bureaux, Farnham Royal, England.

———. 1958. *Bird Hybrids*. Commonwealth Agricultural Bureaux, Farnham Royal, England.

Greenwood, J. J. D. 1974. Effective population numbers in the snail *Cepaea nemoralis*. *Evolution.* **28:** 513–526.

———. 1976. Effective population number in *Cepaea:* a modification. *Evolution* **30:** 186.

Gustafsson, Å. 1951. Mutations, environment and evolution. *Cold Spring Harbor Symp. Quant. Biol.* **16:** 263–281.

Haldane, J. B. S. 1932. *The Causes of Evolution*. Longmans, Green & Co., London.

———. 1957. The cost of natural selection. *J. Genet.* **55:** 511–524.

———. 1960. More precise expressions for the cost of natural selection. *J. Genet.* **57:** 351–360.

Hallam, A., ed. 1973. *Atlas of Paleobiogeography*. American Elsevier, New York.

Hamilton, W. D. 1964. The genetical evolution of social behavior. I and II. *J. Theor. Biol.* **7:** 1–16, 17–52.

Hardin, G. 1960. The competitive exclusion principle. *Science* **131:** 1292–1297.

———. 1972. *Exploring New Ethics for Survival*. Viking Press, New York.

Harding, J., R. W. Allard, and D. G. Smeltzer. 1966. Population studies in predominantly self-pollinated species. IX. Frequency-dependent selection in *Phaseolus lunatus*. *Proc. Nat. Acad. Sci. USA* **56:** 99–104.

Hardy, A. C. 1954. Escape from specialization. In *Evolution as a Process*, ed. by J. Huxley, A. C. Hardy, and E. B. Ford. George Allen & Unwin, London.

Hayes, W. 1968. *The Genetics of Bacteria and Their Viruses*. Ed. 2. John Wiley & Sons. New York.

Heiser, C. B. 1973*a*. *Seed to Civilization*. W. H. Freeman and Co., San Francisco.

———. 1973*b*. Introgression re-examined. *Bot. Rev.* **39:** 347–366.

Hill, J. 1967. The environmental induction of heritable changes in *Nicotiana rustica* parental and selection lines. *Genetics* **55:** 735–754.

Hinton, T., P. T. Ives, and A. T. Evans. 1952. Changing the gene order and number in natural populations. *Evolution* **6:** 19–28.

Hiraizumi, Y., L. Sandler, and J. F. Crow. 1960. Meiotic drive in natural populations of *Drosophila melanogaster*. III. Populational implications of the segregation-distorter locus. *Evolution* **14**: 433–444.

Hovanitz, W. 1953. Polymorphism and evolution. *Symp. Soc. Exp. Biol.* **7**: 240–253.

Hubby, J. L., and R. C. Lewontin. 1966. A molecular approach to the study of genic heterozygosity in natural populations. I. The number of alleles at different loci in *Drosophila pseudoobscura*. *Genetics* **54**: 577–594.

Hubby, J. L., and L. H. Throckmorton. 1965. Protein differences in *Drosophila*. II. Comparative species genetics and evolutionary problems. *Genetics* **52**: 203–215.

Huettel, M. D., and G. L. Bush. 1972. The genetics of host selection and its bearing on sympatric speciation in *Procecidochares* (Diptera: Tephritidae). *Entomol. Exp. Appl.* **15**: 465–480.

Hutchinson, G. E. 1965. *The Ecological Theater and the Evolutionary Play.* Yale University Press, New Haven, Conn.

Huxley, J. S. 1932. *Problems of Relative Growth.* Methuen, London.

———. 1942. *Evolution: The Modern Synthesis.* George Allen & Unwin, London.

———. 1954. The evolutionary process. In *Evolution as a Process,* ed. by J. Huxley, A. C. Hardy, and E. B. Ford. George Allen & Unwin, London.

———. 1958. Evolutionary processes and taxonomy with special reference to grades. In *Systematics of Today,* ed. by O. Hedberg. Uppsala Universitets Årsskrift, Uppsala.

Ingram, V. M. 1963. *The Hemoglobins in Genetics and Evolution.* Columbia University Press, New York.

Ives, P. T. 1950. The importance of mutation rate genes in evolution. *Evolution* **4**: 236–252.

Jain, S. K., and A. D. Bradshaw. 1966. Evolutionary divergence among adjacent plant populations. I. The evidence and its theoretical analysis. *Heredity* **21**: 407–441.

Jepson, W. L. 1909. *A Flora of California.* Vol. 1, Part 1. Associated Students Store, University of California, Berkeley.

Johannsen, W. 1911. The genotype conception of heredity. *Amer. Nat.* **45**: 129–159.

Johnston, J. S., and W. B. Heed. 1976. Dispersal of desert-adapted *Drosophila:* the saguaro-breeding *D. nigrospiracula. Amer. Nat.* **110**: 629–651.

Jones, H. A., J. C. Walker, T. M. Little, and R. M. Larson. 1946. Relation of color-inhibiting factor to smudge resistance in onion. *J. Agric. Res.* **72**: 259–264.

Jones, J. S. 1973. Ecological genetics and natural selection in molluscs. *Science* **182**: 546–552.

Jukes, T. H. 1972. Comparison of polypeptide sequences. In *Darwinian, Neo-Darwinian, and Non-Darwinian Evolution,* ed. by L. M. LeCam, J. Neyman, and E. L. Scott. Berkeley Symposia on Mathematical Statistics and Probability. Proceedings of the Sixth Symposium, Vol. 5. University of California Press, Berkeley.

Keast, A., F. C. Erk, and B. Glass, eds. 1972. *Evolution, Mammals, and Southern Continents.* State University of New York Press, Albany.

Keith, A. 1948. *A New Theory of Human Evolution.* Watts, London.

Kerner, A. 1894–1895. *The Natural History of Plants.* Translation, 2 vols. Blackie & Son, London.

Kerr, W. E., and S. Wright. 1954. Experimental studies of the distribution of gene frequencies in very small populations of *Drosophila melanogaster*. I. Forked. *Evolution* **8:** 172–177.

Kerster, H. W. 1964. Neighborhood size in the Rusty lizard, *Sceloporus olivaceus*. *Evolution* **18:** 445–457.

Kerster, H. W., and D. A. Levin. 1968. Neighborhood size in *Lithospermum caroliniense*. *Genetics* **60:** 577–587.

Kettlewell, H. B. D. 1956. Further selection experiments on industrial melanism in the Lepidoptera. *Heredity* **10:** 287–301.

———. 1973. *The Evolution of Melanism.* Oxford University Press, New York.

Kimura, M. 1968. Genetic variability maintained in a finite population due to mutational production of neutral and nearly neutral isoalleles. *Genet. Res.* **11:** 247–269.

Kimura, M., and T. Ohta. 1971. *Theoretical Aspects of Population Genetics.* Princeton University Press, Princeton, N.J.

——— and ———. 1972. Population genetics, molecular biometry, and evolution. In *Darwinian, Neo-Darwinian, and Non-Darwinian Evolution*, ed. by L. M. LeCam, J. Neyman, and E. L. Scott. Berkeley Symposia on Mathematical Statistics and Probability. Proceedings of the Sixth Symposium, Vol. 5. University of California Press, Berkeley.

King, J. L., and T. H. Jukes. 1969. Non-Darwinian evolution. *Science* **164:** 788-798.

King, M. C., and A. C. Wilson. 1975. Evolution at two levels in humans and chimpanzees. *Science* **188:** 107–116.

Kircher, H. W., and W. B. Heed. 1970. Phytochemistry and host plant specificity in *Drosophila. Recent Adv. Phytochem.* **3:** 191–209.

Koestler, A., and J. R. Smythies, eds. 1969. *Beyond Reductionism: New Perspectives in the Life Sciences.* Alpbach Symposium, 1968. Macmillan, New York.

Kojima, K., and K. N. Yarbrough. 1967. Frequency dependent selection at the esterase-6 locus in *Drosophila melanogaster. Proc. Nat. Acad. Sci. USA* **57:** 645–649.

Kolata, G. B. 1974. !Kung hunter-gatherers: feminism, diet, and birth control. *Science* **185:** 932–934.

Koopman, K. F. 1950. Natural selection for reproductive isolation between *Drosophila pseudoobscura* and *Drosophila persimilis. Evolution* **4:** 135–148.

Kraus, G. 1973. *Homo sapiens in Decline.* New Diffusionist Press, Bedfordshire, England.

Krebs, C. J. 1973. *Ecology: The Experimental Analysis of Distribution and Abundance.* Harper & Row, New York.

Krebs, C. J., M. S. Gaines, B. L. Keller, J. H. Meyers, and R. H. Tamarin. 1973. Population cycles in small rodents. *Science* **179:** 35–41.

Kulp, J. L. 1961. Geologic time scale. *Science* **133:** 1105–1114.

Lack, D. 1944. Ecological aspects of species formation in passerine birds. *Ibis* **86:** 260–286.

———. 1947. *Darwin's Finches.* Cambridge University Press, Cambridge.

Laird, C. D., and B. J. McCarthy. 1968. Magnitude of interspecific nucleotide sequence variability in *Drosophila. Genetics* **60:** 303–322.

Lamarck, J. B. P. de. 1809. *Philosophie Zoologique.* Paris.

———. 1815–1822. *Histoire Naturelle des Animaux sans Vertèbres.* Paris.

Lamotte, M. 1951. Recherches sur la structure génétique des populations naturelles de *Cepaea nemoralis* (L.). *Bull. biol. France Belg.* (suppl.) **35**: 1–238.

———. 1959. Polymorphism of natural populations of *Cepaea nemoralis*. *Cold Spring Harbor Symp. Quant. Biol.* **24**: 65–86.

Laughlin, W. S. 1950. Blood groups, morphology and population size of the Eskimos. *Cold Spring Harbor Symp. Quant. Biol.* **15**: 165–173.

Laycock, G. 1974. Dilemma in the desert: bighorns or burros? *Audubon Magazine*, September 1974.

LeCam, L. M., J. Neyman, and E. L. Scott, eds. 1972. *Darwinian, Neo-Darwinian, and Non-Darwinian Evolution*. Berkeley Symposia on Mathematical Statistics and Probability. Proceedings of the Sixth Symposium, Vol. 5. University of California Press, Berkeley.

Le Gros Clark, W. E. 1960. *The Antecedents of Man*. Quadrangle Books, Chicago.

Leng, E. R. 1960. Long-term selection of corn for oil and protein content. Mimeographed annual report, Illinois Agricultural Experiment Station, Urbana.

Lerner, I. M. 1954. *Genetic Homeostasis*. Oliver & Boyd, Edinburgh and London.

———. 1958. *The Genetic Basis of Selection*. John Wiley & Sons, New York.

Lerner, I. M., and F. K. Ho. 1961. Genotype and competitive ability of *Tribolium* species. *Amer. Nat.* **95**: 329–343.

Levene, H. 1953. Genetic equilibrium where more than one ecological niche is available. *Amer. Nat.* **87**: 331–333.

Levin, D. A. 1970. Reinforcement of reproductive isolation: plants versus animals. *Amer. Nat.* **104**: 571–581.

———. 1971. Plant phenolics: an ecological perspective. *Amer. Nat.* **105**: 157–181.

———. 1972. Low frequency disadvantage in the exploitation of pollinators by corolla variants in *Phlox. Amer. Nat.* **106**: 453–460.

Levin, D. A., and W. L. Crepet. 1973. Genetic variation in *Lycopodium lucidulum*: a phylogenetic relic. *Evolution* **27**: 622–632.

Levin, D. A., and H. W. Kerster. 1967. Natural selection for reproductive isolation in *Phlox. Evolution* **21**: 679–687.

——— and ———. 1968. Local gene dispersal in *Phlox. Evolution* **22**: 130–139.

——— and ———. 1969. Density-dependent gene dispersal in *Liatris. Amer. Nat.* **103**: 61–74.

——— and ———. 1971. Neighborhood structure in plants under diverse reproductive methods. *Amer. Nat.* **105**: 345–354.

——— and ———. 1974. Gene flow in seed plants. *Evol. Biol.* **7**: 139–220.

Levin, D. A., and B. A. Schaal. 1970. Corolla color as an inhibitor of interspecific hybridization in *Phlox. Amer. Nat.* **104**: 273–283.

Levins, R. 1968. *Evolution in Changing Environments*. Princeton University Press, Princeton, N.J.

Lewin, R.A., ed. 1962. *Physiology and Biochemistry of Algae*. Academic Press, New York.

Lewis, H. 1962. Catastrophic selection as a factor in speciation. *Evolution* **16**: 257–271.

Lewis, H., and C. Epling. 1959. *Delphinium gypsophilum*, a diploid species of hybrid origin. *Evolution* **13**: 511–525.

Lewis, H., and M. E. Lewis. 1955. The genus *Clarkia. Univ. Calif. Publ. Bot.* **20**: 241–392.

Lewis, H., and P. H. Raven. 1958. Rapid evolution in *Clarkia. Evolution* **12**: 319–336.

Lewis, H., and M. R. Roberts. 1956. The origin of *Clarkia lingulata*. *Evolution* **10:** 126–138.

Lewontin, R. C. 1955. The effects of population density and composition on viability in *Drosophila melanogaster*. *Evolution* **9:** 27–41.

———. 1970. The units of selection. *Ann. Rev. Ecol. Syst.* **1:** 1–18.

———. 1973. Population genetics. *Ann. Rev. Genet.* **7:** 1–17.

———. 1974. *The Genetic Basis of Evolutionary Change*. Columbia University Press, New York.

Lewontin, R. C., and J. L. Hubby. 1966. A molecular approach to the study of genic heterozygosity in natural populations. II. Amount of variation and degree of heterozygosity in natural populations of *Drosophila pseudoobscura*. *Genetics* **54:** 595–609.

Lillegraven, J. A. 1974. Biogeographic considerations of the marsupial-placental dichotomy. *Ann. Rev. Ecol. Syst.* **5:** 263–283.

Lindsay, D. W., and R. K. Vickery. 1967. Comparative evolution in *Mimulus guttatus* of the Bonneville basin. *Evolution* **21:** 439–456.

Littlejohn, M. J. 1965. Premating isolation in the *Hyla ewingi* complex (Anura: Hylidae). *Evolution* **19:** 234–243.

Loomis, W. F. 1967. Skin-pigment regulation of vitamin-D biosynthesis in man. *Science* **157:** 501–506.

MacArthur, R. H. 1958. Population ecology of some warblers of northeastern coniferous forests. *Ecology* **39:** 599–619.

McVeigh, I., and C. J. Hobdy. 1952. Development of resistance by *Micrococcus pyogenes* var. *aureus* to antibiotics: morphological and physiological changes. *Amer. J. Bot.* **39:** 352–359.

Mangelsdorf, P. C. 1958. Reconstructing the ancestor of corn. *Proc. Amer. Phil. Soc.* **102:** 454–463.

———, ed. 1959. The origin of corn. *Bot. Mus. Leafl. Harvard Univ.* **18:** 329–440.

———. 1974. *Corn: Its Origin, Evolution, and Improvement*. Harvard University Press, Cambridge, Mass.

Mangelsdorf, P. C., H. W. Dick, and J. Camara-Hernandez. 1967. Bat Cave revisited. *Bot. Mus. Leafl. Harvard Univ.* **22:** 1–31.

Mangelsdorf, P. C., and D. F. Jones. 1926. The expression of Mendelian factors in the gametophyte of maize. *Genetics* **11:** 423–455.

Mangelsdorf, P. C., R. S. MacNeish, and W. C. Galinat. 1964. Domestication of corn. *Science* **143:** 538–545.

Margulis, L. 1970. *Origin of Eukaryotic Cells*. Yale University Press, New Haven, Conn.

Martin, P. S. 1973. The discovery of America. *Science* **179:** 969–974.

Martin, P. S., and H. E. Wright, Jr., eds. 1967. *Pleistocene Extinctions: The Search for a Cause*. Yale University Press, New Haven, Conn.

Maynard Smith, J. 1958. Sexual selection. In *A Century of Darwin*, ed. by S. A. Barnett. Heinemann, London.

Mayo, O. 1966. On the evolution of dominance. *Heredity* **21:** 499–511.

Mayr, E. 1942. *Systematics and the Origin of Species*. Columbia University Press, New York.

———. 1954. Change of genetic environment and evolution. In *Evolution as a Process*, ed. by J. Huxley, A. C. Hardy, and E. B. Ford. George Allen & Unwin, London.

————. 1957*a*. Species concepts and definitions. In *The Species Problem*, ed. by E. Mayr. American Association for the Advancement of Science, Washington, D.C.

————. 1957*b*. Difficulties and importance of the biological species concept. In *The Species Problem*, ed. by E. Mayr. American Association for the Advancement of Science, Washington, D.C.

————. 1963. *Animal Species and Evolution*. Harvard University Press, Cambridge, Mass.

————. 1969. *Principles of Systematic Zoology*. McGraw-Hill, New York.

————. 1970. *Populations, Species, and Evolution*. Harvard University Press, Cambridge, Mass.

————. 1972*a*. The nature of the Darwinian revolution. *Science* **176:** 981–989.

————. 1972*b*. Sexual selection and natural selection. In *Sexual Selection and the Descent of Man 1871–1971*, ed. by B. Campbell. Aldine-Atherton, Chicago.

Mettler, L. E., and T. G. Gregg. 1969. *Population Genetics and Evolution*. Prentice-Hall, Englewood Cliffs, N.J.

Michener, C. D. 1947. A revision of the American species of *Hoplitis* (Hymenoptera, Megachilidae). *Bull. Amer. Mus. Nat. Hist.* **89:** 257–318.

————. 1975. The Brazilian bee problem. *Ann. Rev. Entomol.* **20:** 399–416.

Miller, A. H. 1955. A hybrid woodpecker and its significance in speciation in the genus *Dendrocopos*. *Evolution* **9:** 317–321.

Millicent, E., and J. M. Thoday. 1961. Effects of disruptive selection. IV. Gene flow and divergence. *Heredity* **16:** 199–217.

Moos, J. R. 1955. Comparative physiology of some chromosomal types in *Drosophila pseudoobscura*. *Evolution* **9:** 141–151.

Mourant, A. E. 1954. *The Distribution of the Human Blood Groups*. Blackwell Scientific Publications, Oxford.

Muller, C. H. 1966. The role of chemical inhibition (allelopathy) in vegetational composition. *Bull. Torrey Bot. Club* **93:** 332–351.

————. 1970. The role of allelopathy in the evolution of vegetation. In *Biochemical Coevolution*. Oregon State University Press, Corvallis.

Nassar, R., H. J. Muhs, and R. D. Cook. 1973. Frequency-dependent selection at the Payne inversion in *Drosophila melanogaster*. *Evolution* **27:** 558–564.

Nei, M., and Y. Imaizumi. 1966. Genetic structure of human populations. I and II. *Heredity* **21:** 9–35, 183–190.

Newell, N. D. 1949. Phyletic size increase—an important trend illustrated by fossil invertebrates. *Evolution* **3:** 103–124.

————. 1967. Revolutions in the history of life. *Geol. Soc. Amer., Spec. Pap.* **89:** 63–91.

Oakley, K. P. 1949. *Man the Tool-Maker*. British Museum (Natural History), London.

Ohta, T., and M. Kimura. 1971. On the constancy of the evolutionary rate of cistrons. *J. Molec. Evol.* **1:** 18–25.

Oparin, A. I. 1962, 1964. *Life: Its Nature, Origin and Development*. Oliver & Boyd, Edinburgh (1962); Academic Press, New York (1964).

Osborn, H. F. 1910. *The Age of Mammals in Europe, Asia and North America*. Macmillan, New York.

————. 1929. The titanotheres of ancient Wyoming, Dakota, and Nebraska. U.S. Geological Survey, Washington, D.C., Monograph 55.

————. 1934. Aristogenesis, the creative principle in the origin of species. *Amer. Nat.* **68:** 193–235.

Owen, D. F. 1961. Industrial melanism in North American moths. *Amer. Nat.* **95:** 227–233.

Paterniani, E. 1969. Selection for reproductive isolation between two populations of maize, *Zea mays* L. *Evolution* **23:** 534–547.

Patterson, B. 1949. Rates of evolution in taeniodonts. In *Genetics, Paleontology, and Evolution*, ed. by G. L. Jepsen, E. Mayr, and G. G. Simpson. Princeton University Press, Princeton, N.J.

Pianka, E. R. 1974. *Evolutionary Ecology.* Harper & Row, New York.

Pilbeam, D. 1972. *The Ascent of Man.* Macmillan, New York.

Pilbeam, D., and S. J. Gould. 1974. Size and scaling in human evolution. *Science* **186:** 892–901.

Pimentel, D., G. J. C. Smith, and J. Soans. 1967. A population model of sympatric speciation. *Amer. Nat.* **101:** 493–504.

Proctor, M., and P. Yeo. 1973. *The Pollination of Flowers.* Collins, London.

Prout, T. 1962. The effects of stabilizing selection on the time of development in *Drosophila melanogaster. Genet. Res.* **3:** 364–382.

Race, R. R., and R. Sanger. 1962. *Blood Groups in Man.* Ed. 3. Blackwell Scientific Publications, Oxford.

Radinsky, L. 1976. Oldest horse brains: more advanced than previously realized. *Science* **194:** 626–627.

Raven, P. H. 1970. A multiple origin for plastids and mitochondria. *Science* **169:** 641–646.

Raven, P. H., and D. I. Axelrod. 1974. Angiosperm biogeography and past continental movements. *Ann. Missouri Bot. Gard.* **61:** 539–673.

Rejmanek, M., and J. Jenik. 1975. Niche, habitat, and related ecological concepts. *Acta Biotheor.* **24:** 100–107.

Remington, C. L. 1954. The genetics of *Colias* (Lepidoptera). *Adv. Genet.* **6:** 403–450.

Rensch, B. 1960*a.* The laws of evolution. In *Evolution After Darwin*, Vol. 1, ed. by S. Tax. University of Chicago Press, Chicago.

———. 1960*b. Evolution Above the Species Level.* Translation. Columbia University Press, New York.

Rieger, R., A. Michaelis, and M. M. Green. 1976. *A Glossary of Genetics and Cytogenetics: Classical and Molecular.* Ed. 4. Springer Verlag, New York.

Ris, H., and W. Plaut. 1962. Ultrastructure of DNA-containing areas in the chloroplast of *Chlamydomonas. J. Cell Biol.* **13:** 383–391.

Romer, A. S. 1966. *Vertebrate Paleontology.* Ed. 3. University of Chicago Press, Chicago.

Rosin, S., J. K. Moor-Jankowski, and M. Schneeberger. 1958. Die Fertilität im Bluterstamm von Tenna (Hämophilie B). *Acta Genet.* **8:** 1–24.

Ross, H. H. 1957. Principles of natural coexistence indicated by leafhopper populations. *Evolution* **11:** 113–129.

———. 1958. Evidence suggesting a hybrid origin for certain leafhopper species. *Evolution* **12:** 337–346.

Rothschild, M. 1970. Toxic Lepidoptera. *Toxicon* **8:** 293–299.

———. 1975. Remarks on carotenoids in the evolution of signals. In *Coevolution of Animals and Plants*, ed. by L. E. Gilbert and P. H. Raven. University of Texas Press, Austin.

Sakai, K., and K. Gotoh. 1955. Studies on competition in plants. IV. Competitive ability of F_1 hybrids in barley. *J. Hered.* **46:** 139–143.

Sandler, L., and E. Novitski. 1957. Meiotic drive as an evolutionary force. *Amer. Nat.* **91:** 105–110.

Schaeffer, B. 1948. The origin of a mammalian ordinal character. *Evolution* **2:** 164–175.

Schmalhausen, I. I. 1949. *Factors of Evolution: The Theory of Stabilizing Selection.* Translation. Blakiston, Philadelphia.

Schopf, J. W. 1974. Paleobiology of the Precambrian: the age of blue-green algae. *Evol. Biol.* **7:** 1–43.

Selander, R. K. 1970. Behavior and genetic variation in natural populations. *Amer. Zool.* **10:** 53–66.

Selander, R. K., S. Y. Yang, R. C. Lewontin, and W. E. Johnson. 1970. Genetic variation in the horseshoe crab *(Limulus polyphemus),* a phylogenetic "relic." *Evolution* **24:** 402–414.

Sheppard, P. M. 1959. *Natural Selection and Heredity.* Philosophical Library, New York.

Sheppard, P. M., and E. B. Ford. 1966. Natural selection and the evolution of dominance. *Heredity* **21:** 139–147.

Simpson, G. G. 1944. *Tempo and Mode in Evolution.* Columbia University Press, New York.

———. 1949, 1967. *The Meaning of Evolution.* Eds. 1 and 2. Yale University Press, New Haven, Conn.

———. 1951. *Horses: The Story of the Horse Family in the Modern World and through Sixty Million Years of Evolution.* Oxford University Press, New York. Paperback edition, American Museum of Natural History, New York.

———. 1952. How many species? *Evolution* **6:** 342.

———. 1953. *The Major Features of Evolution.* Columbia University Press, New York.

———. 1961. *Principles of Animal Taxonomy.* Columbia University Press, New York.

———. 1967. (See Simpson, 1949, 1967, above.)

———. 1968. Evolutionary effects of cosmic radiation. *Science* **162:** 140–141.

———. 1969. The first three billion years of community ecology. *Brookhaven Symp. Biol.* **22:** 162–177.

———. 1974*a*. The concept of progress in organic evolution. *Social Res.* **41:** 28–51.

———. 1974*b*. Recent advances in methods of phylogenetic inference. *Wenner-Gren Cent. Int. Symp. Ser.* **61:** 1–35.

Simpson, G. G., and W. S. Beck. 1965. *Life: An Introduction to Biology.* Ed. 2. Harcourt, Brace & World, New York.

Snogerup, S. 1967. Studies in the Aegean flora. IX. *Erysimum* sect. *Cheiranthus.* B. Variation and evolution in the small-population system. *Opera Bot.,* No. 14, 1–86.

Soans, A. B., D. Pimentel, and J. S. Soans. 1974. Evolution of reproductive isolation in allopatric and sympatric populations. *Amer. Nat.* **108:** 117–124.

Sokal, R. R., and I. Karten. 1964. Competition among genotypes in *Tribolium castaneum* at varying densities and gene frequencies (the black locus). *Genetics* **49:** 195–211.

Spotila, J. R., P. W. Lommen, G. S. Bakken, and D. M. Gates. 1973. A mathematical model for body temperatures of large reptiles: implications for dinosaur ecology. *Amer. Nat.* **107:** 391–404.

Stanley, S. M. 1973. An explanation for Cope's rule. *Evolution* **27:** 1–26.

———. 1974. Relative growth of the titanothere horn: a new approach to an old problem. *Evolution* **28:** 447–457.

————. 1975. A theory of evolution above the species level. *Proc. Nat. Acad. Sci. USA* **72:** 646–650.

Stebbins, G. L. 1949. Rates of evolution in plants. In *Genetics, Paleontology, and Evolution,* ed. by G. L. Jepsen, E. Mayr, and G. G. Simpson. Princeton University Press, Princeton, N.J.

————. 1950. *Variation and Evolution in Plants.* Columbia University Press, New York.

————. 1969. *The Basis of Progressive Evolution.* University of North Carolina Press, Chapel Hill.

————. 1974. *Flowering Plants: Evolution Above the Species Level.* Harvard University Press, Cambridge, Mass.

Stebbins, G. L., and R. C. Lewontin. 1972. Comparative evolution at the levels of molecules, organisms, and populations. In *Darwinian, Neo-Darwinian, and Non-Darwinian Evolution,* ed. by L. M. LeCam, J. Neyman, and E. L. Scott. Berkeley Symposia on Mathematical Statistics and Probability. Proceedings of the Sixth Symposium, Vol. 5. University of California Press, Berkeley.

Stern, C. 1958. Selection for subthreshold differences and the origin of pseudo-exogenous adaptations. *Amer. Nat.* **92:** 313–316.

————. 1959. Variation and hereditary transmission. *Proc. Amer. Phil. Soc.* **103:** 183–189.

————. 1960, 1973. *Principles of Human Genetics.* Eds. 2 and 3. W. H. Freeman and Co., San Francisco.

Stirton, R. A. 1947. Observations on evolutionary rates in hypsodonty. *Evolution* **1:** 32–41.

Streams, F. A., and D. Pimentel. 1961. Effects of immigration on the evolution of populations. *Amer. Nat.* **95:** 201–210.

Strid, A. 1970. Studies in the Aegean flora. XVI. Biosystematics of the *Nigella arvensis* complex with special reference to the problem of non-adaptive radiation. *Opera Bot.,* No. 28, pp. 1–169.

Stubbe, H. 1960. Mutanten der Wildtomate *Lycopersicon pimpinellifolium* (Jusl.) Mill. *Kulturpflanze* **8:** 110–137.

Sudworth, G. B. 1908. *Forest Trees of the Pacific Slope.* U.S. Department of Agriculture, Washington, D.C.

Sukatchev, W. 1928. Einige experimentelle Untersuchungen über den Kampf ums Dasein zwischen Biotypen derselben Art. *Z. indukt. Abstammungs- Vererbungsl.* **47:** 54–74.

Takhtajan, A. L. 1959. *Essays on the Evolutionary Morphology of Plants.* Translation. American Institute of Biological Sciences, Washington, D.C.

Teilhard de Chardin, P. 1959. *The Phenomenon of Man.* Translation. Harper & Row, New York.

Thoday, J. M. 1958. Natural selection and biological progress. In *A Century of Darwin,* ed. by S. A. Barnett. George Allen & Unwin, London.

————. 1972. Disruptive selection. *Proc. Roy. Soc. London, B* **182:** 109–143.

Thoday, J. M., and T. B. Boam. 1959. Effects of disruptive selection. II. Polymorphism and divergence without isolation. *Heredity* **13:** 205–218.

Thoday, J. M., and J. B. Gibson. 1962. Isolation by disruptive selection. *Nature* **193:** 1164–1166.

—— and ——. 1970. The probability of isolation by disruptive selection. *Amer. Nat.* **104:** 219–230.

Timofeeff-Ressovsky, N. W. 1940. Mutations and geographical variation. In *The New Systematics,* ed. by J. Huxley. Clarendon Press, Oxford.

Turesson, G. 1922. The genotypical response of the plant species to its habitat. *Hereditas* **3:** 211–350.

——. 1925. The plant species in relation to habitat and climate: contributions to the knowledge of genecological units. *Hereditas* **6:** 147–236.

Uzzell, T., and C. Spolsky. 1974. Mitochondria and plastids as endosymbionts: a revival of special creation? *Amer. Sci.* **62:** 334–343.

Waddington, C. H. 1953. Genetic assimilation of an acquired character. *Evolution* **7:** 118–126.

——. 1956. Genetic assimilation of the bithorax phenotype. *Evolution* **10:** 1–13.

——. 1957. *The Strategy of the Genes.* George Allen & Unwin, London.

Walker, E. P. 1964. *Mammals of the World.* 2 vols. The Johns Hopkins University Press, Baltimore.

Walker, J. C., and M. A. Stahmann. 1955. Chemical nature of disease resistance in plants. *Ann. Rev. Plant Physiol.* **6:** 351–366.

Wallace, B. 1966. On the dispersal of *Drosophila. Amer. Nat.* **100:** 551–563.

——. 1968. *Topics in Population Genetics.* W. W. Norton & Co., New York.

Washburn, S. L. 1960. Tools and human evolution. *Sci. Amer.* **203**(3): 62–75.

Washburn, S. L., and R. Moore. 1974. *Ape Into Man.* Little, Brown & Co., Boston.

Watt, W. B. 1969. Adaptive significance of pigment polymorphisms in *Colias* butterflies. II. Thermoregulation and photoperiodically controlled melanin variation in *Colias eurytheme. Proc. Nat. Acad. Sci. USA* **63:** 767–774.

Weismann, A. 1889–1892. *Essays upon Heredity and Kindred Biological Problems.* Translation, 2 vols. Oxford University Press, Oxford.

——. 1893. *The Germ-plasm: A Theory of Heredity.* Translation. Walter Scott, London.

Wells, P. V. 1969. The relation between mode of reproduction and extent of speciation in woody genera of the California chaparral. *Evolution* **23:** 264–267.

Werth, E. 1956. *Bau und Leben der Blumen: Die blütenbiologischen Bautypen in Entwicklung und Anpassung.* Enke Verlag, Stuttgart.

White, M. J. D. 1973. *Animal Cytology and Evolution.* Ed. 3. Cambridge University Press, Cambridge and London.

Whittaker, R. H., S. A. Levin, and R. B. Root. 1973. Niche, habitat, and ecotype. *Amer. Nat.* **107:** 321–338.

Wiener, A. S., and J. Moor-Jankowski. 1971. Blood groups of non-human primates and their relationship to the blood groups of man. In *Comparative Genetics in Monkeys, Apes and Man,* ed. by A. B. Chiarelli. Academic Press, London and New York.

Wilkes, H. G. 1967. *Teosinte: The Closest Relative of Maize.* Bussey Institution, Harvard University, Cambridge, Mass.

——. 1972. Maize and its wild relatives. *Science* **177:** 1071–1077.

Williams, G. C. 1966. *Adaptation and Natural Selection.* Princeton University Press, Princeton, N.J.

——, ed. 1971. *Group Selection.* Aldine-Atherton, Chicago.

——. 1975. *Sex and Evolution.* Princeton University Press, Princeton, N.J.

Williamson, D. L., and L. Ehrman. 1967. Induction of hybrid sterility in nonhybrid males of *Drosophila paulistorum. Genetics* **55**: 131–140.

Wills, C. 1973. In defense of naive pan-selectionism. *Amer. Nat.* **107**: 23–34.

Wilson, E. O. 1971. Competitive and aggressive behavior. In *Man and Beast: Comparative Social Behavior,* ed. by J. F. Eisenberg and W. S. Dillon. Smithsonian Institution, Washington, D.C.

———. 1975. *Sociobiology: The New Synthesis.* Harvard University Press, Cambridge, Mass.

Winter, F. L. 1929. The mean and variability as affected by continuous selection for composition in corn. *J. Agric. Res.* **39**: 451–476.

Wolf, C. B. 1948. Taxonomic and distributional studies of the New World cypresses. *Aliso* **1**: 1–250.

Woodworth, C. M., E. R. Leng, and R. W. Jugenheimer. 1952. Fifty generations of selection for protein and oil in corn. *Agron. J.* **44**: 60–65.

Wright, S. 1931. Evolution in Mendelian populations. *Genetics* **16**: 97–159.

———. 1943*a*. Isolation by distance. *Genetics* **28**: 114–138.

———. 1943*b*. An analysis of local variability of flower color in *Linanthus parryae. Genetics* **28**: 139–156.

———. 1946. Isolation by distance under diverse systems of mating. *Genetics* **31**: 39–59.

———. 1949. Adaptation and selection. In *Genetics, Paleontology, and Evolution,* ed. by G. L. Jepsen, E. Mayr, and G. G. Simpson. Princeton University Press, Princeton, N.J.

———. 1956. Modes of selection. *Amer. Nat.* **90**: 5–24.

———. 1960. Physiological genetics, ecology of populations, and natural selection. In *Evolution After Darwin,* Vol. 1, ed. by S. Tax. University of Chicago Press, Chicago.

Wright, S., and W. E. Kerr. 1954. Experimental studies of the distribution of gene frequencies in very small populations of *Drosophila melanogaster.* II. Bar. *Evolution* **8**: 225–240.

Wynne-Edwards, V. C. 1962. *Animal Dispersion in Relation to Social Behaviour.* Oliver & Boyd, Edinburgh and London.

Index to Authors

Index to Organisms

A number in **boldface** denotes an
illustration of the organism
(or a part of it).

Index to Subjects